Principles of Classical Thermodynamics

Applied to Materials Science

Principles of Classical Thermodynamics

Applied to Materials Science

Didier de Fontaine

University of California, Berkeley, USA

World Scientific

EW JERSEY · LONDON · SINGAPORE · BEIJING · SHANGHAI · HONG KONG · TAIPEI · CHENNAI · TOKYO

Published by

World Scientific Publishing Co. Pte. Ltd.

5 Toh Tuck Link, Singapore 596224

USA office: 27 Warren Street, Suite 401-402, Hackensack, NJ 07601

UK office: 57 Shelton Street, Covent Garden, London WC2H 9HE

Library of Congress Cataloging-in-Publication Data

Names: De Fontaine, Didier, 1931– author.

Title: Principles of classical thermodynamics : applied to materials science / Didier de Fontaine
(University of California, Berkeley, USA).

Description: Singapore ; Hackensack, NJ : World Scientific Publishing Co. Pte. Ltd., [2019] |
Includes bibliographical references and index.

Identifiers: LCCN 2019002090| ISBN 9789813222687 (hardcover ; alk. paper) |
ISBN 9813222689 (hardcover ; alk. paper)

Subjects: LCSH: Thermodynamics. | Materials science.

Classification: LCC QC311 .D37 2019 | DDC 536/.7--dc23

LC record available at https://lccn.loc.gov/2019002090

British Library Cataloguing-in-Publication Data

A catalogue record for this book is available from the British Library.

For any available supplementary material, please visit
https://www.worldscientific.com/worldscibooks/10.1142/10509#t=suppl

This book is dedicated to my wife Danielle for her unflagging support during its preparation, and to our three sons and their families. On a sad note, my wife passed away just as the text was being corrected.

Didier Le Fountain

Foreword

This book is really a combination of class notes, textbook, and review article. It started as class notes for a classical thermodynamics course taught at the UCLA and later at UC Berkeley; that material constitutes most of Part I, devoted to the basic principles of the subject. Part II is of more recent vintage, was partly taught at Berkeley, and is devoted mostly to topics of interest in materials science. It also contains material which is original to this book, as far as I know, the objective being to indicate, in a "review" manner, some of the current trends in the general field of thermodynamics. The presentation throughout is decidedly theoretical, which makes the treatment rather condensed. As a result, the contents will be most relevant to those who are already familiar with basic notions of classical thermodynamics. Those who prefer a more experimental approach have many excellent textbooks at their disposal, some of which are mentioned at the end of the Introduction.

The first chapter of Part I introduces the subject and some of its antecedents, plus the motivation for adopting the Carathéodory treatment of the second law, as presented in Chapter 3. That chapter, which tells us, according to the second law, that some processes in thermodynamics are forbidden, then emphasizes the notion that all equilibrium thermodynamics starts with the (Pfaffian) differential form for the internal energy, with consequent applications arrived at by integration of this form. Manipulations of differential forms, by Legendre transformations for instance, are given in Chapter 4, leading naturally to conditions of phase equilibrium and to chemical potentials. An important chapter is that of solutions (Ch. 7), essential in materials science, less so in solid state physics, chemistry, or mechanical engineering. Part I ends with a superficial introduction to statistical thermodynamics for the purpose of introducing certain applications and also for relating the rather abstract macroscopic world of classical thermodynamics to the more intuitive micro-world which has become so familiar — it was not so in the time of Gibbs.

Part II begins with one of the most important thermodynamics applications in materials science, that of temperature-concentration phase diagrams. These diagrams come in varied and bewildering shapes, while at the same time obeying the rigorous set of directives provided by the Gibbs phase rule. Condensed matter is of course included in the study of phase diagrams, but not, in this book, non-hydrostatic deformations of solids. Solids can be crystals (perfect and imperfect), or glasses (amorphous materials), and subsequent chapters describe their most relevant thermodynamic properties, including point defect equilibria, through chemical equations. As mentioned above, the book is mostly concerned with macroscopic phenomena, and peripherally with microscopic ones, but there is another class: that of mesoscopic ones, less often treated in textbooks. Inclusion of that third scale is warranted as a supplement to a purely Gibbsian treatment of phase uniformity: non-uniformity is indeed the rule in condensed matter, such as that observed in diffuse interfaces, incomplete products of phase transformations and the like, and also appearing in the Landau–Lifshitz–Ginzburg instability theory, an account of which is given in Chapter 15.

Therefore a number of chapters in Part II are concerned with the topic on non-uniformity, which usually necessitates the application of a procedure known as "coarse-graining", whereby concentration averages are taken over small regions of space, thereby making discrete (molecular) matter continuous but non-uniform, hence amenable to treatment by differential equations. One recent consequence of that is the realization that nucleation and growth on the one hand, and spinodal decomposition on the other are but two aspects of the same phase separation mechanism, and can be handled by a single diffusion (differential) equation, or more generally a finite difference equation capable of treating order-disorder in addition to clustering. The final chapter summarizes the book's methodology, so that the reader may wish to to read that chapter first. A set of ten appendices, containing mostly mathematical derivations, completes the book.

To compile these notes, I have made copious use of several well-known textbooks, such as the excellent two-volume work by J. Kestin, that by M. W. Zemansky, both editions of the book by H. B. Callen, the classic *Statistical Physics* by Landau and Lifshitz, the textbook on crystal physics by J. F. Nye, the *Introduction to the Study of Stellar Structure* by S. Chadrasekhar, the monograph *Order and Stability in Alloys* by F. Ducastelle, the books *Phase Transitions in Materials* by B. Fultz, *Chemical Thermodynamics* by C. Lupis, *The Chemistry of Imperfect Crystals* by F. A. Kröger, *The Theory of Phase Transformations in Metal and Alloys* by J. W. Christian, *Phase Transformations* by J. J. Hoyt, *Theory of Structural Transformations in Solids* by A. G. Khachaturyan and *Introduction to Geometry* by H. S. M. Coxeter (all of these and others cited further in the text).

I was fortunate to have learned my thermodynamics from my dissertation adviser at Northwestern, John Hilliard, which allowed me to interact frequently with his friend and collaborator John Cahn. I was also extremely fortunate to have worked, at UCLA and at Berkeley, with remarkable graduate students and post-doctoral fellows, many of whom now hold faculty positions at prestigious universities in the US and abroad. I learned much from them, and I thank them all. My Berkeley colleague Ramamoorty Ramesh read my prose and made valuable suggestions as did Dr. Melanie Lutz who provided very pertinent corrections, while Dr. Alphonse Finel contributed a text on the third law of thermodynamics. I am also grateful to Mathematics Professor Beresford Parlett who pointed out several mistakes in my text. Ms. Wan Wan Liu typed in LaTeX most of Part I, rapidly and accurately, from an early version which had been typed most professionally by Ms. Virginia Wilkins. Thanks are also due to my editor Rachel Seah Mei Hui and her colleagues for their patience and to Mr. Rajesh Babu for help with the LaTeX code at World Scientific Publishing, and also to Mr. Nick Nallick at the Intaglio drafting software. I typed most of Part II in LaTeX, drew all the figures with Intaglio, and created the Index, so that I am totally responsible for any remaining errors. Parts of this book were written when I benefited from grants from the National Science Foundation and the Department of Energy through the Lawrence Berkeley National Lab, and also from a small grant from the U.C. Berkeley Emeritus Fund.

Contents

Chapter 1

Introduction

Die Energie der Welt ist constant.

Die Entropie der Welt strebt einem Maximum zu.[1]

Thermodynamics was born of the Industrial Revolution which historians conventionally place around the end of the eighteenth and the beginning of the nineteenth centuries. Initially, engineers were concerned with getting the maximum amount of "work" out of "heat", two forms of energy which were defined at first in rather fuzzy ways. Indeed it was not originally recognized that both forms of energy were really different kinds of motion, the former (work), roughly speaking, representing external, macroscopic, *ordered* motion, the latter (heat) representing internal, microscopic, *disordered* motion. The reader interested in the history of the subject is referred to the two-volume work by Steven Brush entitled, appropriately enough, *The Kind of Motion We Call Heat* (1976); also of interest is the text by Martin Bailyn (1994) which treats basic principles from a historical perspective, thereby showing how painfully the scientific community arrived at thermodynamic understanding, too often taken for granted today. After 1850, thermodynamics developed into a science with the works of Rudolf Clausius (1822–1888), William Thompson (Lord Kelvin, 1824–1907), James Clerk Maxwell (1831–1879). Ludwig Boltzmann (1844–1906) and J. Willard Gibbs (1839–1903) whose treatise *On the Equilibrium of Heterogeneous Substances*, published in 1875 and 1877 in the *Transactions of the Connecticut Academy of Arts and Sciences*, practically created the field of classical thermodynamics as we know it today.

---

[1] "The Energy of the World is constant. The Entropy of the World tends towards a Maximum". It is with this sweeping statement of Rudolf Clausius that J. Willard Gibbs begins his treatise *On the Equilibrium of Heterogeneous Substances*.

1

Classical — or Gibbsian — thermodynamics, dealing with macroscopic objects, cannot really predict properties of matter, as its equations are actually *identities* indicating how thermodynamic variables must be related to one another given certain postulates which are, in the main, the *Zeroth*, the *First*, and the *Second* laws of thermodynamics. The *Third* law will be treated in a separate chapter. Such is the strength of classical thermodynamics: its results are independent of the particular internal details of matter, thereby resulting in a very compact, very elegant formulation of the theory, free from approximations; it is a "black box", *macroscopic* formulation of the thermal properties of matter. But that strength is also a weakness: the parameters which enter the equations of classical thermodynamics must be obtained from experiment or from first-principles calculations. For a truly predictive theory, the fundamental structure of matter must be taken into account, but the necessary knowledge of atomic theory was not available to Gibbs and to early thermodynamicists. The discrete nature of the structure of fluids and liquids was of course anticipated and allowed Maxwell and Boltzmann to develop *microscopic* formulations of thermodynamics, without however coming to grips with the physics of interacting particles, which had to await the advent of quantum mechanics.

The term "microscopic" thermodynamics can have various meanings: it can refer to the *statistical* mechanics of Maxwell, Gibbs and Boltzmann, as opposed to the *classical* thermodynamics of Kelvin, Clausius and Gibbs which refers to the continuous, macroscopic approach. However, the term "microscopic" may also refer simply to the scale of the system under consideration. Somewhat surprisingly, a macroscopic-continuous-classical treatment, suitably modified, can often remain phenomenologically valid down to almost nanometer scales. The same thermodynamics can be applied as well to systems of huge sizes, for example to the internal equilibrium of stars, as shown in the monograph entitled *Introduction to the Study of Stellar Structure* by Chandrasekhar (1957). This text contains in its first two chapters a general introduction to classical thermodynamics which is one of the clearest and most rigorous that one can find and it has inspired the treatment of the basic laws of thermodynamics given in the present book. Hence, the term "microscopic" is not necessarily equated to "statistical". Several textbooks discuss both the classical and the statistical aspect of the field together, for instance those by Chandrasekhar, Landau and Lifshitz (1958)[2] or Kestin (1966).

The ultimate thermodynamic treatment must of course be based on the atomic nature of matter. But the continuum is a very useful concept due

---

[2] A newer edition delves much deeper into the statistical theory (1980).

to ability to perform analytic calculations and thus to handle much larger systems. Indeed, by adopting the continuum approximation one essentially replaces a system of some 10^{22} degrees of freedom by one consisting of only a few macroscopic variables. But care must be taken in going from an atomistic to a continuum treatment. For example, it is necessary to define properly the concentration c of atoms of a certain type per unit volume by averaging the number of these atoms in a volume δV, large with respect to atoms themselves, but small enough to capture the relevant concentration variation. This procedure was treated very carefully in the work of Langer *et al.* (1975) and of Finel and coworkers (Bronchart *et al.*, 2008), and summarized in Chapter 18 of the present book (see references therein).

Historically, the first thermodynamic systems treated were homogeneous, uniform (of constant composition), continuous fluids, the types of substances which mattered for use in heat engines. An important generalization to "heterogeneous" substances, introduced by Gibbs (1875), gave us the concept of *phase*, i.e. *a portion of matter which is itself homogeneous and uniform, but is in thermodynamic contact with another phase, distinct from the original, but itself also homogeneous and uniform.* Thus was born the concept of *phase equilibrium*, which is the major topic treated in this book. Even more generally, a given phase may present variations in some of its internal variables, whence thermodynamic properties within a given phase now can depend on position, and gradients of thermodynamic variables must be included in the formulation. Early treatments along those lines include those of Landau and Lifshitz (1958) and of Cahn and Hilliard (1958).

Finally, an important distinction must be made between "reversible" and "irreversible" processes (to be defined later). From the foregoing it will be apparent that the field of thermodynamics is extremely broad, so that mostly *equilibrium* phenomena will be presented in this book. What is therefore left out are kinetic and dynamic processes, which are of course of great practical interest, but are not governed by laws having the same degree of rigor and generality as those of equilibrium (reversible) thermodynamics. The interested reader may consult the book by de Groot and Mazur (1962) or that by Nicolis and Prigogine (1977) for a thorough treatment of irreversible phenomena. Nevertheless, a later chapter in this book will be devoted to kinetics.

The first chapters introduce, as they must, the basic principles and laws of thermodynamics. The formulation adopts the axiomatic treatment developed initially by Carathéodory (1909), and continuators. The traditional method of making use of idealized heat engines has its advantages, of course, and physicist Gabriel Weinrich (1969) states that:

The traditional formulation — complete with heat engines and cycles – is amusing and has a great deal of charm; to a large degree, it gives the subject its personality.

Be that as it may however, here is how Chandrasekhar introduces the axiomatic approach of mathematician Constantin Carathéodory, which he prefers, in the first chapter of his book on stellar structure:

The reasons for including this chapter are twofold: first, there exists no treatise in English which gives Carathéodory's theory[3]; and second, on the writer's view, Carathéodory's theory is not merely an alternative, but elegant, approach to thermodynamics but is the only physically correct approach to the second law.

But how did this treatment originate? Physicist Max Born tells the following story from his days in Cambridge in his autobiography (1978):[4]

I had nothing to do, and I spent most of my time lying in a punt or canoe on the Cam and reading Gibbs, whom I now began to understand. From this sprang an essential piece of progress in thermodynamics — not by myself, but by my friend Carathéodory. I had tried to understand the classical foundations of the two theorems, as given by Clausius and Kelvin; they seemed to me wonderful, like a miracle produced by a magician's wand, but I could not find the logical and mathematical root of these marvelous results. A month later I visited Carathéodory in Brussels where he was staying with his father, the Turkish ambassador, and told him about my worries. I expressed the conviction that a theorem expressible in mathematical terms, namely the existence of a function of state like the entropy, with definite properties, must have a proof using mathematical arguments which for their part are based on physical assumptions or experiences but clearly distinguished from these. Carathéodory saw my point at once and began to study the question. The result was a brilliant paper, published in *Mathematische Annalen* (1909), which I consider the best and clearest presentation of thermodynamics.

Chandrasekhar wrote his text on Carathéodory's axiomatic method in 1936, but today that approach is not often taken, particularly since the Carathéodory's axiom has been shown to be equivalent to the traditional Kelvin and Plank statements of the second law (Pippard 1957, Landsberg 1964).

---

[3]at the time that Chandrasekhar wrote his text.

[4]I am indebted to Melanie Lutz for having communicated to me this text of Max Born.

Hence, the traditional and the mathematical approaches can now be combined, as shown for example by Zemansky (1968), a formulation which will be followed here in Sec. 3.3.

Materials science, to a great extent, is the study of *alloys* so that a basic parameter to be dealt with is the concentration, or number of atoms or molecules (or electric or magnetic dipoles) in solution per unit volume or per mole. Consequently, a whole chapter, which closes the first part of the text, is devoted to solutions. The second part, covering mostly applications, includes the study of temperature-concentration phase diagrams, chemical reactions and interfaces. Less common topics include a brief introduction to statistical thermodynamics, non-uniform systems, order-disorder transformations and glasses. Despite announced restrictions, later chapters introduce notions of kinetics, such as nucleation and spinodal decomposition, and more importantly a more recent treatment which combines the two traditional approaches.

The final chapter summarizes the book's methodology, so that the reader may wish to to read that chapter first. The text is subdivided in two parts: Part 1 describes fundamentals, Part 2 describes some applications of Classical Thermodynamics. A set of ten appendices, containing mostly mathematical derivations, completes the book.

References

Bailyn, M. (1994), *A Survey of Thermodynamics*, AIP Press, New York.

Brush, S. G. (1976), *The Kind of Motion We Call Heat*, North-Holland.

Born, M. (1978), *My Life, Recollections of a Nobel Laureate*, Taylor & Francis; Translated from original German edition (1975).

Cahn, J. W. and Hilliard, J. E. (1958), *Journal of Chemical Physics*, **28**, 258.

Carathéodory, C. (1909), *Mathematische Annalen*, **67**, 355.

Chandrasekhar, S. (1957), *Introduction to the Study of Stellar Structure*, Dover; Original edition by the University of Chicago Press (1939).

Clausius, R. (1865), *Mechanische Wärmetheorie*.

de Groot, S. R. and Mazur, P. (1962), *Non-Equilibrium Thermodynamics*, North-Holland.

Gibbs, J. W. (1875), *On the Equilibrium of Heterogeneous Substances*. *Trans. Connecticut Acad.*, pp. 108-248; (1877), *ibidem*, pp. 343-524. See for example (1993), *The Scientific Papers of J. Willard Gibbs*, Ox Bow Press.

Kestin, J. (1966), *A Course in Thermodynamics*, Volumes I and II, Blaisdell.

Landau, L. D. and Lifshitz, I. M, (1958), *Statistical Physics*, Addison-Wesley.

Landau, L. D., Lifshitz, I. M. and Pitaevskii, L. P.(1980), *Statistical Physics*, Third Edition, Pergamon Press.

Landsberg, P. T. (1964), *Nature*, **201**, 485.

Nicolis, G. and Prigogine, I. (1977), *Self-Organization in Non-Equilibrium Systems; From Dissipative Structures to Order Through Fluctuations*, John Wiley & Sons.

Pippard, A. B. (1957), *Elements of Classical Thermodynamics*, Cambridge University Press.

Weinrich, G. (1969), *Fundamental Thermodynamics*, Addison-Wesley.

Zemansky, M. W. (1968), *Heat and Thermodynamics*, *Fifth Edition*, McGraw-Hill Book Co., NY.

Part I

Basic Thermodynamics

Chapter 2

Thermodynamic Systems

In keeping with the concept of "Classical (or Gibbsian) Thermodynamics", we shall henceforth consider only the continuum approximation which, as mentioned in the introductory chapter, irons out internal dynamics and also the very structure of matter: for instance, instead of actual atomic (or molecular) elements of types A, B, ..., one considers small volumes of material, centered at a given point x, the average composition at the point in question yielding the *composition* $c(x)$, which is a continuous function of position. Such a procedure is generally referred to as one of "coarse-graining", and can be used for other *densities*, besides the concentration c (or c_i with $i = 1, \dots n$ if more than one type of atom or molecule is present). In that way, we can represent, in a continuum manner, matter of *non-uniform* composition c varying from point to point. Conversely, if the composition profile is perfectly flat, the material is said to be *uniform, i.e. homogeneous*. In that way, since volume averages are taken, the "points" can be chosen arbitrarily, hence not confined to lattices, and can be separated by arbitrary distances, thereby saving much computer time in numerical modeling. An example of that feature will be given in Chapter 18.

Ordinary matter may also contain distinct portions whose properties differ markedly from one another, although in each separate portion variables such as composition c are either constant or vary in a continuous manner. Such distinct regions, which can in principle be separated from the others mechanically, were called *phases* by Gibbs, mentioned in the present text on page 3 (in italics), although the definition given there enlarges somewhat on that of Gibbs in that the phases themselves may have non-uniform compositions. Thermodynamic systems containing more than one phase are denoted as *heterogeneous*. Examples of phases are the liquid and solid manifestations of the same chemical substance, or solid mixtures of two or more chemical species appearing in different crystal structures or, as a special case, regions of a solid material having same physical properties but whose crystal lattices are oriented differently with respect to one another. In the latter case, one does not usually speak of different *phases*, but of different crystalline *grains*. Neighboring phase regions are separated by interfaces, which are also heterogeneities which must be taken into account in a full thermodynamic treatment. If the characteristic dimensions of the various phases are large ($\geq 100\,\text{nm}$) however, then their surface to volume ratios are small and interfacial contributions to thermodynamic quantities become vanishingly small in the limit of large volumes.

Having cataloged these various complications which the real world presents to us, we are ready to describe the simplest possible case: that of material subjected to hydrostatic stress only, a fluid for instance, of large volume, so that the effect of interfaces may be neglected, of uniform composition, and consisting of only a single phase. The resulting thermodynamic system may appear to be highly impoverished, but it suffices to introduce some of the main concepts of classical thermodynamics, and is indeed the only one treated in many elementary expositions of the subject. Also, such a simple system represents correctly a large volume of homogeneous, uniform fluid, which is after all, what was needed in the early days of the empirical investigation of heat engines. Such experimental situations can be described in a continuum, macroscopic formalism, but only after necessarily introducing novel concepts to take into account the neglected discrete nature of the real world.

2.1   Types of work

To the neophyte, the expression $P\,\mathrm{d}V$ seems to represent the essence of classical thermodynamics. It isn't; in fact it has very little to do with it, being derived from classical continuum mechanics, which is itself arrived at formally through coarse graining as mentioned above. To sketch briefly how this is done, consider at first the elementary work dw done by a force $\boldsymbol{f}$ acting to displace a point particle by $\mathrm{d}\boldsymbol{s}$ along a path S, as shown in Fig. 2.1, illustrating the concept of *mechanical work*. From classical mechanics, we have the familiar scalar product formula $dw = \boldsymbol{f} \cdot d\boldsymbol{s}$ for the elementary work done *by* the force on the particle. Note the use of the slanted symbol d for an arbitrarily small quantity, rather than the straight d for a true differential. The total work from point a to point b along the path is then given by the line integral

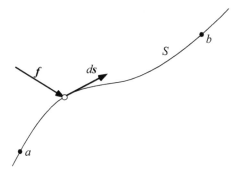

Figure 2.1: Particle acted upon by a force $\boldsymbol{f}$ and constrained to a path S.

$$w_{a \to b} = \int_{S(a)}^{S(b)} \boldsymbol{f} \cdot \mathrm{d}\boldsymbol{s} \,. \tag{2.1}$$

The finite amount of work done by the force will of course depend on the path itself. Strictly, one must again appeal to the the continuity hypothesis — here in a sort of one-dimensional coarse graining — to allow the small quantity $d\boldsymbol{s}$ to become the mathematical differential $\mathrm{d}\boldsymbol{s}$.

Now consider a homogeneous elastically stressed body whose internal stresses and strains are given at each point $\boldsymbol{x}$ of the continuum by the symmetric tensors σ_{ij} and ϵ_{ij}, respectively. Stresses of course are the coarse-grained equivalent of actual quantum mechanical forces acting on atoms, ions and molecules (see for example Landau and Lifshitz 1995). The components of the stress tensor have units of force per unit area and those of the strain tensor have units of displacements per unit length. Now let a small variation of the applied forces on the boundary of the solid, of volume V, cause incremental strains $d\epsilon_{ij}$. Imagine a small cube of volume $\mathrm{d}v(\boldsymbol{x})$ about point $\boldsymbol{x}$ and calculate, by means of Eq. (2.1), the work done by the local forces on displacing the faces of the cube according to the varied strain tensor. It can now be shown (see for example Landau and Lifshitz (1995) or Voorhees and Johnson (2005)), by summing contributions from all elementary volumes, that the incremental work dW is given by the volume integral

$$dW = \int_V \sum_{ij} \sigma_{ij}(\boldsymbol{x}) \, d\epsilon_{ij}(\boldsymbol{x}) \, \mathrm{d}v \,. \tag{2.2}$$

Major simplification occurs if the elastic body is acted upon only by hydrostatic stresses, as is the case for fluids at rest. The stress tensor now reduces to a single diagonal element according to $\sigma_{ij} = \sigma \, \delta_{ij}$, where the Kronecker symbol has been used.[1] The unique stress component σ, constant for a uniform fluid, is none other than the negative of the uniform pressure P, having units of energy per unit volume. The change of sign is conventional: stress is considered to be positive if it is tensile, whereas pressure is considered to be positive if it is compressive. Thus, in the hydrostatic situation, Eq. (2.2) reduces to

$$dW = -P\,dV \qquad (2.3)$$

which is the well-known expression for hydrostatic work given in all elementary treatments of the thermodynamics of fluids. More complete derivations are given in the texts mentioned above.

As another example, let us follow basically the same procedure by starting again from the arrangement depicted in Fig. 2.1, but where now the particle is a point electric charge following a path within an electric field $\boldsymbol{\mathcal{E}}$. Next consider a continuous distribution of charges in a dielectric material to arrive at the incremental electrostatic work expression (see for example Jackson (1962)) or Landau and Lifshitz (1995))

$$dW = \int_V \sum_{i=1}^{3} \boldsymbol{\mathcal{E}}(\boldsymbol{x})\,d\boldsymbol{\mathcal{D}}(\boldsymbol{x})\,dv \qquad (2.4)$$

where $\boldsymbol{\mathcal{E}}$ and $\boldsymbol{\mathcal{D}}$ are the electric field and dielectric displacement vectors, respectively. In like manner we arrive at the incremental magnetic work

$$dW = \int_V \sum_{i=1}^{3} \boldsymbol{\mathcal{H}}(\boldsymbol{x})\,d\boldsymbol{\mathcal{B}}(\boldsymbol{x})\,dv \qquad (2.5)$$

where $\boldsymbol{\mathcal{H}}$ and $\boldsymbol{\mathcal{B}}$ are the magnetic field and magnetic induction vectors, respectively. It is seen that expressions (2.1), (2.2), (2.3), (2.4) and (2.5) are all products of an *intensive* (or field) quantity times the differential of an *extensive* quantity (or density). Such is the general form of work differentials which will be encountered time and again in this book. For short, we shall adopt the general symbols X (for "X-tensive") and Y (for "Yn-tensive") for those two types of variables. Extensive quantities possess an additive property, in the sense that, in energy expressions, volumes, displacements, ... add up cumulatively, whereas intensive (field) variables, such as pressure or temperature or electric and magnetic fields do not. Nonetheless, normalized extensive quantities, thus densities, while not retaining that additive property, still retain the essential characteristic of extensive variables, which is that they appear after the differential sign in the corresponding expressions

---

[1] $\delta_{ij} = 1$ if $i = j$, 0 otherwise.

Table 2.1: Examples of conjugate pairs of intensive and extensive variables (some variables to be defined later).

Type	*Intensive*	*Extensive*	*Work*
Generic	Y	X	$Y \, dX$
Thermal	T	S	$T \, dS$
Mechanical	$-P$	V	$-P \, dV$
Chemical	μ_k	N_k	$\sum_k \mu_k \, dN_k$
Elastic	σ_{ij}	ϵ_{ij}	$\sigma_{ij} \, d\epsilon_{ij}$
Electric	$\mathcal{E}$	$\mathcal{D}$	$\mathcal{E} \, d\mathcal{D}$
Magnetic	$\mathcal{H}$	$\mathcal{B}$	$\mathcal{H} \, d\mathcal{B}$
Interface	σ	A	$\sigma \, dA$

for incremental work. Whenever possible, we shall use upper case letters for extensive quantities themselves, and lower case for normalized extensive quantities, i.e. densities. However, context should be the real guide, rather than pedantic notation. Note that an expression such as $X \, dY$ could be a valid thermodynamic differential, but it is not a work term.

A pair of variables which occur in incremental work expressions such as $Y \, dX$ are known as *conjugate variables*, and play a fundamental role in many branches of physics and related fields. In the mathematical treatment of equilibrium thermodynamics, extensive and intensive variables are actual *point* variables, which means that their values depend only on their point of application in thermodynamic space, not on past history, nor "on how we got there". For example, pressure P must be regarded as belonging to the thermodynamic system itself, rather than to the outside world. This is important: P is then the actual pressure of a fluid inside a container, not the outside pressure applied to a piston connected to the container; that pressure P' is generally different from P because of the forces of friction applied inevitably to the piston head. A summary of conjugate variables is given in Table 2.1; definition of symbols are given later.

2.2 Types of equilibria

Any scientific investigation must necessarily focus on some definite portion of the world at large. In thermodynamics, that portion of interest is designated by the general term *system*, bounded by walls or partitions which separates it from the outside world or from its subsystems. The *state* of a system is determined by the measurements of its collective $\{X\}$ and $\{Y\}$ point variables, hence are properly called *state variables*. In order to perform a meaningful measurement, the system's variables must be observed not to change during a time which is long compared to the time of observation. Such is the definition of *equilibrium*, but it is then predicated upon the patience of the observer. Actually, a system may be in a state of *metastable* equilibrium, meaning that, after a much longer wait, perhaps by an appropriate "jiggling" of the system (from thermal vibrations, for instance), it may eventually reach a lower minimum of energy or potential.

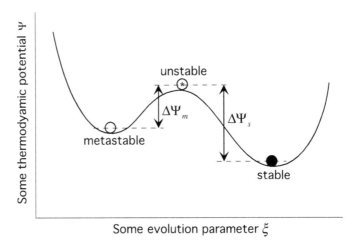

Figure 2.2: Illustration of stable, metastable and unstable equilibria.

Figure 2.2 illustrates the situation schematically by means of the familiar mechanical analogy of a massive ball on hilly terrain: stable equilibrium is indicated by a filled circle located in the deepest potential well, and metastable equilibrium is indicated by an open circle in a local minimum, the two minima being separated by a maximum in the potential curve at the starred point, denoting unstable equilibrium. It follows that, in an analytic treatment of equilibrium, the criterion of setting to zero the derivative of the potential curve with respect to some representative parameter, here denoted by the symbol ξ — later to be identified as an *order parameter* —

does not suffice: the distinction between maximum and minimum must be determined additionally by examining the sign of the second derivative at the extrema, positive for a minimum, negative for a maximum. Hence, to distinguish between stable and metastable minima, one must examine the totality of the potential curve, a much more elaborate proposition. Also, for multicomponent systems, the stability surface must be expanded to second order and the signs of the eigenvectors at the extrema must be examined.

If the system (ball) is in a state of metastable equilibrium (empty circle), how does it (the system) "know" that this equilibrium is not the stable one, that it must eventually jump into the filled circle position and stay there? It doesn't know! But assuming a continual jiggling of the whole hill-and-valley arrangement (a metaphor for atomic vibrations), balls in either potential well will occasionally jump over the energy barrier ($\Delta\Psi$) and fall into the other well. Once there, it can jump out again but the frequency of successful jumps will differ for the two wells; the energy required to jump successfully from the metastable well ($\Delta\Psi_m$) is seen to be less than that required from the stable well ($\Delta\Psi_s$), so that there will be less jumping out of the stable well than out of the metastable one in a given time interval. Thus the situation of the filled circle will surely represent a more stable state than the one represented by the open circle. That is how, while having knowledge of only its immediate environment, a material system can seek stable equilibrium in the apparent absence of information concerning global topography. And yet, thermodynamic systems have no consciousness.[2]

The curve in Fig. 2.2 with its two minima is found frequently in physics, chemistry, or/ and materials science and usually goes by the name of "double-well potential", for obvious reasons. Such a curve will be encountered in this book in Fig. 9.1 in the context of the regular solution model, Fig. 10.4 in the context of glass formation, Fig. 14.1 in the context of interfaces, Fig. 15.3 in the context of the Landau rules, Fig. 16.4 in the context of ferroelectric transitions, and Fig. 18.6 in the context of phase separation.[3]

---

[2]The process just outlined bears pondering: in general, if the only way to explain a certain event requires the attribution of "consciousness" or a "new force" — not yet discovered by scientists — to a material system, it means invariably that the event was ill-observed or that the proposed explanation was ill-logical. That argument also tells us, counterintuitively, that order (here, phase separation) can arise from disorder (local vibrations), as will be examined further in Chapter 10 on topological disorder or Chapters 17 and 18 on phase transformations.

[3]The double-well potential, or at least its extension in higher dimensions, has also been featured in the *Inflationary Universe* cosmology, see for example A. H. Guth *The Inflationary Universe*, Addison-Wesley Publishing Co. (1997).

2.3   Types of evolution

Obviously thermodynamic systems do evolve, and although we are not concerned in this book with kinetics *per se*, we should ask how it may be possible to monitor changes of state by readings of appropriate measuring devices. Three cases are of interest, as illustrated for a two-variable thermodynamic coordinate system in Fig. 2.3: (a) a spontaneous, discontinuous change of state from point a to point b, (b) a *quasi-static* process, *continuous* process, and (c) a *reversible* process.

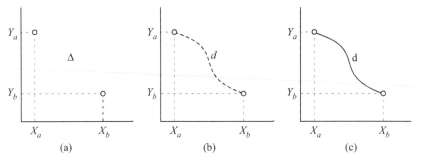

(a)   (b)   (c)

Figure 2.3: Three types of processes: (a) discontinuous, irreversible; (b) quasi-static, irreversible; (c) continuous, reversible.

Irreversible process

In case (a), the state variables of the system change from their values measured at point a, in a state of equilibrium, perhaps so maintained by appropriate constraints, to those measured at point b, also at equilibrium (perhaps metastable). If there exists some thermodynamic function F of the coordinates X and Y, then its change in going from a to b is given by

$$\Delta F = F(X_b, Y_b) - F(X_a, Y_a)$$

since a path cannot be indicated in a thermodynamic diagram such as that of Fig. 2.3, even in principle, at least in this coordinate system. Indeed no path was sketched in part (a) of this figure, the simple reason being that the state variables cannot be monitored during the process; there may not even be unique values of these macroscopic variables during the evolution of the system, possibly subjected to volume oscillations, pressure gradients, turbulence and the like. Usually, we shall reserve the symbol Δ for discontinuous changes of state, by convention taking the final values of the variables minus the initial ones. For such processes, an expression such as $dW = Y \, dX$ makes no sense because differentials cannot be defined for discontinuous changes, nor can one perform "integration along a path", since there is no identifiable path.

Quasi-static process

In case (b) of the above figure, a path has been indicated symbolically, consisting of (theoretically) infinitely small changes in the values of the thermodynamic variables, since it is now possible to determine the point variables at (X_i, Y_i), then at the next point (X_{i+1}, Y_{i+1}), and so on. Ideally, one should make measurements for states as close as possible to equilibrium ones, going from one point in thermodynamic space to the next by successively removing constraints one tiny bit at a time. A quasi-static incremental work expression may thus be written as $Y \, dX$, using the slanted symbol d for "a small quantity", as was done in Section 2.1. Between quasi-equilibrium states, each mini-process may well be irreversible, meaning that some friction or turbulence may take place, thereby causing dissipation to the outside world.

Reversible process

Case (c) of Fig. (2.3) is an idealization: a reversible process, referring as it does to processes which are quasi-static and which can be performed from a to any point b along the path or in the reverse direction from b to a in such a way that, after returning to the starting point, both the system *and the outside world find themselves in exactly the same state as the original one.* The italicized clause in the definition guarantees that there shall be no net changes occurring in this particular cyclic process, for either the system *or its surroundings.* To be sure, for any *cyclic* process (same starting and ending points), a change such as $\Delta F(X, Y)$ must obviously be zero since the state variables have identical values at coinciding points, but cyclic processes generally involve dissipative effects which alter the state of the outside world. To emphasize the nature of incremental work in the reversible case, it is customary to write differential expressions in state variables with a straight "d" for the mathematical differential symbol. Thus, we write the following expression for *reversible* incremental work involving several variables:

$$dW = \sum_i Y_i \, dX_i \,. \tag{2.6}$$

The summands in expression (2.6) are just the building blocks for the general fundamental differential form of classical thermodynamics, from which practically all else follows, as we shall see.

2.4 Types of partitions

In this section, we shall introduce various ways of separating thermodynamic systems (or subsystems) from the surroundings and from one another. In defining appropriate partitions, or walls, we shall be careful *not* to use the concept of *heat* in the definitions; the reason is simple, "heat" has not yet been defined, as it is a notion which is not encountered formally in classical mechanics. For that reason, the "H-word" will be used in quotation marks for the remaining portion of the present chapter

Three types of partitions are of interest:

Adiabatic partition

If a system is entirely enclosed within adiabatic walls, then by definition, the only way that the state of the system can be changed is by performing work, a concept which is known to mechanics. An idealized system so enclosed is represented in Fig. (2.4). Shaded areas denote adiabatic partitions. The types of mechanical work shown are the work of a piston compressing a fluid, stirring the fluid by cranking a paddle-wheel, or mechanically generating an electrical current passing through a resistive device.

Diathermic partition

If two subsystems [α and β in Fig. 2.4] are collectively enclosed within adiabatic walls, but mutually separated by a *diathermic* partition (shown as a heavy solid line in the figure), then *by definition*, it is found that, at equilibrium, a certain definite relation must hold between the variables describing the two subsystems. Such a relation may be given as a functional (F) form between appropriate state variables of the two subsystems:

$$F(X^\alpha, Y^\alpha, X^\beta, Y^\beta) = 0 . \tag{2.7}$$

A relation of this type is known as an *equation of state*. To anticipate, an equation of state for a system of two ideal gases, each present in the same amount, is:

$$P^\alpha V^\alpha - P^\beta V^\beta = 0 .$$

If relations of this type are satisfied, the two (or more) subsystems are said to be in *thermal equilibrium*. "X and Y meters" are indicated symbolically in Fig. 2.4, but it is by no means required that extensive and intensive variables be paired: all that is required is that there be enough independent variables to fix the state of the combined system. A simple example, defined by (Carathéodory, 1909), is one whose state can be completely specified by its extensive variables plus one intensive variable conjugate to one of these. In common parlance, diathermic walls allow transfer of heat and adiabatic walls do not, but the concept of "heat" will be defined only in the next chapter.

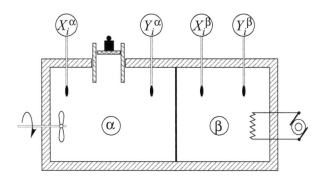

Figure 2.4: Adiabatic system with two subsystems, α and β separated by a diathermic partition. Upper "gadgets" are probes, ones on the sides are "mechanical work" engines (even though the one on the right appears to perform electrical work, as in the next figure).

Permeable partition

Adiabatic and diathermic walls are special cases of *impermeable* walls. But there can also exist *permeable* walls which allow matter to pass through. In fact, there also exist *semipermeable walls*, or partitions, which are ideally "transparent" to one species of matter, but impermeable to others. The state of subsystems will now depend also on the quantity and type of matter present, hence equations of state will have to be enlarged to include amounts of substances, in the form of extensive variables N_i denoting numbers of molecules or moles. Chemists would dearly love to have semipermeable membranes which would greatly facilitate the separation of different chemicals, but nature has not been kind, as it does not provide many sufficiently selective semipermeable membranes. In biology, though, semipermeable partitions constitute an essential element of life itself. Permeable partitions are by necessity diathermic as well.

With these considerations, we are now ready to define types of systems.

2.5   Types of Systems

Isolated system

An isolated system is simply one enclosed entirely by adiabatic walls and which can exchange no work with the outside world. An isolated system is shown schematically, and uninterestingly, in Fig. 2.5, leftmost panel. It can, of course exchange no "heat".

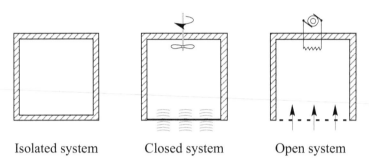

Isolated system   Closed system   Open system

Figure 2.5: Isolated, closed and open systems. An example of adiabatic system is shown in Fig. 2.4.

Adiabatic system

An adiabatic system is one which is entirely enclosed within adiabatic walls but which can *exchange work with the surroundings.* That case was illustrated in Fig. 2.4 for the combined system. Note that although the resistor of Fig. 2.4 is getting warm when current is applied, no "heat" is flowing to the combined system from the outside. Matter in system β may be getting warmer, but this heating up is caused by the work generated outside the adiabatic system. The composite system of this figure is represented as thermally insulated from the outside world, but it also includes internally a diathermic partition (heavy line), so that the individual subsystems α and β are no longer adiabatically enclosed; they are said to be *closed systems*, as defined below:

Closed system

A closed system in one which is entirely enclosed by impermeable walls, a finite portion of which must be of diathermic nature. The central panel of Fig. 2.5 shows such a system schematically. Finally:

Open system

An open system is one which is enclosed by walls, a finite portion of which must be permeable or semipermeable. The rightmost panel of Fig. 2.5 illustrates this case. The notion of types of systems, though idealized is of capital importance in thermodynamics as it allows one to work on some small portion of the world, i.e. on some "system", whose interactions with the surroundings are specified by its "envelope", defined by its partitions, which are the system's boundary conditions. Two subsystems were encountered in Fig. 2.4. There can of course be many more in practice. The notations α and β were used with a purpose, it is the one which will be used most frequently to denote *phases* in heterogeneous systems. In that sense, the two subsystems depicted in Fig. 2.4 can be considered as enclosing macroscopic phases which can exchange work and "heat" with each other, but not matter. Normally, unless constrained not to do so, perhaps because the required kinetics are too sluggish, phases do exchange matter; in that case, the diathermic partition shown in the figure must then be replaced by a (semi)permeable one, or in the notation of Fig. 2.5, it must be "full of holes". In another, seemingly absurd generalization, which will be encountered in the Part II portion of this book, this α and β notation of phases will be extended to sublattices in ordered systems; in that case, the "phases" are actually interpenetrating.[4]

2.6 Reservoirs, Probes

A *probe* is a hypothetical system so small that, when brought into contact with the system of interest, it does not sensibly alter the state of the latter. Typically, probes are used to measure intensive quantities: pressure, temperature, chemical potential... Usually, probes are devices which display changes in their extensive quantities in response to changes in intensive variables brought about the probe's coming to equilibrium with the system of interest. Examples of probes are: a load cell for stress, a thermometer for temperature, a "little cylinder" in a van't Hoff box for chemical potentials μ_i, where the index i denotes a particular atomic or molecular component in a mixture.

A *reservoir* is, conversely, a system so large that it can be used to alter the state of the system of interest with which it is brought into contact

---

[4]Failure to take different types of systems correctly into account has caused serious trouble as misguided philosophers and theologians have at times denied the possibility of order coming out of chaos, invoking the second law of thermodynamics erroneously to closed (open to heat exchanges; recall definition of types of systems, above) or open systems (open to exchange of matter).

without altering its own (the reservoir's) intensive variables. Typical examples are: (a) work reservoirs which can supply (or absorb) infinite amounts of work, (b) thermal reservoirs (or heat baths) which can supply or absorb infinite amounts of heat, (c) chemical reservoirs which can supply or absorb infinite amounts of substances, i.e, chemical species. Reservoir (b) may be used to maintain constant T, (c) may be used to maintain constant μ_i.

2.7   Simple Systems

In most of the early part of this text, *simple systems* are tacitly assumed. Callen (1985, Sec. 1.3) gives the following definition of such systems: these are *systems that are macroscopically homogeneous, isotropic, uncharged, and chemically inert, that are sufficiently large that surface effects can be neglected and are not acted upon by electric, magnetic or gravitational fields.* Later chapters will progressively remove some of these restrictions.

References

Callen, H. B. (1985), *Thermodynamics and an Introduction to Thermo-statistics*, John Wiley & Sons, New York.

Carathéodory, C. (1909), *Mathematische Annalen*, **67**, 355.

Jackson, J. D. (1962). *Classical Electrodynamics*, J. Wiley and Sons, NY.

Landau, L. D. and Lifshitz, I. M. (1995), *Theory of Elasticity*, Oxford, Butterworth-Heinemann, Boston, MA.

Landau, L. D. and Lifshitz, I. M. (1995), *Electrodynamics of Continuous Media*, Oxford, Butterworth-Heinemann, Boston, MA.

Voorhees, P. W. and Johnson, W. C. (2005), *The Thermodynamics of Elastically Stressed Crystals*, in *Solid State Physics*, Ehrenreich, H. and Spaepen, F, editors **59**, pp. 2-201, Elsevier, Academic Press, NY.

Chapter 3

Fundamental Laws

Systems containing a small number of particles require no special laws for their description beyond those of classical mechanics, or, for atomic-size particles, those of quantum mechanics. For collections of very large numbers of particles (10^{23}...), describing the behavior, in time, of all particles is neither feasible nor desirable: only behavior of averages matters. The averaging process introduces *laws*, which may be derived from classical (or quantum) mechanics by statistical processes, or they may be enunciated as *postulates*. The second approach is taken in Classical Thermodynamics. New variables appear, belonging to the collection of particles; these are *absolute temperature* and *entropy* (to be defined), which have no meaning at the individual particle level: there is no such thing as the temperature of one electron, or the entropy of one atom.

Three main laws are required: the *zeroth law* introduces the *empirical temperature* θ, the *first law* introduces the *internal energy* U, and the derived quantity heat Q, the *second law* introduces S and places the temperature on an *absolute scale* T. Another law, the *third law*, concerns the unattainability of the absolute zero of temperature.

It is hardly necessary to stress the importance of this chapter: it contains the description of the first three laws of classical thermodynamics, along with the main mathematical apparatus required to develop a quantitative treatment of applications thereof.

3.1 Zeroth Law

The zeroth law is eminently sensible:

If systems A and C are in thermal equilibrium, and systems B

23

and C are in thermal equilibrium (across), then A and B are themselves in mutual thermal equilibrium,

as shown schematically in Fig. 3.1. This statement can be formulated mathematically (see definition of diathermic partitions):

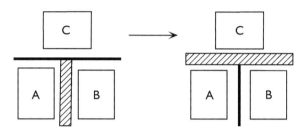

Figure 3.1: Illustration of the zeroth law; shaded bar represents adiabatic partition, full heavy line represents diathermic partition.

$$F^{\alpha\gamma}(X^{\alpha},Y^{\alpha},X^{\gamma},Y^{\gamma}) = 0, \quad F^{\beta\gamma}(X^{\beta},Y^{\beta},X^{\gamma},Y^{\gamma}) = 0\,.$$

If the F-functions are single-valued and well-behaved, one variable can be expressed as a function of the others:

$$Y^{\gamma} = f^{\alpha\gamma}(X^{\alpha},Y^{\alpha},X^{\gamma}), \quad Y^{\gamma} = f^{\beta\gamma}(X^{\beta},Y^{\beta},X^{\gamma})\,.$$

Hence we have the equality

$$f^{\alpha\gamma}(X^{\alpha},Y^{\alpha},X^{\gamma}) = f^{\beta\gamma}(X^{\beta},Y^{\beta},X^{\gamma}) \tag{3.1}$$

which, by the zeroth law, must be true for given α and β, regardless of the nature of γ. Therefore X^{γ} cannot appear, i.e. must cancel from both sides of (3.1), leaving

$$\theta^{\alpha}(X^{\alpha},Y^{\alpha}) = \theta^{\beta}(X^{\beta},Y^{\beta}) = \theta^{\gamma}(X^{\gamma},Y^{\gamma}) = \theta \tag{3.2}$$

where θ^{α} is the function $f^{\alpha\gamma}$, with X^{γ} taken out, then no longer dependent on the nature of γ. Thus, θ is the common value of all those functions, each pertaining to a single system. This common value θ of systems in thermal equilibrium across diathermic partitions is called the *empirical temperature*. Any system, such as C in Fig. 3.1, can then be used as a *thermometer*. A fluid enclosed in a container can thus measure empirical temperatures if properly calibrated as indicated in Fig. 3.2.

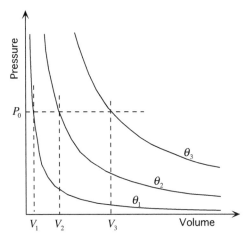

Figure 3.2: At constant pressure P_0, the three volumes V_1, V_2, V_3 can measure the empirical temperatures $\theta_1, \theta_2, \theta_3$ by means of this "thermometer".

3.2 First Law

Joule's statement of the first law is:

> *If an adiabatic system is caused to change from a prescribed initial to a prescribed final state, the work expended is the same for all paths connecting the two states.*

It follows that *adiabatic work*, W^a (work done on an adiabatic system), must be expressed as a difference of two numbers, one corresponding to the initial (1), one to the final state (2).

$$W^a_{1\to 2} = U_2 - U_1 = \Delta U. \tag{3.3}$$

In this equation, U can be given a name: the *internal energy*, which must be a *state function*, or "point function", since its value depends only on the state of the system at the thermodynamic equilibrium point considered, not on past history. The terminology *internal energy* indicates that a body or "system" can possess energy despite being globally at rest, its energy being contained for example in that of interactions between atoms or molecules or that of internal thermal motions. As such, internal energy thus differs from potential or kinetic energies.

Only *differences* of the internal energy matter; hence U implicitly contains an arbitrary additive constant. Thus ΔU can be measured empirically by performing work adiabatically, *i.e* by measuring it by standard techniques of classical mechanics. The internal energy is also additive in the sense that a composite $\alpha+\beta+\gamma+...$ has total energy $U^\alpha+U^\beta+U^\gamma+...$, which can be proved by calculating the work expended in bringing the systems α, β, γ,...individually from some initial reference states to the final states, then comparing to the total (adiabatic) work required to perform the same change on the systems taken collectively.

Now consider a closed but *non-adiabatic* system, and perform work on it so as to bring it from some prescribed initial (1) to prescribed final (2) state. It will be found, in general, that

$$W^a_{1\to2} \neq U_2 - U_1 = \Delta U \tag{3.4}$$

where, again, ΔU must be measured in an independent adiabatic work experiment. Give the difference resulting from Eq. (3.4),

$$Q = \Delta U - W \tag{3.5}$$

a name which *defines* Q as the *heat* transmitted through the diathermic partition of the (non-adiabatic) system. Equation (3.5) is written generally as

$$U = Q + W \tag{3.6}$$

(where Δ has been dropped since the additive constant of U is arbitrary). Equation (3.6) is known as the *mathematical expression of the first law*. Its differential form is

$$dU = dQ + dW \tag{3.7}$$

where dU is an exact differential (since U is a state function) whereas Q along with W must be inexact, since the difference $dU - dQ$ is inexact. Note the subtle but important difference between the straight symbol d (as in dU) and the slanted d (as in dW), the former being a true differential operator, the differential of a point (or state) function, the latter simply denoting a small quantity pertaining to a path-dependent quantity (non-state function). It follows that a system cannot be considered to contain "this much heat Q" or "that much work W", whereas it is meaningful to state that a system, in a given equilibrium state, contains a certain amount of total energy U. In a process, energy transfer can manifest itself as heat, or work, or a combination of both. As for internal energy and work, heat is also additive in composite systems.

3.3 Second Law

The second law is generally regarded (with the law of conservation of energy, or better, of mass-energy) as the most general, most profound, and one of the most mysterious of the laws of the natural sciences.[1] Two apparently trivial and one abstract expression of this law are given below. All three can be shown to be equivalent, along with may other valid formulations. All three are statements about the *impossibility* of performing certain processes.

The Kelvin–Planck statement reads: *No process is possible whose sole result is the absorption of heat from a reservoir and its conversion into work.*

The Clausius statement reads: *No process is possible whose sole result is the transfer of heat from a cooler to a warmer body.*

The Carathéodory statement reads: *Arbitrarily close to any given thermodynamic state there exist states which cannot be reached from it solely by means of adiabatic processes.*

Entropy (S) and (T) are *derived as consequences* of the second law, from which it follows that the second law cannot be formulated in terms of entropy or absolute temperature, as is sometimes done, incorrectly. These two new variables take case of such things as atomic and molecular motions in an average way, without making any attempts at describing these motions, which are generally unobservable. These variables will first appear as the differential expression $T \, dS$ which has the form of a "thermal work term" belonging to a (largely) unseen, but very real microscopic world. The aim of *classical* thermodynamics is to take into account this internal world without, however, attempting to describe it microscopically. It will come as no surprise that deriving the new differential expression $T \, dS$ is not a simple matter. It will be sketched here, whereas the full derivations will be given in Appendix B, following the work of Chandrasekhar in his classic (1957) monograph entitled *Introduction to the Theory of Stellar Structure*, the title of the book indicating the vast scope of classical thermodynamics, extending to the stars of the universe. It therefore follows that adding this magic term $T \, dS$ to a classical differential expression for mechanical, elastic, magnetic, even stellar, ... energies will, as if by magic, transform them into thermodynamic expressions, while keeping microscopic motions "under the table", so to speak, but quite present quantitatively.

The objective is thus to derive new state functions: the *entropy S* and the *absolute* temperature T. For that purpose, we consider *reversible* processes, and show that *reversible adiabats* (adiabatic processes performed

---

[1]The second law has at times been used both to prove and to disprove the existence of God; needless to say, both "proofs" are incorrect.

reversibly) are characterized by the constant value of a new state function, σ, later to be transformed into the entropy itself. The first law for a reversible (R) process can be written in differential form by combining Eqs. (3.7) and (2.6):

$$dU = dQ_R + \sum_i Y_i dX_i . \tag{3.8}$$

For simplicity, consider just two extensive variables X_1 and X_2 (for example volumes V_1 and V_2). The equation of state $F(X_1, Y_1, X_2, Y_2) = 0$ reduces the number of independent variables to just three. Since the internal energy is a state function, U can be used as a state variable, and thermodynamic processes can be plotted in (U, X_1, X_2) space, as in Fig. 3.3.   All possible

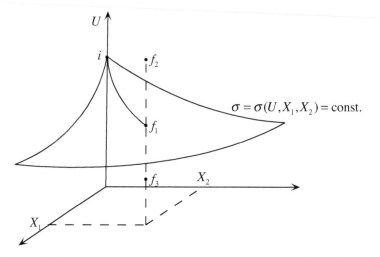

Figure 3.3: Reversible adiabatic surface $\sigma = $ constant in (U, X_1, X_2) space.

reversible adiabats have differential equation $dQ_R = 0$ or $dU = Y_1 dX_1 + Y_2 dX_2$ in the present case. Since X_1 and X_2 are independent variables, a reversible adiabat can be traced by incrementing X_1 and X_2 in an arbitrary way, so that as many reversible adiabats as desired can be constructed, all starting from an initial state i, say. One of these, $i \rightarrow f_1$ is shown in Fig. 3.3 with f_1 having the indicated X_1, X_2 coordinates. It will now be shown, from the second law, that point f_1 is the only point along the (X_1, X_2) vertical which can be reached adiabatically and reversibly from i. Suppose that another point, f_2, was in fact adiabatically accessible from i. For the cycle $i \rightarrow f_1 \rightarrow f_2 \rightarrow i$ (indeed for any cycle) $\Delta U = 0$. By the first law

$U = Q + W$ we have, for different legs of the cycle:

$$
\begin{array}{lll}
i \to f_1 & Q = 0 & \text{original adiabat} \\
f_1 \to f_2 & W = 0 & \Delta U = Q > 0 \\
f_2 \to i & Q = 0 & \text{presumed adiabat}
\end{array}
$$

as represented in Fig. 3.4. For the whole cycle then, $Q > 0$, hence $W < 0$.

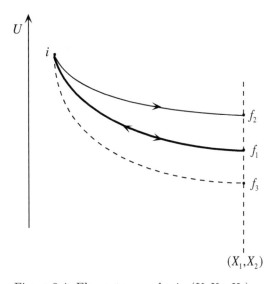

Figure 3.4: Elementary cycles in (U, X_1, X_2) space.

Then, for this cycle, positive Q is extracted along one path of the cycle and completely converted into positive work done by the system $(-W > 0)$, which is forbidden by the Kelvin–Planck statement of the second law. So, any final f_2 "above" f_1 (i.e. with $U_2 > U_1$) cannot exist for reversible paths initiating at i. What about a point "below" f_1, such as f_3? The cycle, $i \to f_1 \to f_3 \to i$, traced clockwise is now possible, since now $Q < 0$ and $W > 0$, i.e. positive work performed on the system is converted into heat given off by the system, which does not conflict with the second law. However, $i \to f_1$, is a *reversible* path, hence the reverse cycle $i \to f_3 \to f_1 \to i$ should be possible, but it is not (see argument concerning "f_2" cycle). It follows that point f_1 on (X_1, Y_1) is *unique* as a possible end-point of a reversible adiabat originating at i.

This conclusion is valid for any end point along the allowed $i \to f_1$ reversible adiabat, and also for any end point f infinitely close to the path. We also expect that any two points separated by δX_1 and δX_2, will be separated in energy by a small quantity of the same order, δU. Therefore, a continuous surface, locus of end points f_1 can be traced through any given initial point i. Furthermore, no two such reversible adiabatic surfaces can intersect; if they did, we could take i along the intersection, and construct an $i \to f_1 \to f_2 \to i$ cycle with f_1 on one surface, and f_2 on another, which is forbidden, as we have just seen. We can now characterize each surface by the equation $\sigma = \sigma(U, X_1, X_2) = $ constant, where σ is continuous and single valued in the indicated variables. Hence, "inversion" is possible:

$$U = U(\sigma, X_1, X_2) \tag{3.9}$$

which can describe internal energy as well as, say, $U(\theta, X_1, X_2)$, where θ is empirical temperature, as in Eq. (3.2).

Let us differentiate Eq. (3.9), or its more general form $U = U(\sigma, X_1...X_n)$:

$$dU = \frac{\partial U}{\partial \sigma} d\sigma + \sum_{i=1}^{n} \frac{\partial U}{\partial X_i} dX_i . \tag{3.10}$$

Now compare to Eq. (3.8). Since the process of total differentiation is *unique*, we must have

$$dQ_R = \frac{\partial U}{\partial \sigma} d\sigma \tag{3.11}$$

and

$$Y_i = \frac{\partial U}{\partial X_i} \qquad (i = 1, 2, \ldots n). \tag{3.12}$$

Equation (3.12) is an important one which equates an intensive variable with the partial derivative of the internal energy with respect to the corresponding extensive conjugate variable, holding all other extensive variables constant. Since Eq. (3.12) holds between state variables, it is universally true, independently of the processes considered, adiabatic or not, reversible or not. A reversible adiabatic process was used here only for the sake of the proof, but the result is path-independent.

For brevity, rewrite Eq. (3.11) as

$$dQ_R = \lambda d\sigma \tag{3.13}$$

where

$$\lambda = \left(\frac{\partial U}{\partial \sigma} \right)_{X_1, X_2} \tag{3.14}$$

written for just two extensive variables X_1 and X_2, as in Eq. (3.9). Equation (3.13) is of the standard "work" form, $Y dX$, but it is not yet the desired result. Thermal equilibrium between two systems is required in order to bring Eq. (3.13) in the required "irreducible" form

$$dQ_R = T dS \tag{3.15}$$

where

$$T = C \exp \left[\int \varphi(\theta) \, \mathrm{d}\theta \right] \tag{3.16}$$

and

$$\mathrm{d}S = \frac{1}{C} \Psi(\sigma) \, \mathrm{d}\sigma. \tag{3.17}$$

In these equations, and as shown in Appendix B, $\varphi(\theta)$ turns out to be a universal function of the empirical temperature θ, and Ψ is some well-behaved function of σ alone. Hence $\mathrm{d}S$ is an exact differential and S is a state function (or variable): the *entropy*. In Eqs. (3.16) and (3.17), C is an arbitrary constant chosen so as to make T positive. It follows that T cannot then become negative but can vanish in the limit of the integral of φ tending to $-\infty$. Since the definition of T contains no *additive* constant (C serves to fix the magnitude of the conventional *degree* of temperature), the singular point $T = 0$ is fixed for all thermodynamic systems, and is unique; it is the *absolute zero* of temperature, setting the zero of the Kelvin scale.

Since S is defined by the differential equation Eq. (3.17), it does contain an arbitrary additive constant. An additional law, the *third law*, is required to fix the zero of entropy. The additive nature of the entropy follows from that of heat Q. It is thus an extensive variable, and T is its conjugate intensive variable.

3.4 Fundamental Energy Differential

By combining Eqs. (3.8) and (3.15) we have the *combined (differential) expression of first and second laws*

$$\mathrm{d}U = T\mathrm{d}S + \sum_{i=1}^{I} Y_i \, \mathrm{d}X_i \tag{3.18}$$

or, more compactly but less "transparently"

$$\mathrm{d}U = \sum_{j=0}^{J} Y_j \, \mathrm{d}X_j \tag{3.19}$$

with

$$T = Y_0 \,, \qquad S = X_0 \,. \tag{3.20}$$

From Eqs. (3.18) or (3.19), the internal energy differential is seen to be expressed as a differential form in the extensive variables $S, X_1, X_2, \ldots X_J$. Hence, the internal energy itself is most basically expressed as

$$U = U(S, X_1, \ldots X_J) \,, \tag{3.21}$$

with entropy S and the X's being the internal energy's *natural variables*. An explicit form of the internal energy U, Eq. (3.21), is to be integrated from the differential form (3.18) by methods to be given in Sec. 3.5 and Appendix A. In formal thermodynamics this differential form dU is absolutely fundamental. Indeed, Gibbs (1875–77) begins his famous treatise with the differential form dU (and also dS), but with explicit notation for the variables X and Y. On comparing Eq. (3.18) to the total differential of $U = U(S, X_1, \ldots X_J)$ we find that

$$T = \left(\frac{\partial U}{\partial S}\right)_{X_j} \tag{3.22}$$

and

$$Y_j = \left(\frac{\partial U}{\partial X_j}\right)_{S, X_{j'}} \quad (j' \neq j), \tag{3.23}$$

Eq. (3.23) being merely Eq. (3.12) in "thermodynamic" notation.

The foregoing discussion indicates how a differential energy expression in mechanics, electromagnetism,... may be formally transformed into a differential *thermodynamic* energy expression simply by adding the term $T\,dS$. Of course, integration must then follow.

Examples

For a closed, single-phase fluid system we have the well-known expression

$$dU = T dS - P dV \tag{3.24}$$

with

$$-P = \left(\frac{\partial U}{\partial V}\right)_S \tag{3.25}$$

where P is the hydrostatic pressure (positive when directed inwards), and

$$T = \left(\frac{\partial U}{\partial S}\right)_V . \tag{3.26}$$

For a closed, single-phase, uniformly stressed solid:

$$dU = T dS + \sigma_{ij} d\epsilon_{ij} \tag{3.27}$$

with

$$\sigma_{ij} = \left(\frac{\partial U}{\partial \epsilon_{ij}}\right)_{S, \epsilon'_{ij}} \tag{3.28}$$

where, as in Eq. (2.2), σ_{ij} is a component of the stress tensor and ϵ_{ij} is the corresponding component of the strain tensor integrated over the volume over which the strain is regarded as uniform. In Eq. (3.28), and in what is to follow later, summation is implied over repeating Cartesian subscripts i, j=1,2,3.

Important caveat

Upon comparing Eqs. (3.7) and (3.18) it does *not* follow that TdS can be equated with dQ, nor $\sum Y_j \mathrm{d}X_j$ with dW, despite the statement in Sec. 2.3 that these were the general expression for heat and work. Actually, those expressions are correct, but, in *irreversible* processes, dQ and dW are heat and work increments exchanged with the outside world, and are measured externally to the system. Hence, in such processes, the outside intensive variables have values slightly different from those of the corresponding inside variables. For example, if friction is involved, the outside pressure P required to cause compression of a fluid must be slightly larger than the inside pressure P in order to overcome frictional forces. The measured work is the outside one $dW = -PdV$ (note the straight "d" for "differential"), different from the corresponding expression appearing in Eq. (3.24). Hence, equating of work terms in Eqs. (3.7) and (3.18) in general is not allowed for irreversible processes.

3.5   Mathematical Interlude

To many, the mathematics of classical thermodynamics appears to differ from that of familiar calculus: the notation appears to be cumbersome and such apparently non-standard concepts as "inexact differentials" are introduced. The purpose of the present section is to reassure the reader that no contradictions arise.

Types of differentials

Let x be an independent variable. We commonly write $y = f(x)$ which is interpreted as "y is the value of the function f evaluated at the point x". Note that y and f have different meanings, hence have different symbols. For reasons of economy, this convention is not respected in thermodynamics; it is considered more important to emphasize the physical meaning of a variable rather than strict mathematical orthodoxy. Thus, in thermodynamics we might write for example $V = V(P)$, meaning "the volume V is a function of pressure P".

The differential of x is written $\mathrm{d}x$ and is a numerically undefined quantity tending to zero in the limit. It follows that any expression containing differentials must be *integrated* before actual values can be assigned to the expression in question. Now consider the differential expression $f(x)\,\mathrm{d}x$. Let us give this evanescent quantity a name (or symbol): $dw = f(x)\,\mathrm{d}x$. Note carefully that the left hand side of this equation does not necessarily imply that dw stands for the differential of some function w, the compact symbol dw, taken as a whole, merely standing for some infinitesimal quantity. It is for that reason that the "d" symbol in the expression dw is written in mathematical convention as an *italic* letter, whereas it is written with a straight "d" in the true differential expression $\mathrm{d}x$, as mentioned previously. To obtain the (finite) increment of w between $x = a$ and $x = b$ say, one must of course integrate:

$$w_{a\to b} = \int_a^b f(x)\,\mathrm{d}x, \qquad (3.29)$$

where the integral clearly depends not only on the limits a and b, but also on the nature of the function f. We could also have written $dw = y\,\mathrm{d}x$ from which it is quite obvious that the required integration cannot be performed without specifying the function $y \equiv f(x)$. A differential form which cannot be integrated without specifying the integrand is called *inexact* (as in dw, above). Some differential expressions in two independent variables, x and y say, can be integrated directly, without specifying the path of integration. The next subsection and also Appendix A give the conditions under which that somewhat surprising property holds true. But first we must investigate how traditional thermodynamic notation handles partial differentials.

Consider two independent variables, x and y, and a function of these, $f(x, y)$ (not the same f as before), suitably differentiable, taking value u at point (x, y). The same u values might be obtained after performing a change of variables $y \to z$, for example, yielding

$$u = f(x, y) = g(x, z),$$

g being, in general, quite a different function from f. In both cases, the total differentials of these functions are written

$$\mathrm{d}u = \frac{\partial f}{\partial x}\,\mathrm{d}x + \frac{\partial f}{\partial y}\,\mathrm{d}y, \qquad (3.30a)$$

$$\mathrm{d}u = \frac{\partial g}{\partial x}\,\mathrm{d}x + \frac{\partial g}{\partial z}\,\mathrm{d}z. \qquad (3.30b)$$

Now suppose that the variables x, y and z represent the thermodynamic variables P (pressure), V (volume) and T (temperature), respectively. The total differentials of functions $U = U(P, V)$ and $U = U(P, T)$, in thermodynamic formulation corresponding to Eqs. (3.30a) and (3.30b), will then look like this:

$$\mathrm{d}U = \left(\frac{\partial U}{\partial P}\right)_V \mathrm{d}P + \left(\frac{\partial U}{\partial V}\right)_P \mathrm{d}V, \qquad (3.31a)$$

$$\mathrm{d}U = \left(\frac{\partial U}{\partial P}\right)_T \mathrm{d}P + \left(\frac{\partial U}{\partial T}\right)_P \mathrm{d}T. \qquad (3.31b)$$

Contrary to proper mathematical usage, the symbol U in Eqs. (3.31a) and (3.31b) here stands for three different things: the value of a function, a given function of P and V, and a generally different function of P and T. In mathematical convention, it is clear that

$$\frac{\partial f}{\partial x} \neq \frac{\partial g}{\partial x}$$

and that the variable to be held constant in the differentiation is the one appearing in the other derivative. In thermodynamic convention, which privileges physical meaning over mathematical symbolism, how is one to indicate that partials differ, as in

$$\left(\frac{\partial U}{\partial P}\right)_V \neq \left(\frac{\partial U}{\partial P}\right)_T ,$$

without indicating explicitly which variables are being held constant in the partial differentiation? Note that although the subscript V in the first partial indicates "at constant V", the derivative itself $(\partial U/\partial P)_V$ is still a function of both P and V, just as the derivative $\partial f/\partial x$ is a function of both x and y. The cumbersome notation of partial derivatives is the price one has to pay in thermodynamics for a more physically relevant notation.

Cyclic partial formula

Consider n variables x_i related by the equation

$$F(x_1, x_2, x_3, \ldots x_n) = 0 . \tag{3.32}$$

Whether or not the variables are independent, the total differential of F is given by

$$dF = F_{,1}\, dx_1 + F_{,2}\, dx_2 + \ldots + F_{,n}\, dx_n = 0 , \tag{3.33}$$

with obvious shorthand notation for the partial derivatives

$$F_{,i} \stackrel{\text{def}}{=} \frac{\partial F}{\partial x_i} , \qquad (i = 1, 2 \ldots n) . \tag{3.34}$$

To obtain the partial derivative of x_1 with respect to x_2 say, holding all other variables constant, divide both sides of Eq. (3.33) by dx_2 and set all other dx_i equal to 0 to obtain

$$\frac{\partial x_1}{\partial x_2} = -\frac{F_{,2}}{F_{,1}} . \tag{3.35}$$

Likewise we find

$$\frac{\partial x_2}{\partial x_3} = -\frac{F_{,3}}{F_{,2}}, \quad \ldots, \quad \frac{\partial x_n}{\partial x_1} = -\frac{F_{,1}}{F_{,n}}. \tag{3.36}$$

We now form the product of all partial derivatives given by the left hand sides of Eqs. (3.35) and (3.36), yielding

$$\frac{\partial x_1}{\partial x_2} \frac{\partial x_2}{\partial x_3} \cdots \frac{\partial x_n}{\partial x_1} = (-1)^n \tag{3.37}$$

since all the $F_{,i}$ cancel and the negative sign is taken n times.

In elementary texts, only three variables are generally considered and Eq. (3.37) reduces to

$$F(x, y, z) = 0,$$

with the analogue of Eq. (3.37) written simply as

$$\frac{\partial x}{\partial y} \frac{\partial y}{\partial z} \frac{\partial z}{\partial x} = -1.$$

Differential forms

The expression

$$dw = y \, dx \tag{3.38}$$

is an example of a differential form in two independent variables x and y; but the most general case of such a form is

$$dw = X(x, y) \, dx + Y(x, y) \, dy \tag{3.39}$$

where X and Y are suitably well-behaved functions of x and y.[2] Although differential (3.38) is surely not integrable if some path in the (x, y) coordinate plane is not given, the general form (3.39) may be integrable, in which case dw is indeed the total differential du of some function $u = f(x, y)$. A fundamental theorem gives the condition for integrability:

The necessary and sufficient condition for the integrability of the differential form (3.39) in two independent variables is that

$$\frac{\partial X}{\partial y} = \frac{\partial Y}{\partial x}. \tag{3.40}$$

---

[2]Exceptionally in this section, the upper-case variables X and Y do not represent the usual *extensive* and *intensive* quantities introduced previously.

(a) Necessary Condition

Assume integrability. Then Eq. (3.39) must represent the total differential of some function $f(x,y)$. Since differentiation with respect to independent variables is unique, by equating the coefficients of the independent differentials in Eqs. (3.39) and (3.30a) we obtain

$$X = \frac{\partial f}{\partial x}, \qquad Y = \frac{\partial f}{\partial y}.$$

By differentiating the first one of these equations by y and the second one by x, and since, for well-behaved functions, the order of differentiation may be permuted, we have

$$\frac{\partial X}{\partial y} = \frac{\partial^2 f}{\partial y \partial x} = \frac{\partial^2 f}{\partial x \partial y} = \frac{\partial Y}{\partial x}$$

thereby proving the necessary condition (3.40).

(b) Sufficient Condition

The proof of the sufficient condition is more involved, and is found in Appendix A, which also gives a standard formula for the integral of a differential form in two variables, to be used several times in this book. It is:

$$f(x.y) = \int_a^x X(\alpha, y)\, d\alpha + \int_b^y Y(a, \beta)\, d\beta \qquad (3.41)$$

in which α and β are dummy variables of integration, the lower limit of the first integral a is *an arbitrary but fixed constant* and the value of the second integral at the lower limit b accounts for the constant of integration. Note the placement of the arbitrary parameter a as an argument of Y in the second integral. We have thus proved that (in abbreviated notation)

$$X\, dx + Y\, dy \quad \text{is an exact differential}$$

and

$$\frac{\partial X}{\partial y} = \frac{\partial Y}{\partial x} \quad \text{is true}$$

are completely equivalent statements. The theorem and the integration formula (3.41) readily generalize to multivariable cases, as shown in Appendix A.

Formula (3.41), rarely appearing in Thermodynamic texts, is fundamental as it is the one required to integrate so-called Pfaffian differential forms. Moreover, the criterion of integrability given by Eq. (3.40) produces the well-known Maxwell equations. Also, the fact that a state function can be integrated over an arbitrary path in thermodynamic space is embodied in the arbitrariness of the constant a in the first integral of Eq. (3.41). As for the lower limit b, it indicates that the value of the unknown function at some fixed (experimental) point must be known in order to determine the function completely.

Formula (3.41) appears rather abstract, but is in fact of simpler application than the procedure given in most thermodynamics textbooks. To illustrate that statement, a simple example of its application is given in Appendix A, along with the more traditional method shown to be clearly more involved. Also illustrated in the example is the property, proven in the present theorem, that the integrated result is independent of the path of integration, which in turn demonstrates that integrated forms $f(x, y)$, under condition (3.39), are *point* functions, depending only on the point (x, y) itself, not on how that point was approached, *i.e.* not on the "history" of the process.[3]

3.6   Thermodynamic Integration

We are now in a position to make the following sweeping statement: *Doing thermodynamics amounts to integrating the fundamental differential form (3.18)*.

The integral can be written down following an extension of Eq. (3.41):

$$U(S, X_1...X_I) = \int_a^S T(s, X_1...X_I) \ \mathrm{d}s + \int_b^{X_1} Y_1(a, x_1, ...X_I) \ \mathrm{d}x_1 +$$

$$\cdots + \int^{X_I} Y_I(a, b, ...x_I) \ \mathrm{d}x_I + C. \quad (3.42)$$

The integration would be straightforward were it not for the fact that the functional form of the intensive variables:

$$T = T(S, X_1, ...X_I) \,, \qquad Y_i = Y_i(S, X_1, ...X_I) \,, \ \forall i \qquad (3.43)$$

is usually not known. If values T^o, Y_j^o are known in some reference (or initial) state, then a multivariable Taylor's expansion might be attempted to evaluate $T, Y_i \ldots$ at neighboring $S, X_i \ldots$ points:

$$Y_j = Y_j^o + \sum_{k=1}^\infty \frac{1}{k!} \left(\sum_{i=0}^I \delta X_i \frac{\partial}{\partial X_i} \right)^k Y_j^o. \qquad (3.44)$$

---

[3]The idea of providing the reader with a simple example was given to me by Dr. Melanie Lutz. The example given here is a somewhat more elaborate one than the one suggested by her. Consulting Appendix A is highly recommended.

Usually, the expansion is terminated after the first few terms, often the very first term:

$$Y_j = Y_j^o + \sum_{i=0}^{I} \delta X_i \left(\frac{\partial Y_j^o}{\partial X_i} \right), \tag{3.45}$$

thereby providing linear *constitutive equations*, often insufficiently accurate, however. Later we shall see the use of more extensive expansions, for instance in the Landau expansion (see Chapter 15).

Examples

For a simple fluid, the internal energy expressed as a function of its natural variables is

$$U = U(S, V). \tag{3.46}$$

In the linear approximation, we write

$$T = T^o + \left(\frac{\partial T}{\partial S} \right)^o \delta S + \left(\frac{\partial T}{\partial V} \right)^o \delta V \tag{3.47}$$

and

$$P = P^o + \left(\frac{\partial P}{\partial S} \right)^o \delta S + \left(\frac{\partial P}{\partial V} \right)^o \delta V. \tag{3.48}$$

It is customary to take the reciprocal of the partials $(\frac{\partial S}{\partial T})$, $(\frac{\partial V}{\partial T})$, $(\frac{\partial V}{\partial P})$,...which can be given physical meaning as *specific heat*. In general, specific heat is defined by:

$$C = \frac{dQ_R}{dT}. \tag{3.49}$$

More conveniently, *molar* specific heats can be defined by normalizing to unit quantity of one mole (or one cm^3, etc...). Lower-case symbols are then used for the densities. The *specific* heat at constant volume is then given by

$$c_V = \left(\frac{\partial q_R}{\partial T} \right)_V = \left(\frac{\partial u + P \partial v}{\partial T} \right)_V = \left(\frac{\partial u}{\partial T} \right)_V = T \left(\frac{\partial s}{\partial T} \right)_V. \tag{3.50}$$

Hence,

$$\left(\frac{\partial T}{\partial s} \right)^o = \frac{T^o}{c_V}. \tag{3.51}$$

c_V is often regarded as constant over a limited temperature range. More generally, thermodynamic tables give values of coefficients of polynomials representing specific heats for various substances. A more useful parameter is the *specific heat at constant pressure*:

$$c_P = \left(\frac{\partial q_R}{\partial T} \right)_P. \tag{3.52}$$

Another useful material parameter is the *expansivity*: at constant entropy it is

$$\beta_S = \frac{1}{V} \left(\frac{\partial V}{\partial T} \right)_S \tag{3.53}$$

so that

$$\left(\frac{\partial T}{\partial V} \right)^o = \frac{1}{V^o \beta_S}. \tag{3.54}$$

A more useful parameter is the *isobaric expansivity*

$$\beta_P = \frac{1}{V} \left(\frac{\partial V}{\partial T} \right)_P. \tag{3.55}$$

Pressure derivatives are evaluated with the help of other material parameters such as the adiabatic compressibility

$$\kappa_S = -\frac{1}{V} \left(\frac{\partial V}{\partial P} \right)_S \tag{3.56}$$

so that

$$\left(\frac{\partial P}{\partial V} \right)^o = -\frac{1}{V^o \kappa_S}. \tag{3.57}$$

A more useful parameter is the isothermal compressibility[4]

$$\kappa_T = -\frac{1}{V} \left(\frac{\partial V}{\partial P} \right)_T. \tag{3.58}$$

These and other parameters are measurable quantities, and can be used to derive simple, empirical expressions for thermodynamic functions. These expressions may then be inserted in integration formulas such as that given above, Eq. (3.41). It is very important to distinguish expansions such as those of Eqs. (3.47) and (3.48) and the definition of the differential of a function such as T, for example:

$$dT = \left(\frac{\partial T}{\partial S} \right) dS + \left(\frac{\partial T}{\partial V} \right) dV. \tag{3.59}$$

---

[4]This notation is not universally adopted: it's not everyone's kappa-T.

The former, Eqs. (3.47) and (3.48), denote linear approximations for finite difference $\delta T \equiv T - T^\circ$ and likewise for δP, with the *constant* partial derivatives evaluated *at* the fixed point S°, V°, whereas the latter, Eq. (3.59), is the *exact* differential expression for the function T (same for S), the associated partials being still functions of S and V, in this example.

Principle of Increase of Entropy

Since entropy S (or the auxiliary function σ) is a good state variable, it is possible to represent thermodynamic processes in $\{S, X_1, X_2\}$ space. Then Fig. 3.3 transforms into the graph below (Fig. 3.5). From considerations of

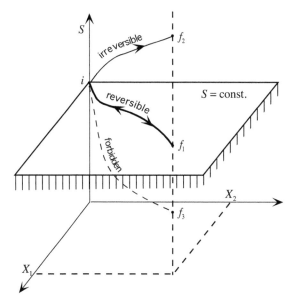

Figure 3.5: Constant entropy surface; reversible (f_1), irreversible (f_2) and forbidden (f_3) *adiabatic* paths.

Sec. 3.3, it is clear that any point f_1 can be reached from i by a reversible adiabatic path lying entirely in the horizontal plane $S = $ constant. Thus, the differential equation of reversible adiabats is $dS = 0$, or $dQ_R = 0$. Consider the cycle $i \to f_2 \to f_1 \to i$, constructed so that only $f_2 \to f_1$ is non-adiabatic. Then $i \to f_2$ must be an irreversible adiabat, according to the arguments set forth in Sec. 3.3. Over path $f_2 \to f_1$ we have $dX_1 = dX_2 = 0$ so that $dU = TdS$, and since $T > 0$, dU and dS must have the same sign. It was shown previously that, in Fig. 3.3, points above $\sigma = $ constant, such as f_2, were adiabatically accessible (by irreversible means) from i. Likewise here, in Fig. 3.5, points *above* $S = $ constant, such as f_2, are adiabatically accessible from i (by irreversible means). Points *below*, such as f_3, are adiabatically *inaccessible*. We thus arrive at the important conclusion:

> *In an adiabatic process (or system), the entropy cannot decrease. It increases in an irreversible adiabatic (no heat or material exchange) process, it stays constant in a reversible adiabatic process.*

In *non-adiabatic* systems, the entropy can perfectly well be made to *decrease*.

Clausius Theorem

Consider an arbitrary quasi-static path $a \rightarrow b$ and a small segment of it $i \rightarrow f$, constructed in such a way that one portion $(i \rightarrow j)$ be reversible and such that δQ_R along it be equal to the original δQ. Then, $j \rightarrow f$ must be an irreversible adiabatic path $(\delta Q_A = \delta Q_{j \rightarrow f} = 0)$. The entropy change δS must be the same for the original and the composite paths, since S is a state function.

Hence we have

$$\delta Q = \delta Q_R = T \delta S_R \tag{3.60}$$

(where δS_R means the change in S over path $i \rightarrow j$). Also

$$\delta S_A \geq 0 \tag{3.61}$$

(where δS_A means the change in S over the path $j \rightarrow f$), since the entropy cannot decrease in an adiabatic (A) process. Hence the total δS is given by

$$\delta S = \delta S_R + \delta S_A = \frac{\delta Q_R}{T} + (\text{positive}) = \frac{\delta Q}{T} + (\text{positive}) \tag{3.62}$$

that is

$$\delta S \geq \frac{\delta Q}{T}. \tag{3.63}$$

This last inequality if often referred to as the "mathematical statement of the second law." For the finite process $a \rightarrow b$, one can add up all infinitesimal contributions (*eq. 46*) to obtain in the limit, by integration

$$S_b - S_a \geq \int_a^b \frac{dQ}{T}. \tag{3.64}$$

For a closed path $(a = b)$,

$$\oint \frac{dQ}{T} \leq 0. \tag{3.65}$$

Such is the mathematical statement of the Clausius Theorem.

"Chemical Work"

Define a *van't Hoff reaction box* (Fig. 3.6) consisting of a major compartment containing a mixture of N substances and a set of small cylindrical compartments separated from the main chamber by appropriate semipermeable membranes in such a way that each small cylinder contains one and only one of the substances of the mixture. Semipermeable membranes which allow the passage of only one component of the mixture and none other are an idealization.[5] Allow the *system* (main chamber) to receive or give off heat dQ_R in a reversible manner, and let the main piston (P) and the auxiliary adiabatic pistons $(p_i, p_j, \ldots, \ i, j = 1, \ldots N)$ perform reversible work on the total (system and auxiliary) apparatus. We assume that the little cylinders have volumes v_i much smaller than V, so that introduction or removal of matter into or from the main chamber does not markedly affect the intensive variables of the system. The fundamental differential

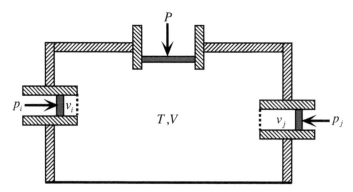

Figure 3.6: Van't Hoff reaction box.

form Eq. (3.18), featuring the symbolic "Work" term $\sum Y dX$ must now include a contribution from the action of the small cylinders which add or remove species N_i to the main chamber by means of the small cylinders, so that the amended fundamental form must now read

$$dU = TdS - PdV + \sum_{i=1}^{N} \mu_i \, dN_i + \sum Y dX \qquad (3.66)$$

where the usual PdV term has been shown explicitly. The "chemical term" $\mu_i \, dN_i$ is of that form also, of course, with the coefficient μ_i, called *chemical potential*, as the intensive variable conjugate to the corresponding extensive

---

[5]One example which is generally given is that of a pure Pd metal sheet which allows the passage of H, but of nothing else.

N_i. The chemical potentials have the important defining equation

$$\mu_i = \left(\frac{dU}{dN_i} \right)_{X_{j \neq i}} , \qquad (3.67)$$

or in words:

> *The chemical potential of a substance i, with respect to the in-*
> *ternal energy U of a mixture containing i, is the change in the*
> *potential U at equilibrium when a small amount of i is added*
> *(or removed) reversibly to the mixture, holding the extensive*
> *variables (other than N_i) constant.*

Other potentials may be considered, as will be shown later. There is more on chemical potentials in Sec. 4.7. Even though "substance" variables have now been added, the arguments presented in Section 2.3 are perfectly valid (the derivations do not depend on the number of variables or of subsystems), so that we may still identify dQ_R with TdS in Eq. (3.18), thus resulting, quite generally, in Eq. (3.66). This form is valid whether the differentials (of extensive quantities) are, or are not, independent. The possible interrelation of variables will play an important role, however, when equilibrium conditions are derived.

Chapter 4

Thermodynamic Equilibria

Recall the fundamental internal energy differential form Eq. (3.66) and the relations between conjugate variables:

$$T = \left(\frac{\partial U}{\partial S}\right)_{X,N_i} \tag{4.1a}$$

$$Y_i = \left(\frac{\partial U}{\partial X_j}\right)_{S,X,N_i} \tag{4.1b}$$

$$\mu_i = \left(\frac{\partial U}{\partial N_i}\right)_{S,X,N_{j\neq i}} \tag{4.1c}$$

where subscripts on the X variables outside the brackets have been left out for simplicity. Equilibrium can be examined by considering Eq. (3.66) and similar dU differential forms, but holding *extensive* variables constant is not practical. Hence it is preferable to change the set of "natural" variables and to define new thermodynamic potentials which are adapted to the particular boundary conditions that are pertinent to the problem at hand. This is generally done by means of *Legendre transformations* (extensively covered in Sec. 8.1) and below in Sec. 4.3. Other "big-name" results will be covered here: Euler integration, the Gibbs–Duhem relation, the Maxwell relations, the Gibbs phase rule.

No new fundamental principles are introduced in this chapter, but those "big-name" derivations are essential for a quantitative treatment of classical thermodynamics; but not only of thermodynamics, as Legendre transformations are those which relate the Lagrangian and Hamiltonian functions of classical mechanics, for example. Also, the Maxwell relations result from the integration conditions of differential forms in general. The name of Gibbs is abundantly featured, as it is he who was responsible for the elegant framework of classical thermodynamics, which is what this book is all about. In particular it was he who first introduced the famous *Gibbs phase rule* which lies at the heart of heterogeneous classical thermodynamics.

4.1 Euler Integration

For brevity, write $Y\,\mathrm{d}X$ and $\mu\,\mathrm{d}N$ collectively for the sums appearing in Eq. (3.66). Since the internal energy $U = U(S, X, N)$ must be a homogeneous function of degree one in its natural (extensive) variables, we have, by Euler's theorem (see Appendix C):

$$U = TS + YX + \mu N. \tag{4.2}$$

An alternate proof to the formal ones given in the Appendix may be obtained by giving proportional increments to S, X (V, for instance) and N so that only the *size* of the system is altered, but *not* its intensive variables. Then the differential form (3.66) can be integrated immediately, with T, $Y(-P)$, μ constant, with $S, X, N, \ldots$ increasing from 0 to the desired final values. In a sense, only a change of scale is taking place. Equation (4.2) then follows immediately. Of course, "Euler integration" applies only to open systems. If V were to decrease in a *closed* system, P and T would normally be altered.

4.2 Gibbs–Duhem

Equation (4.2) can be differentiated totally (whether or not its extensive variables are independent) to yield

$$\mathrm{d}U = T\,\mathrm{d}S + S\,\mathrm{d}T + Y\,\mathrm{d}X + X\,\mathrm{d}Y + \mu\,\mathrm{d}N + N\,\mathrm{d}\mu. \tag{4.3}$$

Comparing Eqs. (4.3) and (3.66) gives the *Gibbs–Duhem* (G-D) *equation*:

$$S\,\mathrm{d}T + X\,\mathrm{d}Y + N\,\mathrm{d}\mu = 0 \tag{4.4}$$

where now the intensive variables are the differentiated ones. This differential form, when integrated, yields a relationship between the intensive variables of a system *via* the corresponding extensive variables. Equation (4.4) is often used at constant T, P (or Y) so that, with

$$c_i = \frac{N_i}{N}, \tag{4.5}$$

where N is the total amount of components, the G-D takes the more familiar form

$$\sum_{i=1}^{n} c_i \, d\mu_i = 0 \,, \qquad (4.6)$$

very important in solution thermodynamics. In particular, its integral provides a relationship between chemical potentials, for instance allowing the determination in a binary system of μ_A from μ_B, or equivalently one activity from the other, as illustrated later in Fig. 7.4.

4.3 Legendre Transformation

With $Y_0 \equiv T, X_0 \equiv S$ (for generality), consider a thermodynamic potential U (not necessarily the internal energy) function of extensive variables $X_i (i = 0, 1, 2, \ldots n)$: $U = U(X_0, X_1, X_2, \ldots X_n)$. As before, whether the extensive variables X are independent or not, the total differential of the potential U is given by

$$dU = \frac{\partial U}{\partial X_0} dX_0 + \frac{\partial U}{\partial X_1} dX_1 + \frac{\partial U}{\partial X_2} dX_2 + \cdots + \frac{\partial U}{\partial X_n} dX_n \,, \qquad (4.7)$$

which may also be written as

$$dU = \sum_{i}^{n} Y_i \, dX_i \,, \qquad (4.8)$$

where, also as before, the Y_i variables are the intensive ones, conjugate to the extensive ones X_i, and numerically equal the partial derivatives of the potential with respect to the X_i, as per Eq. (3.12).

For certain applications, it may be advantageous to replace one or more X variables in a potential by Y variables, thus exchanging extensives by as many intensives as required. For example, experimentally it is generally simpler to perform an experiment at constant pressure P, instead of constant volume V. To accomplish this, for the pair of variables X_1 and Y_1 say, we define a new potential

$$\Phi = U - X_1 Y_1 \,. \qquad (4.9)$$

Differentiating this new function considered as a function of Y_1 and all the X_j, *except* X_1, we obtain the total differential of the new function:

$$d\Phi = \frac{\partial \Phi}{\partial X_0} dX_0 + \frac{\partial \Phi}{\partial Y_1} dY_1 + \frac{\partial \Phi}{\partial X_2} dX_2 + \cdots + \frac{\partial U}{\partial X_n} dX_n \,.$$

Also, from Eq. (4.9) we have

$$d\Phi = \frac{\partial U}{\partial X_0}dX_0 + \frac{\partial U}{\partial X_1}dX_1 + \frac{\partial U}{\partial X_2}dX_2 + \cdots - X_1 dY_1 - Y_1 dX_1 .$$

By comparing these latter two expressions we find, since the $Y_1 dX_1$ terms cancel, for the total differential of $\Phi = \Phi(X_0, Y_1, X_2, \ldots X_n)$:

$$d\Phi = Y_0\, dX_0 - X_1 dY_1 + Y_2 dX_2 + \ldots \tag{4.10}$$

with the coefficients of the differentials given by

$$Y_j = \frac{\partial \Phi}{\partial X_j} = \frac{\partial U}{\partial X_j} \quad (j \neq 1) \tag{4.11}$$

and

$$\frac{\partial \Phi}{\partial Y_1} = -X_1 . \tag{4.12}$$

It therefore follows that, after a Legendre transformation, the derivatives of the new function remain numerically equal to the corresponding ones of the old, except for that variable (X_1 in the present example) with respect to which the transformation is being performed. For that one, the new derivative, with respect to an intensive variable, is just equal to the negative of its conjugate extensive variable. Successive Legendre transformations may be performed so as to replace, as natural variables, as many of the extensive variables by conjugate intensive variables as desired. Note that the Legendre transformation is not limited mathematically to variables which are extensive and intensive; in classical and quantum mechanics they are position and momentum variables, but the X and Y variables must be each other's conjugates. In classical mechanics, for example, Legendre transformations allow the passage of the Lagrangian to the Hamiltonian formulations, and vice versa. Note also that Legendre transformations must be performed on functions which are convex with respect to their extensive variables, and the transformed function must be concave with respect to its new natural variables.[1]

Examples

Consider Eq. (4.2)

$$U = TS - PV + \mu N, \qquad U = U(S, V, N) \tag{4.13}$$

with the familiar $-PV$ replacing the symbolic YX and μN replacing the corresponding sum over components. We can form new thermodynamic functions by Legendre transformations as follows:

---

[1] I thank Dr. Wenhao Sun for reminding me of this important property.

(a) Enthalpy H

$$H = U + PV, \qquad H = H(S, P, N) \tag{4.14}$$

$$dH = TdS + VdP + \mu\,dN \tag{4.15}$$

$$T = \left(\frac{\partial H}{\partial S}\right)_{P,N}, \quad V = \left(\frac{\partial H}{\partial P}\right)_{S,N}, \quad \mu = \left(\frac{\partial H}{\partial N}\right)_{S,P}. \tag{4.16}$$

(b) Helmholtz Free Energy F

$$F = U - TS, \qquad F = F(T, V, N) \tag{4.17}$$

$$dF = -SdT - PdV + \mu\,dN \tag{4.18}$$

$$S = -\left(\frac{\partial F}{\partial T}\right)_{V,N}, \quad P = -\left(\frac{\partial F}{\partial V}\right)_{T,N}, \quad \mu = \left(\frac{\partial F}{\partial N}\right)_{T,V}. \tag{4.19}$$

(c) Gibbs (free) Energy G (or Free Enthalpy, see footnote on page 65)

$$G = G(T, P, N) \tag{4.20}$$

$$G = H - TS = F + PV = U - TS + PV = \sum \mu_i N_i \tag{4.21}$$

$$dG = -SdT + VdP + \mu\,dN \tag{4.22}$$

$$S = -\left(\frac{\partial G}{\partial T}\right)_{P,N}, \quad V = \left(\frac{\partial G}{\partial P}\right)_{T,N}, \quad \mu = \left(\frac{\partial G}{\partial T}\right)_{V,P}. \tag{4.23}$$

(d) Omega Potential (or Grand Potential)

$$\Omega = \Omega(T, V, \mu) \tag{4.24}$$

$$\Omega = U - TS - \mu N = F - G = -PV \tag{4.25}$$

$$d\Omega = -SdT - PdV - N\,d\mu \tag{4.26}$$

$$S = -\left(\frac{\partial \Omega}{\partial T}\right)_{V,\mu}, \quad P = -\left(\frac{\partial \Omega}{\partial V}\right)_{T,\mu}, \quad N = -\left(\frac{\partial \Omega}{\partial \mu}\right)_{T,V}. \tag{4.27}$$

Again, the index "i" has been left out from the "chemical" terms in some of them. Many more "potentials" may be derived by Legendre transformations involving other types of X and Y variables. One more Legendre transformation would be the one yielding the "potential" $\Phi = \Phi(T, P, \mu)$,[2] whose total differential is

$$d\Phi = -SdT + VdP - Nd\mu \equiv 0$$

---

[2]facetiously dubbed the "Omicron Potential" by John Cahn.

which is, by Eq. (4.4), simply the Gibbs–Duhem relation, and is identically zero.

In these examples it is seen that, starting from the internal energy *potential* (or function), which is defined as a function of extensive natural variables, successive Legendre transforms replace extensive variables by intensive ones, so that in particular, energy is replaced by *free* energies, Helmholtz or Gibbs, some of whose natural variables are intensive, such as T or P. The practical advantage of switching to free energies is that, experimentally it is far simpler to impose on a system conditions of constant temperature or pressure, than constant entropy (adiabatic system) or volume (isochoric system).

4.4   Maxwell Relations

From the integrability condition applied to the internal energy differential come such potentials as (formally)

$$\left(\frac{dY_i}{dX_j}\right)_{X_{i\neq j}} = \left(\frac{dY_j}{dX_i}\right)_{X_{j\neq i}} \quad (i \neq j = 0, 1, \dots). \tag{4.28}$$

After application of Legendre transformations we can also derive expressions such as, for example,

$$\left(\frac{dX_1}{dX_i}\right)_{Y_i, X_{i\neq j}} = \left(\frac{dY_i}{dY_1}\right)_{X}. \tag{4.29}$$

More explicitly, the integrability condition applied to the examples of the previous subsection produce the Maxwell that follow:

$$-\left(\frac{\partial S}{\partial P}\right)_{T,N} = \left(\frac{\partial V}{\partial T}\right)_{P,N} \tag{4.30a}$$

or

$$\left(\frac{\partial V}{\partial N_i}\right)_{P,N_{j\neq i}} = \left(\frac{\partial \mu_j}{\partial P}\right)_{T,N} \tag{4.30b}$$

or

$$\left(\frac{\partial \mu_i}{\partial N_j}\right)_{N_{i\neq j},T,P} = \left(\frac{\partial \mu_j}{\partial N_i}\right)_{N_{j\neq i},T,P} \tag{4.30c}$$

or

$$\left(\frac{\partial N_i}{\partial T}\right)_{V,\mu} = \left(\frac{\partial S}{\partial \mu_i}\right)_{T,V,N}. \tag{4.30d}$$

It therefore seems unnecessary to devise mnemonic methods to remember sets of Maxwell relations; knowledge of the "potentials" of the previous subsection suffices.

4.5 Evolution Criteria

Let a system be in contact (diathermic, semi-permeable...) with a very large reservoir, with the system itself plus reservoir being enclosed by rigid adiabatic walls. Let there also be a device capable of performing adiabatic work W^a on the system itself (see Fig. 4.1). Variables for the system are

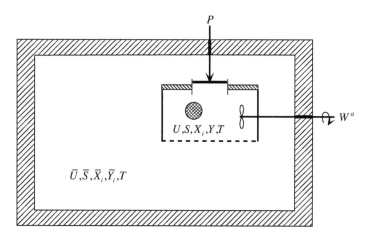

Figure 4.1: Thermodynamic system enclosed in a large reservoir (defined by overbarred variables). Circular feature represents a fluctuation, to be described in Sec. 17.1.

designated: U, S, X_i, T, Y_i; for the reservoir: $\bar{U}, \bar{S}, \bar{X}_i, \bar{T}, \bar{Y}_i$ $(i = 1, 2, \ldots)$. By the statement of the first law we have

$$\Delta U_{tot} = W^a = \Delta U + \Delta \bar{U}. \tag{4.31}$$

For the reservoir (or "surroundings") we have

$$d\bar{U}_i = \bar{T}d\bar{S} + \sum \bar{Y}_i d\bar{X}_i, \tag{4.32}$$

since in a large reservoir the intensive variables are to remain sensibly constant. Integration of Eq. (4.32) from an initial to a final state gives

$$\Delta \bar{U} = \bar{T} \Delta \bar{S} + \sum \bar{Y}_i \Delta \bar{X}_i \tag{4.33}$$

where $\Delta U = U_f - U_i$ etc. The boundary conditions on the total system (adiabatic) impose

$$\Delta X_i + \Delta \bar{X}_i = 0 \quad \text{(all i)}, \qquad \Delta S + \Delta \bar{S} \geq 0. \tag{4.34}$$

Hence, by Eqs. (4.31) and (4.33)

$$W^a = \Delta U_{tot} = \Delta U + \bar{T} \Delta \bar{S} + \sum \bar{Y}_i \Delta \bar{X}_i \tag{4.35}$$

and so, by Eqs. (4.35) and (4.34)

$$W^a \geq \Delta U - \bar{T} \Delta S - \sum \bar{Y}_i \Delta X_i. \tag{4.36}$$

The right-hand side of Eq. (4.36) is known as the "availability function." It is a hybrid quantity featuring variables of both the reservoir (intensive) and system (extensive). The physical interpretation of Eq. (4.36) is as follows: The minimum amount of work which is needed to bring a system from some prescribed initial to some prescribed final state is given by the availability function, under conditions of constant $\bar{T}$ and $\bar{Y}_i$ (all i) of the surroundings. This function has been used for everything from the efficiency of large power plants to the derivation of evolution criteria.

For infinitesimal changes, the left-hand side of Eq. (4.36) can be written

$$\bar{\zeta}\,\delta\xi = \delta U - \bar{T}\delta S - \sum \bar{Y}_i\,\delta X_i \tag{4.37}$$

where we have set
$$\delta W^a \geq \bar{\zeta}\,\mathrm{d}\xi, \tag{4.38}$$

a hybrid work differential containing the differential of the new state function ξ, called the *affinity* function by de Donder (1936). The "affinity" is used in studies of irreversible thermodynamics to measure the "degree of advancement of the thermodynamic process."

In a *spontaneous* process, Eq. (4.36) becomes, in the special case of hydrostatic open systems:

$$\Delta U - \bar{T}\Delta S + \bar{P}\Delta V - \sum \bar{\mu}_i\,\Delta N_i \leq 0. \tag{4.39}$$

Under certain types of boundary conditions or constraints it is possible to eliminate the reservoir variables altogether:

(a) Isolated Systems

$$\Delta U = \Delta V = \Delta N_i = 0. \tag{4.40}$$

Then, from Eq. (4.39), $\Delta S \geq 0$, is the correct evolution criterion, as was already known.

(b) Adiabatic Systems

We have, $\Delta N_i = 0$, hence for the reservoir,

$$\Delta \bar{U} = \bar{T}\Delta \bar{S} - \bar{P}\Delta \bar{V}, \tag{4.41}$$

which, according to the argument that heat absorbed by a reservoir is equivalent to a reversible process (with $\Delta \bar{S} = 0$), then gives $\Delta \bar{U} = -\bar{P}\Delta \bar{V}$ and then $-\Delta U = +\bar{P}\Delta V$, finally giving, from Eq. (4.36), $-\bar{T}\Delta S \leq 0$, hence

$$\Delta S \geq 0 \tag{4.42}$$

which is the correct evolution criterion for an adiabatic system, as was already known.

(c) Mechanical System (isentropic, isochoric, closed)

These are systems envisaged generally in classical mechanics:

$$\Delta S = \Delta V = \Delta N_i = 0. \tag{4.43}$$

Then, from Eq. (4.39),

$$\Delta U \leq 0 \tag{4.44}$$

is the correct criterion for a "non-thermal", mechanical system, as was previously known.

(d) Isentropic, Isobaric System

Same as the previous case but with pressure held constant rather than volume: $\Delta S = \Delta N_i = 0$. To maintain constant pressure, set P equal to that of the reservoir, $\bar{P}$. Then we have

$$\Delta U + \bar{P}\Delta V \leq 0. \tag{4.45}$$

The transition from the initial state to the final may be a violent one, for which it is quite impossible to monitor a unique system pressure. Nevertheless, we must have at the initial and final states $P_i = \bar{P}, P_f = \bar{P}$ so that Eq. (4.45) becomes

$$U_f - U_i + P_f V_f - P_i V_i \leq 0 \tag{4.46}$$

or

$$\Delta(U + PV) \equiv \Delta H \leq 0, \tag{4.47}$$

showing that in such cases, it is the enthalpy that decreases. It is important to note that the pressure need not be constant throughout the entire process, but it must be uniform throughout the system and equal to $\bar{P}$ in both initial and final equilibrium states.

(e) Closed, Isothermal, Isochoric System

We have $\Delta V = \Delta N_i = 0$, and, as in the previous subsection $T_i = T_f = \bar{T}$. Then (5.41) yields

$$\Delta U - \bar{T}\Delta S = \Delta(U - TS) \equiv \Delta F \leq 0. \tag{4.48}$$

In words: *In a closed isothermal system at constant volume, spontaneous processes must be accompanied by a decrease of the Helmholtz free energy.* Note the same restrictions on T as were stated for P in the previous subsection.

(f) Closed, Isothermal, Isobaric System

We have $\Delta N_i = 0$ (all i). Then following the same arguments as above:

$$\Delta U - \bar{T}\Delta S + \bar{P}\Delta V = \Delta(U - TS + PV) \equiv \Delta G \leq 0 \tag{4.49}$$

or, *in a closed isothermal isobaric system, spontaneous processes must be accompanied by a decrease of the Gibbs energy G.* For both temperature and pressure, initial and final values must be the same, and uniform throughout.

(g) Open, Isothermal, Isochoric System

$$\Delta V = 0, \quad T_i = T_f = \bar{T}, \quad \mu_{j,i} = \mu_{j,f} = \bar{\mu}_j \text{ (all } j) \tag{4.50}$$

so that

$$\Delta U - \bar{T}\Delta S - \sum \bar{\mu}_j \Delta N_j = \Delta(U - TS - \sum \mu_i \Delta N_i) \leq 0 \tag{4.51}$$

or

$$\Delta(F - G) \equiv \Delta\Omega \leq 0 \tag{4.52}$$

where Ω is the auxiliary thermodynamics potential defined in Eq. (4.25).

4.6 Phase Equilibrium

A system may be made up of subsystems, separated from one another by permeable (or semipermeable) membranes. A particular subsystem may be arbitrarily small, provided the laws of thermodynamics may be applied to it, that is, its surface- to- volume ratio must not be so large as to require the interface between subsystems to be taken into account explicitly.[3] Also, to qualify as a subsystem, it must be distinguishable "macroscopically" from its neighbors by, say, color, composition, crystal structure, symmetry (but not shape), etc..., and must be capable, at least in principle, of being separated mechanically from its neighbors. Such subsystems were called *phases* by Gibbs. We reserve the Greek superscripts $\alpha, \beta, \gamma, \ldots \varphi$ for these, with φ representing the surrounding "phase" or reservoir. Each phase must be homogeneous with respect to its macroscopic variables.

Consider a closed supersystem in a rigid adiabatic enclosure. Let the state of the system be varied slightly, reversibly, from its equilibrium. We have

$$\sum_{\alpha}^{\varphi} X_j^\alpha = X_j = \text{Const.} \qquad (\text{all } j = 0, 1, 2, \ldots) \tag{4.53}$$

and, by the additive nature of the extensive variables,

$$U = \sum_{\alpha}^{\varphi} U^\alpha = \text{Const.} \tag{4.54}$$

Under the given constraints, the proper equilibrium condition is $dU = 0$ and by the additive property of U,

$$dU = \sum_{\alpha}^{\varphi} dU^\alpha = \sum_{\alpha}^{\varphi} \sum_{j=0}^{J} Y_j^\alpha \, dX_j^\alpha = 0. \tag{4.55}$$

However, the variables X_j are not all independent because of the conditions (4.55). Hence Lagrange multipliers must be introduced as explained in Appendix D: Eq. (4.55) must be differentiated, multiplied by arbitrary parameters λ_j,

$$\lambda_j \sum dX_j^\alpha = 0 \tag{4.56}$$

then combined with Eq. (4.55) to yield

$$\sum_{\alpha}^{\varphi} \sum_{j=0}^{J} (Y_j^\alpha - \lambda_j) \, dX_j^\alpha = 0. \tag{4.57}$$

---

[3]The topic of thermodynamics of interfaces will be taken up in Ch. 13.

Now the coefficients of the differentials may be set equal to zero:

$$Y_j^\alpha = \lambda_j \qquad (\text{all } \alpha, \text{ all } j) \tag{4.58}$$

which is, in symbolic notation, the condition for phase equilibrium, written in terms of intensive variables.

In particular, for hydrostatic systems, Eqs. (4.58) give

$$j = 0: \quad T^\alpha = T^\beta = \cdots = T^\varphi$$
$$j = 1: \quad P^\alpha = P^\beta = \cdots = P^\varphi$$
$$\mu_1^\alpha = \mu_1^\beta = \cdots = \mu_1^\varphi$$
$$\cdots$$
$$\mu_n^\alpha = \mu_n^\beta = \cdots = \mu_n^\varphi.$$

Since there are $(n + 2)$ variables, n being the number of concentration variables (equal to the upper limit J on the summation signs) and φ phases, there are $(n+2)(\varphi - 1)$ separate s (number of equal signs). The number of variables is $(n + 2)\varphi$. There is one Gibbs–Duhem equation for each phase, so this adds φ phases. Hence the number of degrees of freedom f, equal to the number of variables minus the number of conditions, is

$$f = (n + 2)\varphi - \varphi - (n + 2)(\varphi - 1),$$

that is

$$f = n - \varphi + 2. \tag{4.59}$$

This is the famous Gibbs Phase Rule. As stated, it is far from being a universal one. As has been mentioned, each phase must be *homogeneous*, no gradient is tolerated, hence no position-dependent fields are allowed. Also, there can be no *coupling* between phase variables, as would exist if stress fields, electric fields..., were present, or if phase boundaries were sufficiently curved that capillarity effects were to become important. Also, there can be no chemical reactions between species; if there are, Eq. (4.59) must be generalized. For example, there may exist linear constraints between amounts of species originating from the requirement of conservation of lattice sites or conditions of local electroneutrality for ionic substances. Such "constrained equilibria" will be taken up in Chapters 11 and 12.

4.7   Chemical Potentials

Chemical potentials, first encountered in connection with the schematic van't Hoff reaction box (Fig. 3.6), are defined more generally, but equivalently, by:

$$\mu_i = \left(\frac{\partial U}{\partial N_i}\right)_{S,V,N'} = \left(\frac{\partial H}{\partial N_i}\right)_{S,P,N'} = \left(\frac{\partial F}{\partial N_i}\right)_{T,V,N'}$$

$$= \left(\frac{\partial G}{\partial N_i}\right)_{T,P,N'} \tag{4.60}$$

where N' means all N_j with $j \neq i$. Any one of Eqs. (4.60) can serve as a defining expression, the last one being the most useful in practice.

Consider two adjacent phases, α and β, exchanging species i only. Then, at constant T and P, any spontaneous exchange of i must be such that $\delta G_{tot} \leq 0$. In the present case, all $\delta N_j^\gamma = 0$ for (a) if $\gamma \neq \alpha, \beta$, (b) all γ if $i \neq j$: then

$$\delta G = \delta G^\alpha + \delta G^\beta \leq 0 \tag{4.61}$$

or

$$\mu_i^\alpha \delta N_i^\alpha + \mu_i^\beta \delta N_i^\beta = (\mu_i^\alpha - \mu_i^\beta)\delta N_i^\alpha \leq 0 \tag{4.62}$$

since $\delta N_i^\alpha = -\delta N_i^\beta$ by the conservation of species i. Thus, by (5.57), if

$$\mu_i^\alpha > \mu_i^\beta \rightarrow \delta N_i^\alpha < 0, \tag{4.63}$$

phase β becomes enriched in i. If

$$\mu_i^\alpha < \mu_i^\beta \rightarrow \delta N_i^\alpha > 0, \tag{4.64}$$

phase α becomes enriched in i. In other words, matter "flows down the chemical gradient."

More generally, *partial molar quantities* (PMQ) may be defined as follows: consider a thermodynamic potential Z (standing for U, H, F, G, V,...). By an appropriate change of variables, rewrite Z as a function of variables T, P, and mole numbers N_i, and form the derivatives

$$z_{,i} = \left(\frac{\partial Z}{\partial N_i}\right)_{T,P,N'} \qquad (N' \neq N_i). \tag{4.65}$$

By definition, these derivatives are PMQ's. According to this definition, chemical potentials are "partial molar Gibbs energies."

$$\mu_i \stackrel{\text{def}}{=} g_{,i} = \left(\frac{\partial G}{\partial N_i}\right)_{T,P,N'}. \tag{4.66}$$

To simplify the notation, we now adopt the rule (often used in classical mechanics and other fields) that a subscripted index preceded by a comma signifies taking a partial derivative with respect to i. An operational definition of PMQ's can now be given based on Eq. (4.66): For instance,

> the chemical potential μ_i is the relative change in Gibbs free en-
> ergy when a small amount of substance i is added (or removed)
> reversibly at constant T and P.

For a *pure substance*, rather than a mixture, PMQ's simply reduce to normalized thermodynamic potentials, or "densities," or better yet: molar
quantities

$$z_{,i} \equiv z_i \;\to\; z \tag{4.67}$$

where z is simply $\frac{Z}{N}$, N being the total number of moles in the system.
   In the general case, Euler integration gives

$$Z = \sum_i^n z_{,i} N_i \,, \tag{4.68}$$

where n is the total number of chemical constituents present; this being a
very compact and useful expression.

References

de Donder, Th. (1936) *Thermodynamic Theory of Affinity: A Book of
   Principles*, Oxford University Press.

Chapter 5

Ideal Gases

Historically, gases were the first objects to be investigated thermodynamically, being uniform and homogeneous. Also, gases were of considerable importance to heat engines, the cornerstone of the industrial revolution. Ideal gases were of course the simplest to study as being composed of freely and randomly moving non-interacting (except elastically) atoms and molecules, as soon as it was recognized that matter was composed of these units. The present chapter is devoted to such gases, as an introduction to more complex material.

5.1 Definition

An ideal gas is a hypothetical low-density fluid defined by the equation

$$PV = NRT \quad \text{or, according to Eq. (4.67),} \tag{5.1a}$$
$$Pv = RT \tag{5.1b}$$

where R is the universal *gas constant*,[1] v being the molar volume V/N, P the pressure, and T the *absolute* temperature. Equations (5.1) are the *equations of state* of an ideal gas which can be represented by a surface in (P, V, T) space, as shown in Fig. 5.1, this surface being the locus of PV hyperbolae. For a complete definition, an additional statement is required: That the internal energy of an ideal gas be dependent on temperature alone. Thus:

$$U = U(T) = U(PV) \tag{5.2}$$

---

[1] 8.314 4621 J/mol/degree K or 1.985 8775 cal/mol/degree K.

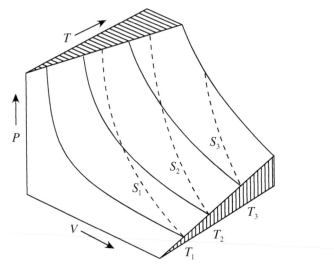

Figure 5.1: Equation of state of an ideal gas plotted in (P, V, T) space (from Zemansky, 1968).

i.e. U depends not on P and V individually, but on the product PV. More basically, the internal energy of an ideal gas can be expressed explicitly as a function of its natural variables S and V.

5.2   Entropy of an Ideal Gas

To calculate the entropy S, integrate $\frac{dQ_R}{T}$. Start from the first law in differential form

$$dQ_R = dU + P\,dV. \tag{5.3}$$

Total differentiation of the energy U in Eq. (5.2) gives

$$dU = U' \cdot (P\,dV + V\,dP) \tag{5.4}$$

where U' is the derivative of U with respect to its argument PV. Then, by Eq. (5.4)

$$dQ_R = (1 + U')P\,dV + U'V\,dP. \tag{5.5}$$

This differential, after division by PV, turns out to be integrable since both derivatives $\partial[(1 + U')/V]/\partial P$ and $\partial(U'/P)/\partial V$ give the same result U'' (second derivative with respect to the product PV), thereby satisfying the integrability criterion Eq. (3.40) (see also Appendix A).

Also,

$$\left(\frac{\partial Q_R}{\partial T}\right)_V = \left(\frac{\partial U}{\partial T}\right)_V + P\left(\frac{\partial V}{\partial T}\right)_V = \frac{\mathrm{d}U}{\mathrm{d}T} \tag{5.6}$$

since U does not depend on V individually. Thus, from the definition of specific heat at constant V (3.49) and (3.50), we have

$$\mathrm{d}U = C_V \mathrm{d}T , \tag{5.7}$$

where C_V must depend on T alone. The entropy differential is then, by Eq. (5.1)

$$\mathrm{d}S = C_V \frac{\mathrm{d}T}{T} + \frac{P}{T}\mathrm{d}V = \frac{C_V}{T}\mathrm{d}T + N\frac{R}{V}\mathrm{d}V . \tag{5.8}$$

Finally, using the standard "integrated differential form" formula (3.41), with $a = 1$, and S_0 as constant of integration (later to be subject to the third law) we find, for the entropy of an ideal gas,

$$S = \int_1^T \frac{C_V(t)}{t}\,\mathrm{d}t + NR\ln V + S_0 ,$$

leading to, if the heat capacity is temperature-independent,

$$S = C_V \ln T + NR\ln V + S_0 . \tag{5.9}$$

Also, from Eq. (5.2),

$$\mathrm{d}Q_R = \mathrm{d}(U + PV) - V\mathrm{d}P = \mathrm{d}U + NR\mathrm{d}T - V\mathrm{d}P$$

so that, according to the definition of the heat capacity at constant pressure,

$$\left(\frac{\partial Q_R}{\partial T}\right)_P \equiv C_P = \frac{\mathrm{d}U}{\mathrm{d}T} + NR.$$

Thus, by Eq. (5.7)

$$c_P - c_V = R. \tag{5.10}$$

Then, since from Eq. (5.8) and $T\mathrm{d}S = C_V\,\mathrm{d}T + P\,\mathrm{d}V$, and also $\mathrm{d}(PV) = P\mathrm{d}V + V\mathrm{d}P$, there follows

$$T\,\mathrm{d}S = C_P\,\mathrm{d}T - V\,\mathrm{d}P, \tag{5.11}$$

hence, by integrating with C_P constant, after division by T, we find

$$S = C_P \ln T - NR\ln P + S_o . \tag{5.12}$$

As above, for this equation we have set $a = 1$ as the lower limit of the integral, and taking care of the constant of integration by setting the lower limit of the integral equal to P_0. This is necessary since the argument of the logarithm must a dimensionless quantity, P_0, say. We summarize the two entropy equations (5.9) and (5.12) thus:

$$S = C_V \ln T + NR \ln V/V_0 \tag{5.13a}$$

and

$$S = C_P \ln T - NR \ln P/P_0, \tag{5.13b}$$

where V_0 and P_0 are the initial volume and pressure, respectively.

Example

Free expansion of an ideal gas. This process is assumed to take place adiabatically, with increasing volume, say from V_0 to V_1, performing no work. We then have $Q = 0$, $W = 0$, hence $\Delta U = 0$, hence T=constant. Therefore, by Eq. (5.13a):

$$\Delta S = NR \ln \left(\frac{V_1}{V_0} \right) > 0. \tag{5.14}$$

As expected, the entropy increases with the ratio of the final to initial volumes.

5.3   Reversible Adiabats

For a reversible adiabat, $dS = 0$, so that differentiating Eqs. (5.13) lead to

$$c_V \, d\ln T = R \, d\ln V, \tag{5.15a}$$
$$c_P \, d\ln T = -R \, d\ln P. \tag{5.15b}$$

Dividing one equation by the other we have

$$\frac{c_P}{c_V} = -\frac{d\ln P}{d\ln V}. \tag{5.16}$$

Define the ratio $\gamma = \frac{c_P}{c_V}$ and assume that it is a constant (a weaker assumption than that of requiring that both c_P and c_V be individually temperature-independent). Integration of the previous equation yields

$$PV^\gamma = Const. \tag{5.17}$$

which is the equation of reversible adiabats for an ideal gas. It follows from the previous equation that $\gamma > 1$. "Reversible isotherms" are given, of course, by $PV = $ const. Isotherms (equilateral hyperbolae) and adiabats are sketched on the state function surface (Fig. 5.1)

5.4 Chemical Potential of an Ideal Gas

The chemical potential of an ideal gas, μ, can be obtained by integrating the Gibbs energy differential form (in a closed system, since a single ideal gas is considered, with $N = 1$)

$$dg = v \, dP - s \, dT \tag{5.18}$$

so that, using the integration formula (3.41) we obtain

$$\mu = g(P, T) = \int_{p^o}^{P} v(p, T) \, dp - \int^{T} s(p^o, t) \, dt + C. \tag{5.19}$$

Now let the last two terms of Eq. (5.19) define the chemical potential in the *standard state* of $p^o = 1$, a fixed but arbitrary pressure:

$$\mu^o = -\int^{T} s(p^o, t) \, dt + C, \tag{5.20}$$

C being a constant of integration to be fixed by the initial conditions. Finally, after substituting Eqs. (5.1b) and (5.20)) into Eq. (5.19) we have

$$\mu = \mu^o + RT \ln \frac{P}{p^o}. \tag{5.21}$$

This formula is a fundamental one for "solution thermodynamics." It is important to recognize that μ^o, from Eq. (5.20), depends on temperature alone.

Chapter 6

Single-Component Equilibrium

This chapter is concerned with phase equilibria in pure (*i.e.* one-component) systems. With graphical help, it is shown that equations of state can, in principle, be obtained from the knowledge of Gibbs free energy functions alone. The practical difficulty lies in determining the free energy itself.[1]

6.1 Phase Transformations

A *real* single-component system, unlike an idealized gas, can undergo transitions to condensed phases, *i.e.* to liquids and solids, because of molecular interactions. During a phase transformation, T and P are usually held constant, so the "potential" of interest is the normalized Gibbs energy: $g(P, T)$, or chemical potential μ, which, for single-component substances is the same thing.

By the additive nature of G we have, for two phases α and β in presence,

$$G = G^\alpha + G^\beta = g^\alpha N^\alpha + g^\beta N^\beta . \tag{6.1}$$

For a small transfer of material (δN) at constant P and T

$$\delta G = g^\alpha \delta N^\alpha + g^\beta \delta N^\beta . \tag{6.2}$$

---

[1]There is still no complete agreement concerning the name of the so-called Gibbs function: most common for the function G is "Gibbs free energy", but increasingly the shorter term "Gibbs energy" is proposed; the French often use the very descriptive term "free enthalpy" (because that's what it is).

For this closed system $\delta N^\beta = -\delta N^\alpha$ we have

$$\delta G = (g^\alpha - g^\beta)\delta N^\alpha. \tag{6.3}$$

Since, in a spontaneous process, $\Delta G \leq 0$ at constant T, P, the difference of chemical potentials $\Delta\mu = g^\alpha - g^\beta$ and δN^α must have opposite signs, as discussed in Sec. 4.5. So, for $g^\alpha > g^\beta \rightarrow \delta N^\alpha < 0$, hence the α phase must transfer material to the β phase which has lower free energy, and conversely. At equilibrium, $\Delta\mu = 0$, *i.e.* $g^\alpha = g^\beta$ and a small virtual change δN^α does not modify the thermodynamic potential, to first order.

The free energy curves for α and β can in principle be extended beyond their domains of stability (dashed lines), and the transfer δN^α may be represented graphically (see Fig. 6.1). The intersection defines the thermodynamic equilibrium transition temperature T_0. These curves are seen to intersect cleanly so that derivatives on the right and on the left of T_0 are different. This characterizes a *first-order transition*, the word *transition* being reserved to transformations occurring at equilibrium. The word *transformation* will then denote any change of phase, at or away from equilibrium, reversibly or not.

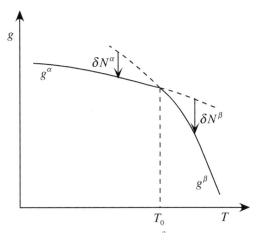

Figure 6.1: Gibbs energy curves g^α and g^β at a first-order transition T_o.

6.2 Derivatives and Discontinuities

For each phase we have the basic differential form (5.18),

$$\mathrm{d}g = -s\,\mathrm{d}T + v\,\mathrm{d}P, \tag{6.4}$$

hence

$$-s = \left(\frac{\partial g}{\partial T}\right)_P, \qquad v = \left(\frac{\partial g}{\partial P}\right)_T \qquad (6.5)$$

and, from results of Sec. 3.6,

$$c_P = T\left(\frac{\partial s}{\partial T}\right)_P = -T\left(\frac{\partial^2 g}{\partial T^2}\right)_P \qquad (6.6a)$$

$$\kappa_T = -\frac{1}{v}\left(\frac{\partial v}{\partial P}\right)_T = -\frac{1}{v}\left(\frac{\partial^2 g}{\partial P^2}\right)_T. \qquad (6.6b)$$

At equilibrium two-phase temperature–pressure points (T_o, P_o), the actual g-curve cannot be differentiated. Instead, one must consider left and right derivatives and their differences. Hence at (T_o, P_o) we have:

$$\left(\frac{\partial g^\beta}{\partial T}\right)_P - \left(\frac{\partial g^\alpha}{\partial T}\right)_P = -s^\beta + s^\alpha = -\Delta s \qquad (6.7a)$$

$$\left(\frac{\partial g^\beta}{\partial P}\right)_T - \left(\frac{\partial g^\alpha}{\partial P}\right)_T = v^\beta - v^\alpha = \Delta v. \qquad (6.7b)$$

Also, for each phase,

$$g^\alpha = h^\alpha - T^\alpha s^\alpha; \qquad g^\beta = h^\beta - T^\beta s^\beta \qquad (6.8)$$

so that, for $T^\alpha = T^\beta = T_o$, $g^\alpha = g^\beta$ (phase equilibrium), we have

$$\Delta h = T_o \Delta s = L \qquad (6.9)$$

which defines the latent heat L. Heat capacities and compressibilities become delta functions at first-order transitions.[2] Since, away from the transition itself, c_P and κ_T must be positive quantities, then the slopes of s and v and curvatures of g are determined in sign. Plots in Fig. 6.2 have been constructed accordingly.

There are also transitions for which the phase change is more gradual, for instance those for which the free energy is still continuous, but for which the first derivatives are continuous as well, though presenting a singularity with an inflection point with vertical tangent at the transition point, as shown in Fig. 6.3(a). It follows that second derivatives of the free energies also diverge at the transition, but not in delta function manner. That behavior is sketched at Fig. 6.3(b), characteristic of a *second-order phase transition* The distinction between first- and second-order transition will be discussed further in Sec. 15.3.

---

[2]In mathematics, the derivative of a Heavyside function, or step function, is a Dirac delta function, the derivative of a discontinuous function.

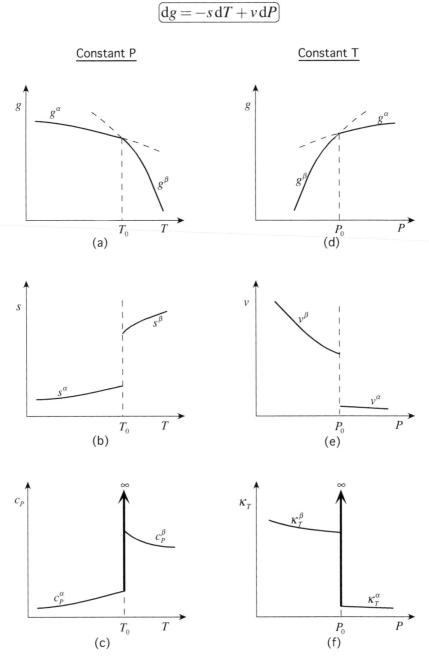

Figure 6.2: Successive derivatives of the Gibbs free energy at constant pressure (P, left column) and at constant temperature (T, right column).

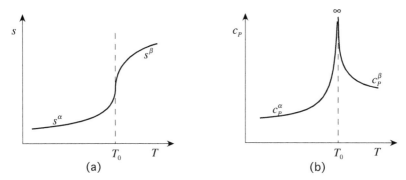

Figure 6.3: First (a) and second (b) derivatives with respect to temperature (at constant pressure) of the free energy for the case of a second-order transition. These two panels correspond respectively to Figs. 6.2 (b) and (c) of the first-order case. Similar behavior is to be found for the volume and compressibility, respectively.

6.3 P–V–T Equation of State

The equation of state of a condensable substance can be plotted on a P–V–T diagram as was done for an ideal gas in Fig. 5.1. Now, of course, the equation of state is represented by a more complicated surface, as shown in the perspective drawing of Fig. 6.4. It is seen that what was the smooth hyperbolic surface of an ideal gas equation of state (PV = constant) is now replaced by a surface containing portions of ruled surfaces.[3] Inside those regions the parallel lines, called "tie lines", link at their extremities the (P, V) coordinates which are those of the two phases in thermodynamic equilibrium, such as gas-liquid, liquid-solid or solid-gas. The tie lines must be parallel to the V axis (extensive variable), i.e. normal to the $P - T$ plane intensive variables).

The plot of the $f(P, V, T) = 0$ surface can be regarded as the loci of isotherms such as that shown on Fig. 6.2 (e) for various temperatures. The vertical portions of such isotherms become the horizontal tie lines of the state function of Fig. 6.4. Take β to represent the gas phase (high-T) and α the liquid (or solid) phase (low-T: condensed phase). For high enough temperatures, isotherms are continuous; at lower temperatures, they are discontinuous (Fig. 6.2). Between these two regimes, there exists a *critical isotherm* (T_c) which exhibits an inflection point with vertical tangent. At the inflection point, the phase transition is of *second-order type*. Figure

---

[3]In geometry, a surface is *ruled* if through every point there exists a straight line that lies in it; in the present case, these are parallel lines.

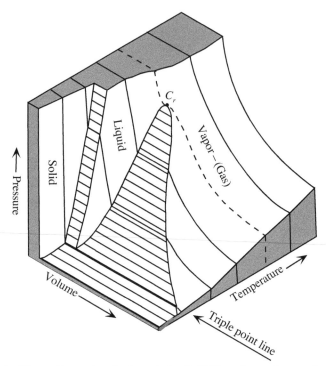

Figure 6.4: Three-dimensional plot of the P-V-T surface of a condensable gas showing two-phase tie lines (from Zemansky, 1968).

6.4 also shows regions of stability of solid phases and attendant two-phase regions. Indeed the term "condensed phase" can mean either liquid or solid (generally crystalline solids). Moreover, solids can take on various crystalline structures for the same pure element. For example, boron, sulphur and plutonium possess a large number of stable crystalline phases for various ranges of temperatures and pressures.[4]

For convenience, state function surfaces are generally plotted in the form of two-dimensional sections, as in Fig. 6.5 which indicate liquid, gas and solid regions respectively, here with symbols α, β, γ respectively. Figure 6.5(b) shows typical isotherms on a PV diagram. At very high temperatures, isotherms are practically the equilateral hyperbolae of ideal-gas type. At lower T, the isotherms distort until T_c is reached, for which a horizontal tangent is expected at the top of the $(\alpha + \beta)$ two-phase region. For $T < T_c$,

---

[4]One well-known example is that of iron which exists both in fcc and bcc forms, with important consequences for the technology of steels. An even better-known example is that of the two forms of carbon: graphite and diamond.

isotherms are seen to pass through the $(\alpha + \gamma)$ two-phase region. At high enough pressures, the gas phase must indeed condense, regardless of temperature. At still lower temperatures, the solid *sublimes*,[5] transforms from

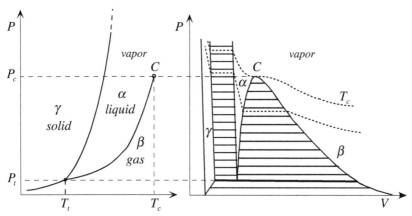

Figure 6.5: PT- (left) and PV-projection (right) diagrams.

the solid phase directly to the gas phase without passing through a liquid state. The three two-phase regions come together at a unique pressure P_{tr} and temperature T_{tr}, which are those of the so-called *triple point*. At the other remarkable point on this graph C, the *critical point*, the transition liquid-vapor is said to be of *second-order*, meaning that the distinction between the two phases (here liquid and gas) vanishes at that point (T_c, P_c). The corresponding P-T diagram is shown in Fig. 6.5 (left). The two-phase $(\alpha + \beta)$ line "dies in mid-air" at C. It is thus possible to go directly from a liquid to a gas phase with no discontinuity at pressures which are high enough. The phase which can be thought of as either a viscous gas or a dilute liquid is generally called "vapor". There is no corresponding critical point at the liquid-to-solid transition. The solid-liquid coexistence curve on the P-T section has positive slope, the usual case in practice. For water, however, and a few other elements such as bismuth, the coexistence curve has negative slope. In that case the solid tends to liquefy when pressure is applied.[6]

---

[5]The author wrote this chapter just two days after NASA announced the discovery of liquid water on the planet Mars (September 28, 2015). The press release also mentioned that, at the short Martian spring, the solid water (containing large concentrations of salts) actually sublimes, the surface temperature and pressure of Mars being extremely low, in accordance with the schematic diagram of Fig. 6.5

[6]It is said that such is the reason that skaters are able to skate on ice, but not on polished steel, for example.

The "phase rule," Eq. (4.59), may be applied to the case of Fig. 6.5(a): since $n = 1$, we have

$$f = 3 - \varphi. \tag{6.10}$$

f	φ
2	1
1	2
0	3

All possibilities for $n = 1$ are listed in the table (φ must be > 0, f must be ≥ 0). The conclusions are: A single-phase region has two degrees of freedom (di-variant, or two-dimensional continuum), a two-phase region has one degree of freedom (mono-variant, or one-dimensional continuum), a three-phase region is reduced to a point (invariant). The application of the rule to Fig. 6.5 (right) requires a bit of caution: It would seem that the two-phase regions also consist of two-dimensional continua. Actually, this is not so, as those regions consist of sets of tie lines, only the extremities of which represent thermodynamic states. In fact, the single-line-regions of Fig. 6.5(left) "open up" to become double-line regions in Fig. 6.5(right), with no thermodynamic states in between the two end points of the tie lines. This "opening up" of two-phase regions takes place quite generally when an intensive variable (say T) is replaced by an extensive variable (say V) as coordinate.

Other diagrams, such as T–S instead of P–V can be constructed in a like manner. Two-dimensional plots can be combined to produce P–V–T three dimensional surfaces such as that of Fig. 6.5. It is seen, therefore, that thermodynamic functions, in particular the state functions of substances, can be derived from the knowledge of the Gibbs free energies $g^\alpha(P, T)$ of the various phases. It suffices, in principle, to obtain the appropriate partial derivatives, and to plot these along with the loci of discontinuities. In practice, one goes generally the other way: from the measurable quantities such as $c_P, \kappa_T, \ldots$ to the Gibbs functions by integration.

6.4   Clapeyron–Clausius Equation

Consider a single-component system. Along a two-phase locus, we have

$$g^\alpha = g^\beta, \quad \text{hence} \quad dg^\alpha = dg^\beta \tag{6.11}$$

or

$$-s^\alpha dT + v^\alpha dP = -s^\beta dT + v^\beta dP. \tag{6.12}$$

Therefore,

$$\frac{dP}{dT} = \frac{\Delta s}{\Delta v} = \frac{L}{T\Delta v}, \tag{6.13}$$

which is one form of the Clapeyron–Clausius equation,[7] from which it follows that the slope $\frac{dP}{dT}$ of the two-phase coexistence line is fixed by the sign of the latent heat. For substances which expand on melting, the slope is positive (the usual case). For those which contract on melting (H_2O, for example), the slope is negative. Since the volume always increases in going from a condensed to a gas phase, along with an increase of entropy, the *vaporization* and *sublimation* curves have positive slopes.

An approximate formula can be given for the sublimation. Since $v^\beta \gg v^\gamma$, v^γ can be neglected and $v^\beta \cong \frac{RT}{P}$ (ideal gas approximation, valid since the pressures are typically very low) then, from Eq. (6.13) we have

$$\frac{dP}{dT} = \frac{LP}{RT^2}. \tag{6.14}$$

If latent heat is approximately temperature-independent, Eq. (6.13) may be integrated to give

$$\ln P = -\frac{A}{T} + B \tag{6.15}$$

where $A = \frac{L}{R}$ and B is a constant of integration.

References

Zemansky, M. K. (1968) *Heat and Thermodynamics, Fifth Edition,* McGraw-Hill Book Co., NY.

---

[7]I use here the French nomenclature. In Germany Eq. (6.13) is the "Clausius–Clapeyron" equation.

Chapter 7

Solutions

The following chapter is a particularly "busy" one, involving many physical parameters and equations describing their mutual interactions. The theme of the chapter is the mixing of constituents, one that, to a great extent, distinguishes it from the topic of mechanical thermodynamics, and also to some extent from condensed matter physics. A most important variable in *solution thermodynamics* therefore is the concentration c_i which may be defined as mole fractions or densities of constituents (atoms, molecules, ...). Then

$$c_i = \frac{N_i}{N} \qquad (i = 1, 2, \ldots n), \qquad (7.1)$$

where n is the number of constituents in solution.[1] Densities are correspondingly defined as

$$\rho_i = \frac{N_i}{V} \qquad (7.2)$$

where V is the total volume occupied by all the N_i moles together (or number of atoms, molecules) of species $(i = 1, \ldots n)$ in the solution.

Mole fractions (concentrations in general) c_i are usually represented graphically on multidimensional simplexes[2]: *i.e.* a straight line segment for binaries, an equilateral triangle for ternaries, a regular tetrahedron for quaternaries, etc.... The simplex construction embodies the "sum rule" of concentrations, *i.e.*:

$$\sum_{i=1}^{n} c_i = 1. \qquad (7.3)$$

Hence, there are at most $n - 1$ independent concentrations c_i present.

---

[1] A more sophisticated definition is given by Eq. (9.17) in Ch. 9.

[2] In geometry, a k-simplex is a figure in k-dimensional space formed by k+1 points all linked to each other.

7.1   Chemical Potentials in Solutions

Since experiments on solutions are usually performed at constant T, P, the
Gibbs energy is of paramount importance. Hence, the following definition
of chemical potentials will be used (following the original definition (4.60))

$$\mu_i = \left(\frac{\partial G}{\partial N_i}\right)_{T,P,N'} \tag{7.4}$$

where N' designates all other N_j, $j \neq i$. Explicit expressions for μ_i will
be arrived at by successive steps: (a) pure gas, (b) mixture of gases, (c)
mixture of condensable substances. The chemical potential of an isolated
ideal gas was obtained in Sec. 5.4.

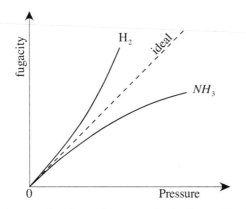

Figure 7.1: Fugacity as a function of pressure for both positive (H_2) and negative
(NH_3) deviations form ideality.

Fugacity of a Real Gas

Here, a real (single-component) gas is regarded as a "perturbed ideal gas."
Thus, define *fugacity* f by the equations

$$\frac{\mathrm{d}f}{f} = \frac{\mathrm{d}P}{P_{id}} \tag{7.5a}$$

$$\lim_{P \to 0} f = P \tag{7.5b}$$

where

$$P_{id} = \frac{RT}{v} \tag{7.5c}$$

by definition. Thus P_{id} is the pressure the gas would have at T, v, if it were ideal. The real pressure P is, generally, different. It is seen that fugacity is defined by a differential equation (7.5a), so that an initial condition (7.5b) must be supplied to complete the definition. The latter equation is reasonable since one expects that a real gas will approach ideal conditions at infinite dilution, as $P \to 0$. Equation (7.5a) may also be written as

$$d \ln f = \frac{v}{RT} \, dp \tag{7.6}$$

which is the more usual defining equation of fugacity. A schematic fugacity diagram is shown in Fig. 7.1, illustrating the fugacity of real gases H_2 and NH_3.

The chemical potential of a pure (single-component) *real gas* is obtained by integrating $dg = -sdT + vdP$. Just as in Sec. 5.4, we have

$$\mu \equiv g(P, T) = \int_{p^\circ}^{P} v(p, T) \, dp - \int_{T_0}^{T} s(p^\circ, t) \, dt \tag{7.7}$$

where p° is a *standard-state* fixed but arbitrary pressure, and T_0 denotes the fixed point temperature at which integration is initiated. Again, as in Sec. 5.4, define the standard state chemical potential by

$$\mu^\circ \equiv g(p^\circ, T) = - \int_{T^\circ}^{T} s(p^\circ, t) \, dt \tag{7.8}$$

so that Eq. (7.7) becomes, by means of Eqs. (7.6) and (7.8)

$$\mu(P, T) = \mu^\circ(T) + RT \ln \frac{f}{f^\circ} . \tag{7.9}$$

Equation (7.9) is often stated as an empirical law, to be justified by experiment. Here, however it was shown that this "law" can be derived quite rigorously by integrating the appropriate differential form according to the method described in Appendix A. Of course one still must find expressions, numerical or algebraic, for the fugacity f, but expression (7.9) can be arrived at rigorously.

Mixture of Real Gases

Formally, a van't Hoff reaction box is used, as in Sec. 3.6, but with the modification that each auxiliary piston is fitted with a shutter which can transform the semipermeable membrane into a diathermic partition, as indicated schematically in Fig. 7.2. Begin by drawing some gas i into the

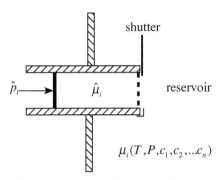

Figure 7.2: Small cylinder in van't Hoff reaction box, equipped with shutter.

auxiliary chamber, shutter open. At equilibrium:

$$\hat{\mu}_i(T, \hat{p}_i) = \mu_i(T, P, c_1 \ldots c_n) \qquad (7.10)$$

where $\hat{\mu}_i$ is the chemical potential of i in the small cylinder, μ_i the chemical potential in the main chamber, $\hat{p}_i$ is the equilibrium pressure on the small piston, P that in the main chamber. Equation (7.10) expresses the phase equilibrium across the semipermeable membrane. Temperatures are also equal, the semipermeable membrane being necessarily diathermic as well.

Now close the (diathermic) shutter and modify the pressure $\hat{p}_i$ to some chosen reference pressure (standard state) $p_i{}^\circ$. Finally return to the equilibrium state (shutter closed). If the process is performed reversibly, when $\hat{p}_i$ is reached, opening the shutter causes no change in the state of the system. Since the small cylinder contains a pure gas, we have, according to the previous subsection, in going from $p_i{}^\circ$ to $\hat{p}_i$ as in Eq. (7.9):

$$\hat{\mu}_i(T, \hat{p}_i) = \hat{\mu}_i(T, p_i^\circ) + RT \ln \frac{\hat{f}_i}{f_i^\circ}. \qquad (7.11)$$

Since chemical potentials of i must be equal across the partition (open to substance i , then, by Eq. (7.10), we must have

$$\mu_i(T, P, c_1, \ldots c_n) = \mu_i^\circ(T) + RT \ln \frac{f_i(T, P, c_1 \ldots c_n)}{f_i^\circ(T)} \qquad (7.12)$$

where $\mu_i^\circ(T)$ has been written for $\hat{\mu}_i(T, \hat{p}_i)$, and where $f_i(T, P, c_1, \ldots c_n)$ is the fugacity f_i expressed as a function of the variables which influence it, *i.e.*, those of the main cylinder (mixture).

According to Eqs. (7.11) or (7.12) then, the auxiliary cylinders of a van't Hoff reaction are seen as probes for measuring chemical potentials μ_i in gas mixtures. It is important to note that, whereas μ_i depends on all the variables of the mixture, including all concentrations, μ_i° depends *only on temperature* for fixed standard state pressure, in the pure state.

Now define *partial pressure* by the equation

$$p_i = c_i P \tag{7.13}$$

where P is the overall pressure. Generally, there is no reason for partial pressure to be equal to the actual "escape pressure" $\hat{p}_i$.

Define the *ideal gas mixture* as one which behaves globally as $PV = NRT$ where P is the overall pressure, V, the total volume and N the total quantity of all gases, *and* for which each individual gas also behaves according to the ideal gas law. Multiplying the above equation by c_i (the concentration of i in the mixture) gives, by the definition of partial pressure (7.13):

$$p_i V = N_i RT. \tag{7.14}$$

Equation (7.14) shows that N_i moles of i alone in V at T, would be equilibrated by pressure p_i on the main cylinder since gas i (and all the other gases) is in this ideal state. Since pressures on either side of a flat semipermeable membrane, perfectly "transparent" to i, must be equal, then p_i (partial pressure) $= \hat{p}_i$ (escape pressure) in the case of *ideal gas mixtures*.

In each auxiliary cylinder we have (for $p_i^\circ = 1$)

$$\hat{\mu}_i(\hat{p}_i, T) = \mu_i^\circ(T) + RT \ln \hat{p}_i \tag{7.15}$$

so that, by the equality of chemical potentials,

$$\mu_i(P, T, c_1...c_n) = \mu_i^\circ(T) + RT \ln p_i \tag{7.16}$$

for an ideal gas mixture.

Mixture of Condensable Substances

Let a condensed-phase mixture contain species i in equilibrium with a gas phase mixture where the partial pressure is p_i. When gas i is drawn through a semipermeable membrane into its auxiliary chamber, and when, as in the previous section, the pressure is increased from $\hat{p}_i$ to the standard one of p_i° (usually one atm) it may well happen that pure i will condense at some T according to the diagram of Fig. 7.3.

Again, integration gives, for isolated i:

$$\hat{g}_i(\hat{p}_i, T) = \int_{p_i^\circ}^{\hat{p}_i} \hat{v}_i \, dp - \int_{T_0}^{T} \hat{s}_i \, dt. \tag{7.17}$$

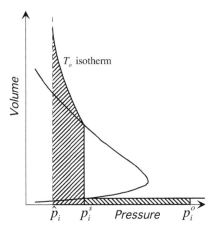

Figure 7.3: Integral of $v\,\mathrm{d}p$ for condensable substance.

The second integral is by definition $\mu_i^\circ(T)$ and the first, over p, must be split up as

$$\int_{p_i^\circ}^{\hat{p}_i} \hat{v}_i(p,T)\,\mathrm{d}p = \int_{p_i^\circ}^{p_i^s} v_i^\alpha\,\mathrm{d}p + \int_{p_i^s}^{\hat{p}_i} v_i^\beta\,\mathrm{d}p \qquad (7.18)$$

where p_i^s is the saturation pressure for pure i at T and α and β are condensed and gas phases respectively. Since the molar volume of the condensed gas remains practically constant at $\bar{v}$ we find, from (7.17) and (7.18),

$$\mu_i(P,T,c_1\ldots c_n) = \hat{g}_i(\hat{p}_i,T) = \mu_i^\circ(T) + \bar{v}_i(p^s{}_i - p^\circ{}_i) + RT\ln\frac{f_i}{f_i^s}. \qquad (7.19)$$

We now make the reasonable assumption that the $v\,\mathrm{d}p$ integral over the condensed phase is very much less than that over the gas phase. Then Eq. (7.19) becomes

$$\mu_i(P,T,c_1\ldots c_n) \approx \mu_i^\circ(T) + RT\ln a_i(P,T,c_1\ldots c_n) \qquad (7.20)$$

where the *activity* a_i has been defined as the ratio of fugacities (approximately equal to the ratio of partial pressures):

$$a_i = \frac{f_i}{f_i^s} \approx \frac{p_i}{p_i^s}. \qquad (7.21)$$

For a *pure condensed* substance i at T in equilibrium with a mixture of gases under total pressure P, the partial pressure p_i is expected to lie very close to the saturation value p_i^s in the absence of all other (relatively inert) gases. Hence,

$$a_i(\text{pure, condensed}) \approx \frac{p_i{}^s}{p_i{}^s} = 1 \qquad (7.22)$$

an approximation much used in chemical thermodynamics.

Note that with $\Delta\mu_i = \mu_i - \mu_i{}^\circ$ we have:

$$a_i = \exp\left(\frac{\Delta\mu_i}{RT}\right) \tag{7.23}$$

so that activity really contains the same information as chemical potentials. The range of activity is more convenient, however:

$$-\infty < \Delta\mu_i \leq 0 \quad \rightarrow \quad 0 \leq a_i \leq 1. \tag{7.24}$$

Change of Standard State

It is often useful to consider standard states other than some fixed p°. For example, the standard state for condensed phases can be taken as the overall pressure P. The new equation for chemical potentials is then

$$\mu_i(T, P, c_1 \ldots c_n) = \mu_i^*(T, P) + RT \ln a_i^*(T, P, c_1 \ldots c_n). \tag{7.25}$$

Furthermore, it is often useful to define the *activity coefficient* γ by the equation

$$a_i = c_i\gamma_i \tag{7.26}$$

so that we have, in abbreviated form

$$\mu_i = \mu_i^* + RT \ln c_i + RT \ln \gamma_i. \tag{7.27}$$

The goal of experimental solution thermodynamics is then to determine activity coefficients γ_i.

7.2 Henry and Raoult Laws

Definition: The majority component is called *solvent*, the others (minority components) are called *solutes*.

It was shown above that for a pure condensed solvent, the activity is practically unity. At the other extreme ($c_i \rightarrow 0$), the activity practically vanishes since, for a substance capable of entering in the condensed-phase solution, $p_i = 0$ in the gas phase if i is absent from the condensed matter. In conclusion:

$$a_i(\text{for } c_i \rightarrow 0) \approx 0, \qquad a_i(\text{for } c_i \rightarrow 1) \approx 1. \tag{7.28}$$

Stronger statements can be made about the neighborhoods of zero solvent and pure solute. In particular, *Henry's law* states that, for very dilute solute c_i (say), the corresponding activity a_1 is proportional to c_1, with proportionality factor γ_1° (Henry's law constant) independent of c, and depending only on the ratios of the other components. That is, for variations

$$\frac{\delta c_2}{c_2} = \frac{\delta c_3}{c_3} = \cdots = \frac{\delta c_n}{c_n} = \delta\lambda_1 \qquad (7.29)$$

then

$$\lim_{\delta\lambda_1 \to 0} a_1 = c_1\gamma_1^\circ, \quad 0 < \gamma_1^\circ < \infty \qquad (7.30)$$

with γ_1° depending only on the ratios of the c_2, c_3... This empirical law can be rationalized as follows: If c_1 is increased from zero, keeping all other N_i constant, then one expects that molecules of type 1 will not at first interact with each other (dilute limit), so that p_1 in the gas phase will, at first, increase linearly with N_1, *i.e.* with c_1.

At the other extreme of almost pure solvent 1, *Raoult's law* holds, stating simply that

$$\lim_{c_1 \to 1} a_1 = c_1. \qquad (7.31)$$

Raoult's law may be proven from Henry's law as follows. First note that $c_i = 1$ defines vertex i of the composition simplex and that $c_i = 0$ defines the face opposite to i (with $n-2$ degrees of freedom). By the Gibbs–Duhem equation (4.6) at constant T, P, we have

$$\sum_{i=1}^{n} c_i \, d\mu_i = 0. \qquad (7.32)$$

Then, by Eq. (7.20), this equation becomes, since the μ_i° depend only on temperature,

$$\sum_{i=2}^{n} {}'c_i \, d\ln a_i = -c_1 \, d\ln a_1, \qquad (7.33)$$

the accent on the summation denoting a restricted sum. For c_1, close to 1, all other concentrations are small, so that Henry's law is expected to hold separately for all $i \neq 1$. Since the variations dc_i ($i \neq 1$) may be regarded as independent, they are chosen, in each term of the left hand side of Eq. (7.33), in such a way that the variations of a_i are taken along the corresponding $d\lambda_i$'s. Thus, by Henry's law,

$$\sum_{i=2}^{n} {}'c_i \, d\ln \gamma_i^\circ c_i = -c_1 \, d\ln a_1 \qquad (7.34)$$

becomes (since in magnitude the sum of all concentration variations must vanish)

$$\sum_{i=2}^{n} {}' \frac{\mathrm{d}c_i}{c_i} = -\mathrm{d}c_1 = -c_1 \, \mathrm{d}\ln a_1 \tag{7.35}$$

or

$$\mathrm{d}\ln c_1 = \mathrm{d}\ln a_1 \tag{7.36}$$

from which $a_1 = kc_1$. But since $a_1(c_1 = 1) = 1$, by Eq. (7.28), Raoult's law (7.31) is proven.

This "law" can be interpreted physically as follows: For almost pure solvent, the addition of solute, of arbitrary nature, in small quantities merely reduces linearly the partial pressure p_i in equilibrium in the gas phase. The properties of the system are thus governed by the solvent. At the other extreme, the nature of the system, for the very dilute "1", depends only on the mixed solvent, *i.e.* on the fixed ratios of "2," "3,"... The domain of validity of Raoult's law depends on the domain of validity of Henry's. Thus, the Raoult domain may be more extended along the simplex faces for which the Henry domain is more extended.

In binary systems, it is convenient to plot activities inside a square. Because of the Gibbs–Duhem equation, curve a_1 follows from that of a_2 and vice versa. Such curves are usually determined experimentally (one of them at least, the other by Gibbs–Duhem integration).[3] Then chemical potentials may be obtained from Eq. (7.20), hence Gibbs energies themselves, hence everything else. If $a_i = c_i$ for the whole range $0 \le c_i \le 1$, the solution is said

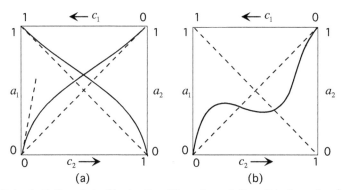

Figure 7.4: Activity plots illustrating Henry's and Raoult's laws for (a) stable and (b) unstable solution.

---

[3] As is the case for chemical potentials, the n activities are related by a Gibbs–Duhem differential equation; one of these activities may thus be obtained from the $n - 1$ others after integration of the differential equation.

to be *ideal* in i. If this is true for all $i = 1, 2 \ldots n$, the solution is "perfect". Activity plots can exhibit extrema, as in Fig. 7.4(b). The extrema are located at vanishing of the first derivative of chemical potentials, hence at points of vanishing of second derivatives of Gibbs energy with respect to concentration squared. The locus of such points defines spinodal lines (see Sec. 18.1).

7.3 Gibbs–Helmholtz Equation

Entropy, volume and chemical potentials may be derived from the Gibbs free energy by means of Eqs. (4.23). The enthalpy can be obtained from the Gibbs–Helmholtz equation, which can be derived thus derived thus:

$$\left(\frac{\partial (G/T)}{\partial T} \right)_P = \frac{1}{T} \left(\frac{\partial G}{\partial T} \right)_P - \frac{1}{T^2} G = -\frac{1}{T^2} (TS + G), \tag{7.37}$$

yielding the Gibbs–Helmholtz equation

$$\left(\frac{\partial (G/T)}{\partial T} \right)_P = \frac{H}{T^2}. \tag{7.38a}$$

By taking partial derivatives with respect to N_i of both sides of this equation one also obtains the partial molar form

$$\left(\frac{\partial (\mu_i/T)}{\partial T} \right)_{P,N'} = -\frac{h_{,i}}{T^2}. \tag{7.38b}$$

7.4 Quantities of Mixing

Partial molar quantities (or PMQs for short) were defined by Eq. (4.65) in Sec. 4.7 as the partial derivatives with respect to N_i of any extensive function (or thermodynamic potential) $Z = Z(Y_0, Y_1 \ldots, N_1, N_2 \ldots N_n)$, function of mole quantities N_i and all other intensive variables ($Y_0 \equiv T, Y_1 \ldots$).

Because of *interactions* between species, "partial molar Z" is in general not equivalent to "molar Z":

$$z_{,i} \neq \left(z_i = \frac{Z_i}{N_i} \right) \tag{7.39}$$

the latter, measured for i in the pure state, *i.e.* in isolation. The difference

$$\Delta z_i = z_{,i} - z_i \tag{7.40}$$

is called *partial Z of mixing* and is a rough measure of the degree of inter-action between i and the solution. Multiply both sides of Eq. (7.40) by N_i and sum over all components gives:

$$\Delta Z_M \equiv \sum_{i=1}^{n} N_i \Delta z_i = Z - Z_L \qquad (7.41)$$

where Euler integration (4.68) has been used. ΔZ_M is by definition the (integral) "Z of mixing" and Z_L is defined as the linear (L) combination

$$Z_L \stackrel{\text{def}}{=} \sum_{i=1}^{n} N_i z_i . \qquad (7.42)$$

Equation (7.40) can be "normalized"

$$\Delta z_M = \sum_{i=1}^{n} c_i \, \Delta z_i = z - z_L . \qquad (7.43)$$

This *molar quantity of mixing* is represented graphically in Fig. 7.5 for a binary case (two constituents, $n = 2$). In general, z of mixing measures the deviation of the thermodynamic potential z from the value it would have in the ideal case.

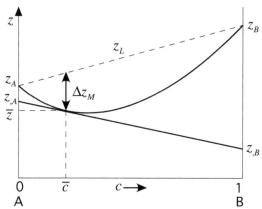

Figure 7.5: Molar (z), partial molar ($z_{,i}$), ideal mixing (z_L) and partial z of mixing (Δz_M). The intercept rule is also illustrated (see Sec. 7.5).

Examples

Useful properties may be derived from the formula for chemical potentials, Eq. (7.25).

Volume of Mixing

Take the partial derivative with respect to P of Eq. (7.25):

$$\left(\frac{\partial \mu_i}{\partial P}\right)_{T,N} = \left(\frac{\partial \mu_i^*}{\partial P}\right)_T + RT\left(\frac{\partial \ln a_i^*}{\partial P}\right)_{T,N}.$$

Now exchange the order of differentiation in the derivative on the left hand side of $\left(\frac{\partial^2 g}{\partial P \partial N_i}\right)$ to obtain

$$\left(\frac{\partial}{\partial N_i}\left(\frac{\partial G}{\partial P}\right)_{T,N}\right)_{T,P,N'} = \left(\frac{\partial v}{\partial N_i}\right)_{T,P,N'} = v_{,i}.$$

Likewise, the molar volume is

$$\left(\frac{\partial \mu_i^*}{\partial P}\right)_T = \left(\frac{\partial g_i}{\partial P}\right)_T = v_i$$

so that

$$\Delta v_i \equiv v_{,i} - v_i = RT\left(\frac{\partial \ln a_i^*}{\partial P}\right)_{T,N'} = RT\left(\frac{\partial \ln \gamma_i}{\partial P}\right)_{T,N'} \qquad (7.44)$$

which defines the *partial volume of mixing*. The integral volume of mixing is obtained from the general formula (7.43) applied to the volume.

If component i behaves ideally, $a_i^* = c_i$, so that the molar volume of mixing vanishes. If this is true for all i, the solution is "perfect" and $\Delta v_n = 0$ so that a plot of v versus concentration is linear, *i.e.*, v coincides with v_L. For crystalline solids, ideality translated in terms of lattice parameter yields *Vegard's law*.

Enthalpy of Mixing

By the Gibbs–Helmholtz equation (7.38) we have, from Eq. (7.25)

$$\left(\frac{\partial(\mu_i/T)}{\partial T}\right)_{P,N} = \left(\frac{\partial(\mu_i^*)/T}{\partial T}\right)_P + R\left(\frac{\partial \ln a_i^*}{\partial T}\right)_{P,N},$$

that

$$\Delta h_i \equiv h_{,i} - h_i = -RT^2\left(\frac{\partial \ln a_i^*}{\partial T}\right)_{P,N} = -RT^2\left(\frac{\partial \ln \gamma_i}{\partial T}\right)_{P,N}. \qquad (7.45)$$

For a perfect solution, all Δh_i vanish and $\Delta H_n = 0$. Thus there is no evolution of heat in case of ideal mixing, hence no latent heat.

Entropy of Mixing

Taking partial derivatives with respect to T in Eq. (7.25), one gets

$$\left(\frac{\partial \mu_i}{\partial T}\right)_{P,N} = \left(\frac{\partial \mu_i^*}{\partial T}\right)_P + R \ln a_i^* + RT \left(\frac{\partial \ln a_i^*}{\partial T}\right)_{P,N}$$

or

$$\Delta s_i = s_{,i} - s_i = -R \ln a_i^* - RT \left(\frac{\partial \ln a_i^*}{\partial T}\right)_{P,N}, \qquad (7.46)$$

for the partial entropy of mixing. If i behaves ideally, the ideal *partial* entropy of mixing is

$$(\Delta s_i)_{id} = -R \ln c_i. \qquad (7.47)$$

If the solution is perfect, the ideal *integral* entropy of mixing is

$$\Delta s_{id} = -R \sum_{i=1}^{N} c_i \ln c_i, \qquad (7.48)$$

which surely does not vanish, but is a universal function, plotted in Fig. 7.6 for a binary solution. Note symmetry about $c = \frac{1}{2}$ where $\Delta s_{id}\left(\frac{1}{2}\right) = R \ln 2$. At $c = 0$ and $c = 1$ (pure states), $\Delta s_{id} = 0$ and has infinite slopes, which has important consequences.

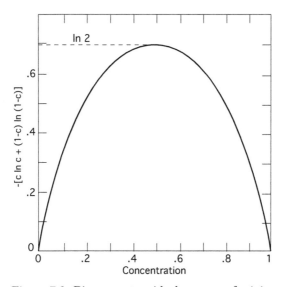

Figure 7.6: Binary system ideal entropy of mixing.

Gibbs Energy of Mixing

Most useful is the Gibbs energy of mixing

$$\Delta g_i = g_{,i} - g_i = \mu_i - \mu_i^* = RT \ln a_i^* = RT(\ln c_i + \ln \gamma_i).$$

By multiplication by c_i and summation over i we obtain:

$$\Delta g_M = RT \sum_i (c_i \ln c_i + c_i \ln \gamma_i) \qquad (7.49)$$

where Eq. (7.26) has been used. Equation (7.49) can be written as

$$\Delta g_M = \Delta g_{id} + \Delta g_{xs} \qquad (7.50)$$

where the ideal Gibbs energy of mixing is given by

$$\Delta g_{id} = RT \sum c_i \ln c_i \qquad (7.51)$$

and, by definition, the *excess* Gibbs energy is the *deviation from ideality*

$$\Delta g_{xs} = RT \sum c_i \ln \gamma_i. \qquad (7.52)$$

The latter quantity may be expressed as

$$\Delta g_{xs} = \Delta h_{xs} - T \Delta s_{xs}. \qquad (7.53)$$

Hence, the only contributions which need to be determined in practice are the excess enthalpy and entropy. Equivalently, "doing solution thermodynamics" means determining activity coefficients, γ_i. Today, fairly reliable analytical models for Δh_{xs} and Δs_{xs} are becoming available, thanks to theoretical models for electronic, elastic, vibrational and configurational contributions, as briefly indicated in Eq. (9.16) in the section on first principles calculations in Chapter 9.

7.5   Intercept Rule

The intercept rule provides a very useful graphical construction for chemical potentials and other partial molar quantities. The proof for n-component systems is usually not given. Here is one version: the basic equation (4.68), normalized, is

$$z = \sum_{i=1}^{n} c_i z_{,i} \qquad (7.54)$$

and can be rewritten

$$\Phi(c_1, c_2 \ldots c_n, z) \equiv \sum c_i z_{,i} - z = 0$$

which represents a surface in n-dimensional space. Whether or not the variables are independent, and whether or not the coordinate system is Cartesian, the tangent plane at point $(\bar{c}_1 \ldots \bar{c}_n, \bar{z})$ to the surface $\Phi = 0$ is given by

$$\sum \frac{\partial \Phi}{\partial \bar{c}_i}(\eta_i - \bar{c}_i) + \frac{\partial \Phi}{\partial \bar{z}}(\zeta - \bar{z}) = 0$$

where η_i and ζ denote coordinates on the tangent plane, and where the derivatives must be evaluated at the tangency point $\bar{c}_i, \bar{z}$. The total differential of Φ is, by (6.52) and the Gibbs–Duhem equation:

$$d\Phi = \sum z_{,i} dc_i - dz = 0$$

so that, at the point in question,

$$\frac{\partial \Phi}{\partial c_i} = \bar{z}_{,i} \quad \text{and} \quad \frac{\partial \Phi}{\partial z} = -1$$

and the equation of the tangent plane becomes

$$\sum \bar{z}_{,i}(\eta_i - \bar{c}_i) - (\zeta - \bar{z}) = 0 \,.$$

Point $(\bar{c}_1, \bar{c}_2 \ldots \bar{z})$ is located on the surface $\Phi = 0$, which satisfies Eq. (7.54), so that the equation of the tangent plane, with coordinates $(\eta_1, \eta_2 \ldots \zeta)$ is

$$\zeta = \sum_{i=1}^{n} \eta_i \bar{z}_{,i} \,. \tag{7.55}$$

Equation (7.55) resembles (7.54) but means something entirely different: Eq. (7.54) is the equation for the molar quantity $z = z(c_1, c_2 \ldots c_n)$, and is *not* linear in the concentration variables since $z_{,i}$ is also a function of the concentrations. However, Eq. (7.55) is linear in the $(\eta_1, \eta_2 \ldots \zeta)$ since $\bar{z}_{,i}$ is a fixed quantity, calculated at the fixed point $(\bar{c}_1, \bar{c}_2 \ldots \bar{c}_n)$; therefore it represents a plane, which was proved to be the tangent plane at the point $\bar{c}_i$.

Consider now one vertex of the composition simplex, say the ith one. At that point, all η's vanish except $\eta_i = 1$. Then Eq. (7.55) yields

$$\zeta(c_i = 1) = \bar{z}_{,i} \,. \tag{7.56}$$

Such is the mathematical expression of the *intercept rule*. It expresses the fact that the intercept of the tangent hyperplane at $c_i = 1$ gives the value of the partial molar quantity at that vertex. The rule is illustrated for $z_{,A}$ and $z_{,B}$ (binary example) in Fig. 7.5 and also in Fig. 7.10.

7.6   Independent Variables

If all constituents are listed in $i = 1, 2 \ldots n$, then

$$\sum_{i=1}^{n} c_i = 1. \tag{7.57}$$

Hence the concentration variables are not all independent. It follows that the derivative $z_{,i}$ can be written formally as

$$z_{,i} = \left(\frac{\partial Z}{\partial N_i} \right)_{P,T,N'} = \frac{\partial (Z/N)}{\partial (N_i/N)} = \left(\frac{\partial z}{\partial c_i} \right)_{P,T,N'}, \tag{7.58}$$

where the symbol N' designates, as above, all components other than i. Equation (7.58) is correct, including its latter form, but in

$$z_{,i} \neq \left(\frac{\partial z}{\partial c_i} \right)_{P,T,c_{i \neq j}} \tag{7.59}$$

the equality is not correct because it is impossible to vary c_i while holding all other c_j's constant, given the sum rule of Eq. (7.57).

It is therefore often useful to express thermodynamic potentials in terms of independent concentration variables, say in the set $(c_2, c_3, \ldots, c_n)$. Let us denote by $\tilde{z}$ the function z expressed in these independent variables. PMQs then are expressed as differences, as shown below:

$$d\tilde{z} = \sum_{i=1}^{n} z_{,i}\, dc_i = z_{,1}\, dc_1 + \sum_{j=2}^{n} z_{,j}\, dc_j = -\sum_{j=2}^{n} z_{,1}\, dc_j + \sum_{j=2}^{n} z_{,j}\, dc_j \tag{7.60}$$

where the differential of (7.57) has been used. Hence

$$d\tilde{z} = \sum_{j=2}^{n} \tilde{z}_{,j}\, dc_j \tag{7.61}$$

where

$$\left(\frac{\partial \tilde{z}}{\partial c_i} \right)_{c_{i \neq 1}} = \tilde{z}_{,i} = z_{,i} - z_{,1}. \tag{7.62}$$

For example, in the case of conservation of species, the proper chemical potential to use is the difference $[\mu_i - \mu_1]$ $(i = 2, 3, \ldots n)$. This potential $\tilde{\mu}_i$ has been termed "diffusion potential" by Larché and Cahn (1985 and papers cited therein) but, by analogy with magnetic systems, we shall call it the *chemical field*.

It is instructive to rewrite the intercept rule in terms of independent variables. So, multiply both sides of Eq. (7.62) by c_i and sum over *all* i to obtain:

$$\sum_{i=2}^{n} {}' c_i \, \tilde{z}_{,i} = \sum_{i=1}^{n} c_i \, \tilde{z}_{,i} = \tilde{z} - z_{,i}, \tag{7.63}$$

whereupon, since $\tilde{z}_1 = 0$, by rearranging we obtain

$$z_{,1} = \tilde{z} - \sum_{i=2}^{n} c_i \, \tilde{z}_{,i} \tag{7.64}$$

where $\tilde{z}_{,i}$, being the coefficient of dc_i in Eq. (7.61), appears as a true partial derivative, with all other variable c_j ($j \neq i \neq 1$) held constant. Equation (7.64) is thus the one to use for the purpose of calculating PMQ's from analytic functions expressed in terms of concentration (rather than "amount") variables. In particular, the chemical potential formula for binary systems is, from Eqs. (7.63) and (7.64),

$$\mu \equiv \mu_2 = g - c \frac{dg}{dc} \tag{7.65}$$

with $c \equiv c_2$. Note use of *total derivative* of $g(c)$ rather than partials. Equation (7.65) could of course have been derived "geometrically" from Fig. 7.5. Equation (7.65) is often referred to as *the intercept rule*, rather than its more general form (7.64). For binary solutions we have, by Eq. (7.62) or (7.65) for the "chemical field",

$$\mu = \mu_2 - \mu_1 \qquad \text{or equivalently} \qquad \mu = \mu_B - \mu_A. \tag{7.66}$$

Note also that Eq. (7.61) is a general property of derivatives and is not limited to extensive functions. Thus, for any function $\Phi(c_1, c_2, \ldots c_n)$ from which c_1 is eliminated by Eq. (7.57), one has

$$\left(\frac{\partial \Phi}{\partial c_i} \right)_{c_{i \neq j \neq 1}} = \frac{\partial \Phi}{\partial c_i} - \frac{\partial \Phi}{\partial c_1} \qquad (i \neq 1) \tag{7.67}$$

where the partials on the right are physically meaningless derivatives taken "at everything else held constant."

7.7 Qualitative Discussion of Mixing

Many important conclusions can be drawn from the form of the equations of solution thermodynamics without actually considering specific models.

Ideal Behavior

At any vertex of the composition simplex, the free energy of mixing Δg_{id} vanishes. This must be so since there is no mixing at a pure component state. Also, mathematically, at $c_1 = 1$, say, all other c's are zero so that

$$\Delta g_{id} = RT \sum_{i=1}^{n} c_i \ln c_i \qquad (7.68)$$

which, for $i \neq 1$ goes as

$$\lim_{c \to 0} c \ln c = \lim_{c \to 0} \frac{\ln c}{\frac{1}{c}} \to -\lim_{c \to 0} \frac{\frac{1}{c}}{\frac{1}{c^2}} = 0, \qquad (7.69)$$

the second equation resulting from l'Hospital's rule. The slopes of actual partial derivatives, such as $\frac{\partial g}{\partial c_2}$, all other $c_3, c_4, \ldots c_n$ being held constant, with c_1 as *dependent variables*, go as $(\mu_2 - \mu_1)$ according to Eq. (7.62). Then, at $c_1 = 1$ (vertex "1"), with

$$\mu_i = \mu_i^* + RT \ln c_i, \qquad (7.70)$$

gives $\mu_i = \mu_i^*$ and $\mu_2 \to -\infty$, etc...., hence the slope $\mu_2 - \mu_1$ tends to minus infinity. This is true for any derivative $\left(\frac{dg}{dc_i} \right)_{i \neq 1}$ at vertex "1".

"Real" Behavior

Limiting behavior of Δg_n (real) can be determined by adding the contribution

$$\Delta g_{xs} = RT \sum_{i=1}^{n} c_i \ln \gamma_i$$

to Δg_{id}. At and near $c_1 = 1, \gamma_1 = 1$ (Raoult's law) and all other $c_i \ln \gamma_i$ ($i \neq 1$) go as $0 \ln \gamma^\circ_1$ so that Δg_{xs} vanishes at all vertices. Hence, in general Δg_{xs} vanishes at all simplex vertices, as expected.

The slopes of partial derivatives, in the "real" case, will also be infinite, unless the contribution of the excess free energy was also infinite and of opposite sign. This, however, is not the case as can be seen by application of Henry's law: for instance

$$\left(\frac{\partial \Delta g_{xs}}{\partial c_2} \right) = RT \sum_{i=1}^{n} \left(\frac{\partial}{\partial c_2} c_i \ln \gamma_i \right)_0, \qquad (7.71)$$

where the subscript zero means that, at vertex "1," all $c_3, c_4 \ldots$ are set equal to zero. The derivatives are thus taken along the simple edge "2-1" and Henry's law holds. Then since $\gamma_1 = 1$ in the vicinity of $c_1 = 1$, the only contribution to the right hand side of Eq. (7.71) is from

$$\left(\frac{\partial}{\partial c_2} c_2 \ln \gamma_2\right)_0 = \left(\frac{\partial}{\partial c_2} - \frac{\partial}{\partial c_1}\right) c_2 \ln \gamma_2 \qquad (7.72)$$

where Eq. (7.67) has been used. In Eq. (7.72), γ_2 may then be replaced by γ_2°, which by (7.30) is neither zero nor infinity, so that Eq. (7.72) goes as $\ln \gamma_2$, which is well-behaved.

Figure 7.4(a) shows an example of positive deviation from ideality, with $\gamma > 1$. Negative deviations, $\gamma < 1$, are also possible, over the whole or a portion of the concentration axis. Near a vertex, Δg_n must then behave, along a simplex edge, as shown in Fig. 7.7(b). The infinite slopes at the

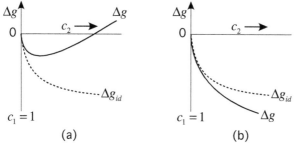

Figure 7.7: Limiting behavior of free energy of mixing for positive (a) and negative (b) deviation from ideality.

origin insure that:

1. Perfectly pure substances will always tend to reduce the free energy by at least slight mixing, *i.e.* or contamination, regardless of the sign of deviation from ideality. This is because, infinitely near a vertex, the mixing entropy (ideal) always predominates over all other contributions.[4]

2. If two free energy curves intersect, it will always be possible to construct a *common tangent* to the two curves, within the $(0,1)$ concentration interval (see Fig. 7.8) thanks to the infinite slopes at the origins. Some general conclusions may also be drawn concerning volume and enthalpy of mixing. By Eqs. (7.44) and (7.45), it is seen that the signs of Δv_i and Δh_i depend both on the sign of the $\ln \gamma_i$ and on the sign of the logarithmic derivatives of γ_i. On physical grounds, it is expected that deviation from ideality (magnitude of $\ln \gamma_i$) will increase with increasing pressure (promotes atomic interactions) and decrease with increasing temperature so

---

[4]Consequently, "purification" of a material beyond the equilibrium concentration is a non-equilibrium process, and calls for such processes as *zone melting*, used by the electronics industry to obtain extremely pure crystals of Si and Ge, for example.

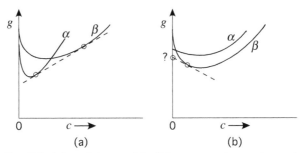

Figure 7.8: Possible (a) and impossible (b) common tangent constructions near pure state.

that one expects

$$\frac{\partial |\ln \gamma|}{\partial P} > 0, \qquad \frac{\partial |\ln \gamma|}{\partial T} > 0. \tag{7.73}$$

Thus, by Eqs. (7.44) and (7.45) we expect $\Delta v_i > 0$ and $\Delta h_i > 0$ for positive deviation and conversely. Thus negative deviation indicates a tendency towards association of unlike species, unlike-bond formation, *i.e.* order, with negative volume of mixing and rejection (evolution) of (latent) heat of mixing.

7.8 Example: Mixing of Hard Spheres

Consider the mixing of two populations of (inert) hard spheres in random close packing, large ones (L) and small ones (s), the small ones being approximately small enough to fit within the interstices of the large ones. Let the two concentrations of spheres in the mixture be $c = c_L$ and $c_s = 1 - c$. With only large spheres ($c = 1$) we have approximately, since the addition of few small spheres leaves the total volume of N spheres invariant to first order,

$$v_{,s}(c = 1) = \left(\frac{\partial V}{\partial N_s} \right)_{c \to 1} = 0$$

and

$$v_{,L}(c = 1) = \left(\frac{\partial V}{\partial N_L} \right)_{c \to 1} = v_L$$

where v_L is the average volume occupied by one large sphere in a dense random packing of exclusively large ones.

At the other extremity of the concentration axis, for $c = 0$, we likewise have

$$v_{,s}(c = 0) = \left(\frac{\partial V}{\partial N_s} \right)_{c \to 0} = v_s$$

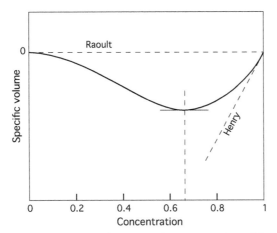

Figure 7.9: Specific volume as a function of concentration of large spheres (c_L) and limiting laws of Henry and Raoult.

and

$$v_{,L}(c = 0) = \left(\frac{\partial V}{\partial N_L} \right)_{c \to 0} = v_L$$

since the addition of one large sphere to a population of exclusively small spheres raises the total volume by roughly the volume of one large sphere, which is somewhat less than v_L.

The specific volume of mixing, given by the previous equation, is plotted against the concentration c of large spheres in Fig. 7.9. Without knowing more about the details of this two-sphere system, not even the ratio of small to large sphere diameters, we can nevertheless construct a plausible plot which turns out to look remarkably like an activity plot for a thermodynamic mixture of two interacting constituents. To show that, we determine the (total) derivative with respect to c of the Δv_M curve

$$\frac{d \Delta v_M}{dc} = -(v_{,s} - v_s) + (v_{,L} - v_L)$$

which, at the two extremities of the concentration axis, yield approximately v_s at $c = 1$ and zero at $c = 0$. In the derivation of the previous equation, the Gibbs–Duhem equation has been used to eliminate the derivatives of the partials. Now the simplest function of c which has the right limiting behavior according to these considerations is the cubic

$$\Delta v_M = -v_s c^2 (1 - c).$$

Maximum (negative) deviation from ideal volume of mixing is found by taking the derivative $d \Delta v_M / dc = -v_s c(2 - 3c)$ which vanishes at $c = 0$

and $c = 2/3$. Hence, the average density of spheres will be highest at the concentration of large spheres approximately equal to $2/3$. Experimental results appear to confirm this result.[5]

One might wonder what causes a mixture of non-interacting spheres to behave in a non-ideal manner. The answer is that the spherical objects are indeed interacting; they do so indirectly through the influence of gravity — which creates the close packing — in an environment in which large and small spheres behave quite differently.[6]

7.9   Barycentric Coordinates

Consider n components distributed among $\varphi (\leq n)$ phases (not necessarily in thermodynamic equilibrium). Denote corresponding extensive quantities by $Z^\alpha, Z^\beta, \ldots$, each containing $N^\alpha, N^\beta, \ldots$ moles (atoms, molecules) of solutions of the n components. Total "Z" is given by

$$Z = \sum_{\alpha=1}^{\varphi} N^\alpha z^\alpha . \tag{7.74}$$

After division by $N = N^\alpha + N^\beta + \ldots$, the average Z of the phase mixture is found to be

$$\bar{z} = \sum_\alpha \nu^\alpha z^\alpha \tag{7.75}$$

with relative amounts $\nu^\alpha = \frac{N^\alpha}{N}$. These parameters ν^α, which obey the relation

$$\sum_\alpha \nu^\alpha = 1 \tag{7.76}$$

are called *barycentric coordinates* (see for example Coxeter, 1961) since Eq. (7.75) is analogous to the formula used to obtain the average mass ($\bar{z}$) of a set of masses of relative masses ν^α (the same Greek symbol ν will also be used in Chapter 11, but with a different meaning and with subscripts instead of superscripts, to designate stoichiometric coefficients in chemical equations).

Likewise, the coordinates of the average concentrations $\bar{c}_i$ can be obtained in terms of these barycentric coordinates as follows: for each i ($i = 1, 2, \ldots, n$),

$$N_1 = \sum_{\alpha=1}^{\varphi} N_i^\alpha \tag{7.77}$$

---

[5]L. J. De Jonge, private communication.
[6]There is more on packing of spheres in Chapter 10, from a geometrical viewpoint.

where, as before, $N_i^\alpha, N_i^\beta \ldots$ are the number of moles of component i in phase $\alpha, \beta, \ldots$. Dividing by N and multiplying each term of the sum by $1 = \frac{N^\alpha}{N^\alpha}, \ldots$, we have, for the ith coordinate of the average concentration,

$$\bar{c}_i = \frac{N^\alpha}{N} \frac{N_i^\alpha}{N^\alpha} + \frac{N^\beta}{N} \frac{N_i^\beta}{N^\beta} + \ldots \tag{7.78}$$

or, after rearranging,

$$\bar{c}_i = \sum_\alpha \nu^\alpha c_i^\alpha \quad (\forall i). \tag{7.79}$$

Because of Eq. (7.76), necessarily, the $\bar{c}_i$'s will sum to unity.

Since average molar Z and average concentrations are all expressed by the same linear relations (7.75) and (7.79) with same barycentric coordinates ν^α, $\bar{z}$ will be found at the intersection of the plane passing through the z^α's and the "vertical" erected at the average concentration point $\bar{c}$ (considered as a vector in concentration space, with the z coordinate on an axis orthogonal to concentration space). This *chord rule* is best illustrated for the case of a two-component (binary) system, as shown in Fig. 7.10, where two arbitrary points (open circles) are placed on Gibbs energy curves g^α and g^β. The average free energy $\bar{g}$ is then situated on the corresponding chord (g^α, g^β) at the intersection with the vertical line drawn from the average concentration $\bar{c}$. Thermodynamic equilibrium requires that (at constant T,

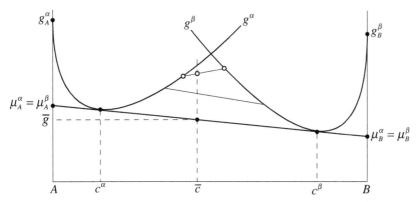

Figure 7.10: Free energy curves for phases α and β and illustration of *chord* and *intercept* rules.

P) the total Gibbs free energy, or average free energy $\bar{g}$, be minimized. This can be accomplished here by dragging the chord (g^α, g^β) as far down as it will go without losing contact with the free energy curves. Obviously, the lowest average free energy will be found on the *common tangent*, thereby providing a simple proof of the *common tangent construction* for phase equilibrium. This construction can be generalized to an arbitrary number n of components (though the geometrical representation may not be simple for $n \geq 3$). By the property of the intercept rule we have, at the common tangent condition,

$$\mu_i^\alpha = \mu_i^\beta = \dots \quad (\forall i) \tag{7.80}$$

which is, of course, the familiar condition of phase equilibrium, given in Sec. 4.6.

It is seen from Fig. 7.10 that the concentration of phases α and β at equilibrium are given by the tangency points c_0^α and c_0^β (only the independent-concentration $c = c_B$ notation is used). The corresponding amounts ν^α and ν^β are given by the binary system version of Eq. (7.79):

$$\bar{c} = \nu^\alpha c^\alpha + \nu^\beta c^\beta \tag{7.81}$$

or, after multiplying the left hand side by $1 = \nu^\alpha + \nu^\beta$ and rearranging, we have

$$(\bar{c} - c^\alpha)\nu^\alpha = (c^\beta - \bar{c})\nu^\beta \tag{7.82}$$

which will be recognized as the familiar *lever rule* which states that the relative amounts of material are inversely proportional to the corresponding "lever arms" measured by concentration differences.

7.10 Regular Solution Model

Real solutions do not behave ideally. Hence, in order to treat practical cases, it is necessary to find approximate models for the excess free energy g^{xs}. The *regular solution model* is the simplest possible approximation available: it features a quadratic form in concentration variables c_i for the excess enthalpy h^{xs} and neglects the excess entropy s^{xs}. A quadratic form is chosen since a linear function of concentrations, g_L, as in Eq. (7.43), can be eliminated from it, as will be presently shown. Moreover, forms of power greater than two should not be used generally since higher powers of concentration tend to interact with the logarithms, possibly giving extraneous extrema in the free energy (see de Fontaine and Hilliard, 1965). The most general quadratic form in c_i, that is $q(c)$, is as follows:

$$q(c) = \sum_{i,j} W_{ij}\, c_i\, c_j \,. \tag{7.83}$$

The elements of the $n \times n$ symmetric matrix W are assumed to be independent of temperature. Because of the concentration conservation relation, it is possible to transform Eq. (7.83) to a strictly non-diagonal form (unlike indices in the interactions W_{ij}). This may be done by replacing one of the c_i's in each diagonal term by a sum of all other concentrations:

$$c_i^2 \rightarrow c_i(1 - \sum_{j \neq i} c_j). \qquad (7.84)$$

The quadratic form may then be written

$$q(c) = \sum_i W_{ii}\, c_i + 2 \sum_{i<j} \Omega_{ij}\, c_i\, c_j \qquad (7.85)$$

with *interaction coefficients* defined by

$$\Omega_{ij} = W_{ij} - \frac{W_{ii} + W_{jj}}{2}. \qquad (7.86)$$

The regular solution model (RS) for the Gibbs free energy may then be written by summing the various contributions given above, with the inclusion of a constant term W_o for generality (it merely fixes the height of the Gibbs function on the g axis):

$$g^{RS}(c, T) = \overbrace{W_o + \sum_i W_{ii}\, c_i}^{g_L} + \overbrace{\underbrace{2 \sum_{i<j} \Omega_{ij} c_i c_j}_{h_{int}} + \underbrace{RT \sum_i c_i \ln c_i}_{-T s_{id}}}^{g_M}. \qquad (7.87)$$

In this equation, the meaning of the terms are indicated by horizontal braces, namely: the linear term $g_L = h_L - T s_L$, the mixing term g_M, consisting of a non-diagonal quadratic term, also called an interaction term h_{int} (regarded as temperature-independent), and the ideal entropy contribution $-T s_{id}$.

The terms used here in Sec. 7.10 have all been defined in Sec. 7.4, above. It is seen that the derivation of the regular solution model, given by Eq. (7.87), rested entirely on classical thermodynamical notions. For the case of crystalline solid solutions, this formula may also be arrived at through elementary statistical mechanical arguments (but need not be). In the statistical derivation, W_{ij} appears as the product $N_0\, r\, w_{ij}$, where w_{ij} is a *pair energy* between atoms of type i and j located on neighboring lattice sites, r is the number of such pairs per site, and N_0 ($\approx 6.0225 \times 10^{23}$ atoms/mole) is Avogadro's number. It follows that the linear term in Eqs. (7.85) or (7.87) is just the sum of the free energies of the fully separated constituents, isolated on the sites of the same lattice as that of the solution. The quadratic term of Eq. (7.87) then represents a true interaction term, with $\Omega_{ij} = N_0\, r\, w_{ij}$

defining the effective pair interaction parameters w_{ij} which, according to Eq. (7.86), may be positive or negative depending upon whether like pairs (i, i), (j, j) are favored over unlike pairs (i, j), or conversely.

The regular solution (or RS) model applied to binary solutions will be discussed in more detail in Sec. 9.4. The method is convenient but purely phenomenological, despite the impression often given that it is derived from statistical bond counting considerations. It is for this reason that, in the present treatment, the regular solution model was given a fully classical thermodynamic derivation. Going beyond the RS model requires a more flexible enthalpy polynomial, offering a few more fitting parameters, but the approximations of the model are not really improved upon. In those RS-like cases, the only configuration variable is the average concentration c, offering the minimal *mean field* approximation. A definite improvement can be achieved by introducing *cluster variables*, first used by Kikuchi (1957). Going beyond that requires appeal to basic physics, including quantum and statistical mechanics, to be briefly presented in Sec. 9.4 under the title "First Principles Calculations".

References

Coxeter, H. S. M. (1961) *Introduction to Geometry*, John Wiley & Sons, Inc., NY.

de Fontaine, D. and Hilliard, J. E. (1965), *Acta metall.*, **13**, 1019.

Larché, F. C. and Cahn, J. W. (1985), *Acta metall.*, **33**, 331.

Chapter 8

Introduction to Statistical Mechanics

The notion of entropy is a difficult one to grasp, particularly in the context of classical thermodynamics. In that approach, entropy is a concept *derived* from the second law, *i.e.* not one *formulating* the second law, as is often stated; the second law merely tells us that certain thermodynamic processes are forbidden. Statistical mechanics introduces entropy in a more physical manner, which makes the subject more approachable, but the mathematics of the field of statistical mechanics (or statistical thermodynamics) quickly becomes more complex. This book is indeed limited to the classical approach, but in later chapters appeal must be made at times to the statistical formulation, so a few simple aspects thereof are introduced in this section.

8.1 Reminder of Basic Results

Before plunging into the microscopic world however, let us recall the main features of the classical formulation as presented in the previous chapters. We begin with the differential form for the internal energy dU as derived in Chapter 4, particularly in Sec. 4.3, that is Eq. (3.18):

$$dU = T\,dS + \sum_{i=1}^{n} Y_i\,dX_i\,, \qquad (8.1)$$

where U is regarded as a function of its *natural extensive* variables $(S, X_1, \ldots X_n)$. Equation (3.18) (or (8.1), which is the same), the *fundamental internal energy differential form*, is the most important equation

in the whole book; it suggests that two basic types of processes contribute
to energy changes: those measured by reversible heat exchanges and those
measured by reversible work performed, "work" being understood in a very
general way, *i.e.* that of hydrostatic, elastic, electrostatic, magnetic, compositional, ... way. The expression "natural variables" in the differential
form (8.1) is defined as referring to the variables which are differentiated
which, in the case of the internal energy, are all extensive ones. The pair
(intensive)/(extensive) in a given differential work term are by definition
conjugate ones (see Table 2.1).

It follows that the simple adjunction of the (inexact) differential $T \, dS$
to a sum of classical work differentials transforms an expression obtained
by macroscopic derivations into one belonging to the discipline of (classical) thermodynamics. It is mainly this apparent sleight-of-hand, which is
usually accepted only reluctantly by beginners, and which makes the subject conceptually difficult. As was mentioned earlier, $T \, dS$ is of the form
(intensive) d*(extensive)*, which is of the form of a classical work differential, but is really a very different animal as its variables T and S must be
arrived at with the help of *postulates* which appear not to be obvious at
first. Note that this adjunction of a mysterious "thermal work" differential
is one that is featured systematically in the several-volume physics treatise
by Landau and Lifshitz: after having derived by classical arguments an energy differential for the subject at hand, the authors add the magical $T \, dS$
contribution to obtain a corresponding thermodynamic differential which
now contains, in an averaged manner, the (hidden) atomic, molecular, electronic, spin,... motions of the microscopic world.[1]

A Legendre transformation involving the strictly "thermal" conjugate
variables T and S (see Sec. 4.3) produces the *free* energy,[2] more useful in
practice than the internal energy itself:

$$F = U - TS \, . \tag{8.2}$$

This is the thermodynamic potential denoted *Helmholtz free energy*, regarded as a function of the natural variables $(T, X_1, \ldots X_n)$. In all generality then, the total differential of the free energy (8.2) is then

$$dF = -S \, dT + \sum Y_i \, dX_i \, , \tag{8.3}$$

where the Gibbs–Duhem relation has been taken into account and where the
X_i and Y_i have their usual meanings of extensive and intensive variables.

---

[1]Here is a little mnemonic trick which can help a novice recall whether it is $T \, dS$
which should be added, or $S \, dT$: think of "tedious"!

[2]The word "free" is not to be construed as related to the expression "no free lunch"
for example, but rather to "free" as in "available".

Various other free energy differential forms can be obtained in a like manner by $X \Leftrightarrow Y$ Legendre transformations.

Whether starting from an internal energy or a free energy, the corresponding differential form must be integrated. Since the forms are integrable, thanks to the Maxwell relations, this can be done formally by means of Eq. (3.41) of Chapter 3 or by one of its generalizations such as Eq. (A.7) in Appendix A. Let us apply this procedure to the Helmholtz free energy differential form [only one conjugate pair, T, X, is used here in Eq. (8.3), for simplicity, X being the volume V, or the concentration c_i treated at some length in the previous chapter, for example]. The integration gives

$$F(T, X) = -\int_a^T S(t, X) \, \mathrm{d}t + \int_b^X Y(a, x) \, \mathrm{d}x . \tag{8.4}$$

The practical problem now is that the integrals cannot be calculated if the state functions $S(t, x)$ and $Y(t, x)$ are not known, as is often the case. That problem is mentioned in Sec. 3.6, where it was suggested to use Taylor's expansions, for instance. Taylor's expansions are usually not satisfactory for functions as $S(T, c)$, essential for treating alloys however, so that model functions are required, such as the regular solution models covered in Sec. 7.10, or derived from "first principles".

Regardless of the approximation used, entropy as a function of the temperature T and the extensive variables X (again X without indices standing for all extensive X_i required) has total differential, in mathematical notation:

$$\mathrm{d}S = \frac{\partial S}{\partial T} \, \mathrm{d}T + \frac{\partial S}{\partial X} \, \mathrm{d}X . \tag{8.5}$$

The entropy partials may be replaced by measurable quantities: thus, in thermodynamic notation,

$$\left(\frac{\partial S}{\partial T} \right)_X = \frac{C_X}{T} \tag{8.6}$$

where C designates a suitable specific heat, and where the subscript X indicates "at constant extensive variables". Also, from the integrability condition of the differential form (8.3) we have the corresponding Maxwell relations

$$\left(\frac{\partial S}{\partial X} \right)_{T,X} = -\left(\frac{\partial Y}{\partial T} \right)_X . \tag{8.7}$$

Equations (8.6) and (8.7) substituted into (8.5) yield a useful expression

$$\mathrm{d}S = \frac{C_X}{T} \, \mathrm{d}T - \left(\frac{\partial Y}{\partial T} \right)_X \, \mathrm{d}X . \tag{8.8}$$

The first contribution to the entropy differential is an essentially "thermal" one, and the second one is an essentially "mechanical" or "chemical" one.

Similar results may be obtained after exchanging the roles of any number of conjugate X and Y variables to obtain corresponding Gibbs energies, for example. Then the entropy must be regarded as dependent on T and on the mixed set Y and some X_i, the latter dependence being ignored in what follows here. We have

$$dS = \frac{\partial S}{\partial T}dT + \frac{\partial S}{\partial Y}dY \quad, \tag{8.9}$$

where the symbol Y might refer to $-P$, for example. Proceeding as above, but using Maxwell relations pertaining to *free enthalpies*, we find

$$dS = \frac{C_Y}{T}dT + \left(\frac{\partial X}{\partial T}\right)_Y dY \tag{8.10}$$

in which the same "thermal" and "mechanical" contributions appear. Equation (8.10), similar to its *extensive* counterpart Eq.(8.8), will be used later in Sec. 8.4.

The configuration-type partials in Eqs. (8.8) and (8.10) may also be expressed as second derivatives since we have (in more convenient "mathematical" notation):

$$\frac{\partial Y_i}{\partial T} = \frac{\partial^2 F}{\partial T \partial X_i}, \tag{8.11a}$$

$$\frac{\partial X_j}{\partial T} = -\frac{\partial^2 \Psi}{\partial T \partial Y_j}. \tag{8.11b}$$

These partials are off-diagonal elements of a Hessian matrix (or simply Hessian)[3] to be used in Chapter 16 for stability analysis. As an example, if X_j denotes volume, the partial derivative in Eq. (8.11b) represents the isobaric thermal expansivity. If X_j is a component of the magnetic field, the partial in Eq. (8.11b) represents the corresponding component of the pyromagnetic (first-rank) tensor field, numerically equal to the same component of the magnetocaloric tensor (see Chapter 16). Specific heat contributions to expressions (8.8) and (8.10) are diagonal elements of the appropriate Hessians.

The notion of "first principles" calculations was mentioned above, and will be described briefly in Chapter 9, as the possibility of obtaining thermodynamic quantities directly from microscopic considerations *i.e.* from

---

[3]In mathematics, a Hessian matrix is a square symmetric matrix of second-order partial derivatives of a (scalar) function.

quantum mechanical calculations, followed by appropriate averaging. The passage from the microscopic to the macroscopic (or mesoscopic) realm through averaging is a subject requiring appeal to *statistical mechanics*, to be given a brief introduction in the next sections.

8.2 Statistical Interpretation

This section is not meant to provide a sound treatment of statistical thermodynamics, as there are excellent textbooks devoted to that, but aims to present arguments informally linking classical (macroscopic) thermodynamics and statistical (microscopic) concepts. We are so accustomed to regarding matter as consisting of aggregates of discrete units that we often forget that this apparently obvious fact was established quite recently in the history of science. It was only in the late 1800s, when the atomistic hypothesis was still not yet universally accepted, that three almost contemporaneous scientists, J. C. Maxwell (1831–1879), J. W. Gibbs (1839–1903) and Ludwig Boltzmann (1844–1906), incorporated the atomic concept into a full theory of thermodynamics. Then, given the discrete nature of the world, it became possible to *derive* — with a few axioms — the macroscopic laws of matter, previously regarded as postulates, from suitable averages taken over large numbers of particles.

Averages

We therefore have two types of description of matter or states: the macroscopic one, which was that examined heretofore in classical thermodynamics, referring to *macrostates*, which are characterized fully by a few state variables such as $U, V, S, F, G, \dots$, and one referring to arrangements of microscopic units (atoms, molecules, elements, particles...), referred to as *microstates*. Basically, there are two ways of taking averages: (a) start from a material system (solid, liquid, gas) and average over microstates as a function of time, or (b) take averages as a function of extent, *i.e.* of space, actually over replicas of the system in question. It would appear that those two methods are equivalent, but they are not necessarily so. In fact if this equivalence holds, the system is said to be *ergodic*. The difficult question of whether a given system is or is not ergodic will not be considered here. Let us take the temporal average viewpoint (a). Consider a thermodynamic system characterized by its macroscopic extensive variables, U, V, N_i or its intensive ones P, T, μ_i or a combination of both, and apply suitable conditions which can turn the system into an isolated, adiabatic, or closed (diathermic) one. Open systems will not be treated explicitly, so that the number of atomic or molecular species N_j in the system will be constant for all j. The important distinction between those systems was illustrated schematically in Fig. 2.5.

Adiabatic systems

Let the system be enclosed by *rigid* adiabatic walls so that V and N (standing for all of the N_i) be constant as well as the energy E.[4] The particular arrangements of microscopic particles (let us retain that generic term) may also change however, as we allow internal motions to take place, leading to different *microstates*, the number of which, Ω, will generally be huge, if the number N itself is large. Note that the symbol N does not necessarily indicate that one type of particles or elements are present; there may be several, in numbers $N_1, N_2, \ldots,$ with $N = \sum N_j$. *In such an isolated system, with fixed $V, N_i,$ and E, it is a fundamental principle of statistical thermodynamics that all microstates (under those conditions) are equally probable at equilibrium.* I Boltzmann had the insight to express the (macroscopic) entropy S of an isolated system as a function of the number Ω of microstates available to it, thus

$$S = f(\Omega). \tag{8.12}$$

For the function f, Boltzmann adopted the logarithmic one

$$S = k \log W, \tag{8.13}$$

where k is, quite appropriately, known as the Boltzmann constant (as will be shown later), often written today as k_B, and W is the present Ω. That formula was engraved on Boltzmann's tomb in Vienna. The Boltzmann formula tells us that entropy in an isolated system is a monotonic function of the number Ω (or W) of microstates, so that such systems will naturally evolve towards macrostates with the largest number of microstates.

It is interesting to note that the logarithmic functional form adopted by Boltzmann can actually be derived logically from the very weak starting assumption (8.12). To show that, it is necessary to appeal to two systems, A and B, with entropies S_A and S_B, as was done in the macroscopic (classical) case examined in Appendix B. Let S be the entropy of the combined system $A + B$. Then, by the additive property of the entropy, we have[5]

$$S = S_A + S_B. \tag{8.14}$$

For the number of microstates, or multiplicities, or *complexions* we have

$$\Omega = \Omega_A \Omega_B \tag{8.15}$$

---

[4]For historical reasons, the internal energy in classical thermodynamics is often designated by the symbol U, but does not differ from the energy generally designated by E in statistical thermodynamics.

[5]The derivation is from Kestin (1968).

since to each microstate of A there are Ω_B of B, and conversely. From Eqs. (8.14) and (8.15) we have, for the combined system, the functional equation

$$f(\Omega_A \Omega_B) = f(\Omega_A) + f(\Omega_B) \,. \tag{8.16}$$

The logarithmic function $S = k \ln \Omega$ (where k is an arbitrary constant) has the desired property, *i.e.* it is a *sufficient* condition.[6] But we must show that it is the *only* such function, *i.e.* that it is a *necessary* condition. By taking the partial derivative of both sides of Eq. (8.16) with respect to Ω_A we arrive at

$$\Omega_B \, f'(\Omega_A \Omega_B) = f'(\Omega_A) \,, \tag{8.17}$$

where the accent, or "prime" on the function symbol indicates derivation with respect to the argument of the function. Taking the partial derivative with respect to Ω_B of the latter equation then yields

$$\Omega_A \Omega_B \, f''(\Omega_A \Omega_B) + f'(\Omega_A \Omega_B) = 0 \tag{8.18}$$

or

$$\Omega \, f''(\Omega) + f'(\Omega) = 0 \,. \tag{8.19}$$

Hence, the functional equation (8.16) has been transformed into a differential equation for f. The unique solution for this latter equation is $f(\Omega) = k \ln \Omega + C$, as can be easily ascertained. Thus we recover the Boltzmann equation (8.13), although it remains to be shown that the constant k is indeed the Boltzmann constant k_B and that the integration constant C may be set equal to zero. That will be done in following sections. We also recover the notion that in an adiabatic system equilibrium is reached for maximum entropy, or maximum "complexion" Ω.

Optimal Distribution

We now introduce the concept of a *distribution* ν which is the set of numbers n_i^ν, or *occupation numbers*, which denote the number of particles occupying a given energy level i, or *quantum-energy-level* to emphasize the fact that the energy of such levels should be calculated, if at all possible, by quantum mechanical means.[7] For N large, the spacing of quantum levels will be very small, so small that we can also consider the occupation numbers as continuous variables when appropriate. In Fig. 8.1 the energy levels are shown equidistant for the purpose of illustration, but in general need not be. The meaning of the different panels will be given later. It is useful

---

[6]In physics texts, the natural logarithmic function is generally indicated as "log", but as "ln" in chemistry and engineering texts. From now on, we shall adopt the latter convention.

[7]Examples of calculations are mentioned further in Sec. 9.4.

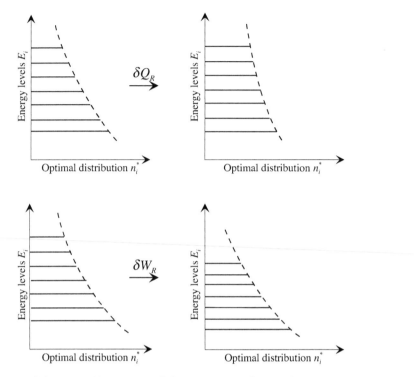

Figure 8.1: Schematic illustration of the two types of virtual changes: reversible heat exchange δQ between upper left system and one at upper right where mostly occupation numbers are altered, and reversible work δW taking place between lower left system and the one at lower right, where mostly spacing between levels are altered (after Kestin, 1968).

to gather the definitions of the various concepts used in this section into a table.[8] The number of ways Ω_ν in which N particles can be distributed over m levels with prescribed occupation numbers n_i^ν is given by

$$\Omega_\nu = \frac{N!}{n_1^\nu!\, n_2^\nu!\, n_3^\nu! \ldots n_m^\nu!}, \qquad (8.20)$$

which is the way that N distinguishable objects can be distributed on m levels with given number n_i on each level, regardless of their order. If there were no distinct levels, then the total number would be simply the number of permutations $N!$. But in the present case, that would be overcounting,

---

[8]Kestin's nomenclature is mostly followed.

Table 8.1: Definitions of notation used in this section.

Symbol	Definition
N	Number of particles
E	Total energy of system
ϵ_i	Energy of quantum-level i
m	Level with largest energy considered
ν	Index of distribution
n_i^ν	Occupation number on level i in distribution ν
Ω_ν	Number of ways of forming a given distribution ν
$\mathcal{Z}$	Partition function
$\alpha, -\beta$	Lagrange multipliers
k_B	Boltzmann's constant $= 1.38 \times 10^{-23}$ Joules/$^\circ$K

since permutation on levels is permitted without changing the overall distribution; hence we must divide by the factorial of the number of particles on each level, bearing in mind that $0! = 1! = 1$. Since the system is adiabatically enclosed, equilibrium will be attained for that distribution ν which maximizes Ω_ν, provided that constraints (8.21) are obeyed

$$G_1 \equiv \sum_{i=1}^{m} n_i^\nu - N = 0 \quad \text{(a)} \quad \text{and} \quad G_2 \equiv \sum_{i=1}^{m} n_i^\nu \epsilon_i - E = 0 \quad \text{(b)}. \quad (8.21)$$

Thus, we must solve a problem of constrained extremum, for which the general case is described fully in Appendix D. We choose to maximize the logarithm of Ω rather than Ω itself because it enables us to use Stirling's approximation for the logarithm of factorials: $\ln n! \cong \ln n - n$, valid for large n. That procedure is allowed since the logarithm is a monotonically increasing function of its argument. The method of Lagrange multipliers applied to the present case tells us to solve the system of equations D.10 with the function to be optimized being

$$\mathcal{F} = N \ln N - N - \sum_{i=1}^{m} (n_i \ln n_i - n_i) \quad (8.22)$$

(where a script F has been used instead of the plain F so as not to cause confusion with the Helmholtz free energy F and where the index ν has been dropped). The constraints $G_1 = 0$ and $G_2 = 0$, given by Eqs. (8.21 a) and (8.21 b) respectively, are to be multiplied by appropriate Lagrange multipliers.

Use of Eqs. (D.10) (in Appendix D) requires taking the partial derivatives with respect to n_i indicated by a prime (′) of the functions $\mathcal{F}, G_1$, and G_2, which we label $\mathcal{F}'_i, G'_1$ and G'_2 for short, yielding

$$\mathcal{F}'_i = 0 - [\ln n_i + \frac{n_i}{n_i} - 1] = \ln n_i \tag{8.23}$$

and

$$G'_1 = 1 \quad \text{and} \quad G'_2 = \epsilon_i \,.$$

Combining these equations with the two Lagrange multipliers, which we take here to be α and $-\beta$, we find the required expressions for the formula (D.10)

$$-\ln n_i + \alpha - \beta \epsilon_i = 0 \quad (i = 1, 2, \ldots, m) \tag{8.24}$$

and by taking antilogs we find

$$n_i^* = e^\alpha e^{-\beta \epsilon_i} \,, \tag{8.25}$$

which gives the required expression for the occupation number n_i^* of the ith level in the optimal distribution (Ω^*). We now use the constraint (8.21a) itself: by summing the equilibrium occupation numbers (8.25) we obtain (dropping the asterisk)

$$\sum_{i=1}^{m} n_i = e^\alpha \sum_{i=1}^{m} e^{-\beta \epsilon_i} = N$$

hence

$$e^\alpha = \frac{N}{\mathcal{Z}} \tag{8.26}$$

with the all-important *partition function*[9] defined by

$$\mathcal{Z} = \sum_{i=1}^{m} e^{-\beta \epsilon_i} \,, \tag{8.27}$$

featuring the famous *Boltzmann factor* $\exp(-\beta \epsilon_i)$.

Let us now imagine performing a small virtual change δ of the optimal distribution Ω:

$$\delta \ln \Omega = -\sum_{i=1}^{m} \ln n_i \, \delta n_i = -\sum_{i=1}^{m} (\alpha - \beta \epsilon_i) \, \delta n_i \tag{8.28a}$$

$$= -\alpha \delta \sum_{i=1}^{m} n_i + \beta \delta \sum_{i=1}^{m} \epsilon_i \, n_i = 0 + \beta \, \delta E \tag{8.28b}$$

---

[9] *Zustandsumme* in German, *i.e.* *sum-over-states*; hence the letter "Z", and in "script" font so as to avoid confusion with the Helmholtz function.

where we have used the constraints (8.21), noting that E is varied in the process but that N is not. The second term in Eq. (8.28b) is valid because the energies ϵ_i are considered fixed for this *virtual* change. Dividing through by $\beta\,\delta E$ produces the following partial derivative at constant extensive variables X (entropy excluded), for example V, which are those which would alter the values of the energy levels ϵ_i:

$$\left(\frac{\partial \ln \Omega}{\partial E}\right)_X = \beta. \qquad (8.29)$$

In Eq. (8.29) it is permissible to write

$$\left(\frac{\partial\, k \ln \Omega}{\partial E}\right)_X = \left(\frac{\partial S}{\partial E}\right)_X \qquad (8.30)$$

since $dE = TdS + YdX$ and hence $\left(\frac{\partial E}{\partial S}\right) = T$ at constant X. In Eq. (8.30) the entropy S was identified with $k \ln \Omega$ through the arguments of the subsection on adiabatic systems, with the proportionality constant k left provisionally undefined. Comparison with simple systems for which the calculations can be carried out exactly indicate that the proportionality constant is indeed equal to Boltzmann's constant k_B. Finally we find the important result:

$$\beta = \frac{1}{k_B T}. \qquad (8.31)$$

The value of Boltzmann's constant, given in Table 8.2, of course depends on the "size" of the degree. Normally it is the Celsius degree which is used, which places 100 degrees C between the freezing and boiling of water at atmospheric pressure.

Correspondence with Classical Thermodynamics

We have seen how Boltzmann was able to relate macroscopic and microscopic concepts for the entropy in the case of adiabatic systems. We now look for correspondence in the case of other thermodynamic quantities, which will necessitate the introduction of closed (non-adiabatic) systems. It must be emphasized that such a program to unite the macro- and microscopic worlds spans a huge range of orders of magnitude. It is therefore to be expected that the arguments presented will be mostly heuristic ones. To justify them, some simple practical systems should be examined; however, in this introductory treatment, only the general framework of the derivations will be given.

Since the partition function is central to statistical thermodynamics, we shall start with its total differential, or better, the differential of its logarithm as a function of the natural variables β and the $\epsilon_i (i = 1, 2, \ldots, m)$:

$$\mathrm{d}\ln \mathcal{Z} = \frac{\partial \ln \mathcal{Z}}{\partial \beta}\mathrm{d}\beta + \sum_{i=1}^{n} \frac{\partial \ln \mathcal{Z}}{\partial \epsilon_i}\mathrm{d}\epsilon_i . \tag{8.32}$$

Expressing the partials explicitly requires the calculation of

$$\frac{\partial \ln \mathcal{Z}}{\partial \beta} = -\frac{E}{N} \quad \text{and} \quad \frac{\partial \ln \mathcal{Z}}{\partial \beta} = \beta\frac{n_i^*}{N} . \tag{8.33}$$

Inserting back into the original differential form gives

$$\mathrm{d}\ln \mathcal{Z} = -\frac{E}{N}\mathrm{d}\beta - \frac{\beta}{N}\sum_{i=1}^{n} n_i^*\mathrm{d}\epsilon_i , \tag{8.34}$$

where the normalization N is required because the left hand side of the equation refers to a single particle, whereas E and the *sum* refer to the whole system.

Correspondence with classical thermodynamics necessitates bringing the differential form (8.34) to one that resembles the fundamental form (3.18), which is the heart and soul of classical thermodynamics. The left hand side of Eq. (8.34) suggests the differential of entropy (by Boltzmann's equation), so that we should make the first term on the right hand side resemble the differential of energy E. For that to happen, it is necessary to perform the following substitution $E\,\mathrm{d}\beta = \mathrm{d}(E\beta) - \beta\,\mathrm{d}E$. Thus, after performing what is essentially a Legendre transformation, setting $\beta = \frac{1}{kT}$, and multiplying through by N we obtain

$$\mathrm{d}\left[Nk \ln \mathcal{Z} + \frac{E}{T}\right] = \frac{\mathrm{d}E}{T} - \frac{1}{T}\sum_{i=1}^{n} n_i^*\mathrm{d}\epsilon_i \tag{8.35}$$

or, after rearranging

$$\mathrm{d}E = T\,\mathrm{d}\left[Nk \ln \mathcal{Z} + \frac{E}{T}\right] + \sum_{i=1}^{n} n_i^*\mathrm{d}\epsilon_i . \tag{8.36}$$

Equation (8.36) is of the form introduced very early in these notes, namely Eq. (3.18) or Eq. (8.1)

$$\mathrm{d}U = T\,\mathrm{d}S + \sum_{i} Y_i\,\mathrm{d}X_i \tag{8.37a}$$

or

$$\mathrm{d}U = \delta Q_R + \delta W_R , \tag{8.37b}$$

where the total energy U is the same as E, the subscript R denotes reversible changes, and the variables X_i and Y_i are extensive and intensive classical variables used previously, in particular in connection with Eq, (8.30), above.

These considerations suggest the following identifications:

$$dU = dE \tag{8.38a}$$

$$\delta Q_R = T\,dS \tag{8.38b}$$

$$\delta W_R = \sum Y dX \quad (\text{forexample} : -P\delta V), \tag{8.38c}$$

leading to

$$S = Nk_B \ln \mathcal{Z} + \frac{E}{T} \tag{8.39}$$

where we can now confidently write k_B for the Boltzmann constant, and from which we have the important relation for the Helmholtz free energy:

$$F = E - TS = -Nk_B T \ln \mathcal{Z}. \tag{8.40}$$

We have now derived statistical versions of such classical thermodynamic concepts as internal energy, entropy, Helmholtz free energy and temperature, but other quantities, such as Gibbs free energy and chemical potentials can also be derived. For that, open systems must be considered, but the derivations will not be presented however since the mathematical techniques to be employed are quite similar. The statistical approach has significant advantages as it provides physical insight into the problems studied. For example, the reversible change in the total energy E, in Eq. (8.36) consists of two contributions, one acting mainly on the population of energy levels (bracketed term), and one acting mainly on the placement of the levels themselves. We say "mainly" because such a separation of contributions would be only valid if these two would be completely independent. The case of independence is illustrated schematically in Fig. 8.1: the upper panels represent reversible heat transfer δQ_R where only occupation numbers n_i change, and the lower panels represent reversible work exchanged, δW_R symbolically, resulting in alteration of levels themselves.

Discussion

Going from a differential form to one with finite quantities only requires integration, with resulting constants of integration. We seem to have neglected that here. That is true, and that is why identifications of microscopic to macroscopic variables are introduced as "suggestions", to be made plausible by further examinations of simple cases (which we leave to appropriate textbooks in the field of Statistical Mechanics). Likewise, the practice of identifying sums term-for-term is usually not justified (*i.e.* as assuming that, if $A + B = C + D$ then it is true that $A = C$ and $B = D$, which it normally isn't) and is done here in the same practical way. In truth, finding the correct correspondence between micro- and macro-realms is not guaranteed by any exact procedures, but rests partly on physical intuition. The difficulty is that the two realms in question are so different in orders of magnitude that identification of the two is difficult to justify theoretically.

Another source of approximation is replacement of the full distribution by its largest value Ω^*, thereby resulting in the neglect of fluctuations in the optimal distribution. However, this approximation is normally acceptable since, for large number of particles N, the Ω^* distribution is known to be sharply peaked, practically a delta function. Nonetheless, as will be seen in Sec. 15.6, that approximation will produce incorrect critical exponents in second-order transitions.

It is often stated that entropy is a measure of disorder. That is not generally true because the concept of "order" is to some extent a subjective one. As an example, Callen (1985) suggests the following amusing analogy:

> A neatly built brick wall is evidently more ordered than a heap of bricks. [...] Unfortunately the "heap of bricks" may be the prized creation of a modern artist, who would be outraged by the displacement of a single brick!

Still, the linking together of entropy and disorder is tempting for the following reason: if we decide that an *ordered* arrangement of particles is one that can be described by a few number of rules (such as "all white particles on the left, all black ones on the right"), and that a *particular disordered* arrangement is one which cannot be described by a few simple rules, then there will obviously be a far greater number of micro states labeled "disordered" than ones labeled "ordered". Thus, by the Boltzmann formula, the probability of finding a system in a disordered state will be greater than that of finding it in an ordered one. Since the value of the logarithmic function increases monotonically with its argument, entropy S will increase spontaneously as the system of particles evolves towards disordered states. However, as Callen's example shows, the notion of linking entropy to disorder is only valid if the concept of "order" has been defined properly for the system at hand. In other words, identifying disorder and entropy is correct only if the very notion of order, or disorder, is *defined* by the Boltzmann formula itself, it which case the "law" stating that entropy promotes disorder is nothing but a tautology; it is not a law of thermodynamics. Counterintuitive examples will be given in Sec. 10.1.

Finally it was stated in the introduction to this section on Statistical Interpretation that two descriptions of matter were available: the *macro-*

scopic and the *microscopic*. Actually, there is a third one: the *mesoscopic* one which, as the name implies, is intermediate between the other two. Section 14.1 will provide an introduction to this approach. At present suffice it to say that it combines the continuous nature of the macroscopic description and the discrete nature of the microscopic one. That approach is particularly useful in phase diagram and phase field calculations, as will be seen.

8.3 The Third Law of Thermodynamics

In the early 1900's, galvanic cell methods had been perfected for carrying out chemical reactions nearly reversibly. It was then noticed that, as the operating temperature was lowered, free energy (or free enthalpy) changes for isothermal reactions, denoted "Δ" in Eqs. (8.41), such as chemical reactions, or phase changes at equilibrium, tended to approach the corresponding internal energy (or enthalpy) changes. The following property thus appeared to hold universally:

$$\lim_{T \to 0} (\Delta U - \Delta F) = 0 \quad \text{or} \quad \lim_{T \to 0} (\Delta H - \Delta G) = 0 \qquad (8.41)$$

as the temperature approached absolute zero. In this section, the symbol "lim" will always denote the limit as T approaches absolute zero.

Of course, at $T = 0$, Eqs. (8.41) hold trivially since, for isothermal processes $\Delta F = \Delta U - T \Delta S$ and similarly for ΔG. Observation indicated, however, that the energy differences in Eqs. (8.41) (label them both ΔD, for short) reached their limiting value of zero with vanishing slope as with the full curve of Fig.8.2 rather than with finite slope (dashed curve). Hence, a statement stronger than (8.41) can be made:

$$\lim \frac{\partial \Delta D}{\partial T} = \lim \frac{\partial (T \Delta S)}{\partial T} = 0 \qquad (8.42)$$

or

$$\lim \left(\Delta S + T \frac{\partial \Delta S}{\partial T} \right) = 0. \qquad (8.43)$$

The latter expression is verified for $\lim \Delta S$ *not* equal to zero only if ΔS and its temperature derivative multiplied by T approach the same non-zero limit, thus implying $\frac{\partial \Delta S}{\partial T} \to \infty$ as $T \to 0$. Since evidence ruled out this case, Nernst and his followers were forced to conclude $\Delta S \to 0$ for $T \to 0$. A new law of thermodynamics, the *Third Law*, was therefore proposed (for which Nernst was awarded a Nobel prize in 1920), one formulation of which is (A. B. Pippard, *Classical Thermodynamics*, p. 51):

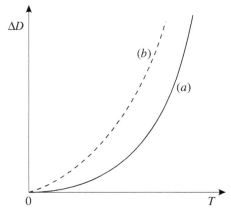

Figure 8.2: The difference ΔD (or $T \Delta S$) must approach absolute zero with zero slope, as in curve (a), rather than with finite slope, as with the dashed curve (b).

As the temperature tends to zero, the magnitude of the entropy change in any reversible process tends to zero.

It is important to point out what the third law says and does not say. First, it is not required that the thermodynamic change of state denoted by the symbol "Δ" ($\equiv$ initial $\rightarrow$ final) be an actual reversible process. All that is required is that the two states (initial, final) be connected, in principle, by some reversible path. From the definition of the term "reversible," it would appear that the two states in question must be equilibrium ones. This is true in a limited sense only: at low temperatures, many types of disorder may be "frozen in," in other words, certain degrees of freedom normally available to a thermodynamic system at elevated temperatures may become unavailable due, for example to sluggish reaction kinetics. Hence, by the term "equilibrium," implied by the term "reversible," we merely mean a state of minimum energy with respect to those degrees of freedom still available at low temperature.

With that understanding, then the third law may be applied to glasses, crystals with disordered distributions of atoms on lattice sites, crystalline substances containing dislocations, grain boundaries, and other disordered configurations. Even perfect elemental single crystals will normally contain random distributions of isotopes and "zero-point" electronic motion. In all such cases, the residual entropy resulting from frozen-in disorder is implicitly subtracted out when taking the difference $\Delta S = S(final) - S(initial)$.

Another statement that the third law does not make is that the entropy of thermodynamic systems should vanish at absolute zero. Entropy, being an extensive quantity, contains implicitly an arbitrary additive constant.

However, since specific heats are positive for any stable substance, S must decrease monotonically with T, so that entropy must reach its minimum value at $T = 0$, at least for those portions of S which depend on available degrees of freedom. Also, we may assume that contributions to entropy coming from defects such as frozen-in vacancies, dislocations and interfaces will be small, simply because the relative amount of such defective material will be small. Accordingly, we may, by convention, set $S = 0$ for all perfect crystals at $T = 0$, the term "perfect" designating the elemental crystals or stoichiometric compounds free from compositional disorder. Finally note that, as a consequence of the foregoing, it is not correct to state, as is sometimes affirmed, that the third law disallows disorder at absolute zero. Disorder is not forbidden at $T = 0$, but the entropy of such states does not vanish.

Some authors (for example Landau and Lifshitz, 1958) refer to the third law as the "Nernst Theorem" which states:

> In the limit of absolute zero, the entropy is equal to Boltzmann's constant times the natural logarithm of the degeneracy of the ground state, (the number of microstates at the lowest energy level, g_o),

its mathematical formulation being

$$\lim_{T \to 0} S = k_B \ln g_o \qquad (8.44)$$

(see derivation in Appendix F. In the classical context, however, the "Nernst principle" must be regarded as a "law" of thermodynamics). For non-degenerate systems (the usual case) the entropy does indeed vanish at absolute zero. An application of Eq. (8.44) is described in Sec. 9.5 in the context of temperature–concentration phase diagrams at low temperature.

The third law has important consequences for properties of materials at low temperature. Thus, from the Maxwell equation

$$\left(\frac{\partial V}{\partial T} \right)_P = - \left(\frac{\partial S}{\partial P} \right)_T \qquad (8.45)$$

it follows that, for isentropic processes, the isobaric thermal expansivity β (not to be confused with the variable $1/k_B T$ of the previous section) must tend to zero as $T \to 0$:

$$\lim \beta = \lim \frac{1}{V} \left(\frac{\partial V}{\partial T} \right)_X = 0. \qquad (8.46)$$

Specific heats ("at constant something," X) must likewise tend to zero at low temperatures with, as mentioned above, the temperature derivative of

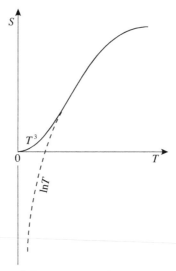

Figure 8.3: The behavior of the entropy at low temperature specific heat cannot be extrapolated from high temperature; for solids one expects a T^3 dependence.

entropy remaining finite. It should then be possible to expand the specific heat about $T = 0$ as $C_X = KT^m$, m being the exponent of the leading term, and K being independent of temperature. The following integral must converge for $T \to 0$:

$$\int^T \frac{C_X}{t}\, dt = \int^T Kt^{m-1}\, dt = \frac{K}{m}T^m \tag{8.47}$$

leading to the condition $m > 0$. Specific heats must then tend to zero as a positive power for T. For solids, the Debye model predicts $m = 3$, in good agreement with low-temperature measurements. For metals, the electronic contribution, according to the Sommerfeld model, gives $m = 1$. For superfluids, however, C exhibits an exponential decay with temperature. At the other extreme of very high temperature, the specific heat of solids at constant volume approaches the Dulong and Petit limit of $3R$, at least for non-metals. For most substances, the rule $c_V = 3R$ is already well obeyed around room temperature. At higher temperatures entropy should then vary as $\ln T$, as for an ideal gas with constant specific heat [see Eq. (5.9)]. Taking both high and low temperature limits into account results in $S(T)$ curves similar to the one shown schematically in Fig. 8.3. It is seen that extrapolating the $C_v = $ const. behavior, acceptable at high temperature, would make S tend to $-\infty$ at very low temperature. The third law shows that such extrapolations of familiar high-temperature properties is invalid:

quantum effects come into play to avert an "entropy catastrophe".

8.4   The Unattainability of Absolute Zero

Since $T = 0$ appears to be a singular point in thermodynamic space, it may
be surmised that reaching it poses a difficult practical problem. How one
might go about doing this practically is illustrated schematically in Fig. 8.4:
Two $S(T)$ curves, such as those of Fig. 8.4, are shown for two different values

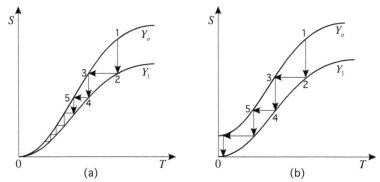

Figure 8.4: Two entropy curves as a function of T for different values of some intensive variable. In case (b), the zero of temperature may be reached (incorrectly)
in a finite number of steps; in case (a) (correct), it cannot.

Y_0 and Y_1 of some intensive quantity, such as pressure. Starting from initial
point 1, one might first extract heat isothermally and reversibly to point
2, then cool adiabatically to point 3, then repeat these processes to points
4,5,6... Since pairs of points 1-2, 3-4, 5-6, *etc.*, are on successive isotherms,
the differences $\Delta S_{1\to2}$, $\Delta S_{3\to4}$,...must become progressively smaller, by
virtue of the third law, until the curves of Y_0 and Y_1 finally merge at
$T = 0$. It is therefore clear that the absolute temperature can never be
reached by a finite number of such processes. If the third law were violated
as in Fig. 8.4(b), it would be perfectly possible to reach absolute zero by a
finite number of steps.

Consider the entropy differential (8.10) in which the substitution $X \to$
$V, Y \to -P$ is performed. We have (per mole):

$$\mathrm{d}s = \frac{c_P}{T}\mathrm{d}T - \left(\frac{\partial v}{\partial T}\right)_P \mathrm{d}P . \tag{8.48}$$

Step $1 \to 2$ could be achieved by increasing pressure reversibly until some
higher pressure P_1 is reached, with the substance to be cooled placed

in diathermic contact with a reservoir at temperature T_1. According to Eq. (8.48) if the thermal expansivity is positive (as is usually the case), the entropy must continually decrease during the isothermal process. The original pressure may then be restored by decompressing adiabatically and reversibly to reach point 3 back on the P_0 isobar. During process $2 \rightarrow 3$, since c_P is positive, Eq. (8.48) shows that, with $ds = 0$, temperature must continually decrease.

In practice, successive expansions and adiabatic compressions cannot attain very low temperatures. To proceed to lower temperatures, it is necessary to resort to so-called *adiabatic demagnetization*. The same figure (8.4) can be used and the same differential (8.10), but now with the substitution $X \rightarrow \mathcal{M}$ (magnetization) and $Y \rightarrow \mathcal{H}$ (magnetic field), producing:

$$ds = \frac{c_{\mathcal{H}}}{T} dT + \left(\frac{\partial \mathcal{M}}{\partial T} \right)_{\mathcal{H}} d\mathcal{H}. \tag{8.49}$$

Curve $\mathcal{H}_0$ may refer to zero applied field, $\mathcal{H}_1$ to non-zero. Step $1 \rightarrow 2$ represents isothermal magnetization during which, by Eq. (8.48), entropy decreases since the pyromagnetic coefficient $\frac{\partial \mathcal{M}}{\partial T}$ is negative. Physically, the ordering of elementary magnetic dipoles (spins) in the solid decreases the configurational entropy. Step $2 \rightarrow 3$ is performed by then insulating the substance and decreasing the magnetic field to zero ($\mathcal{H}_0$). With $ds = 0$, (8.49) indeed shows that, for this process, we must have $dT < 0$. In this way, temperatures of the order of a fraction of a Kelvin may be reached. Lower temperatures still may be reached by using nuclear magnetism. The lowest temperature yet achieved is about 15 microdegrees Kelvin.

The question may be raised as to how to provide the required reservoirs at successively lower temperatures $T_1, T_3, T_5, T_7, \ldots$. In principle, one starts with the largest amount of material possible at point 1, assuming that a T_1 reservoir be available. After reaching point 3 by an adiabatic process, a fraction of the substance is magnetized (or compressed) isothermally, using the remainder of the material as reservoir at T_3. The same procedure is used in step $5 \rightarrow 6$: a still smaller amount of material is "ordered" (magnetized or compressed), with the remainder of the material used as a reservoir at temperature T_5, *etc.*

Of course, only idealized processes have been described here; real cooling experiments are very difficult to carry out. Nevertheless, the basic idea is always the same: cooling is achieved by zig-zag processes which operate alternately on *thermal and configurational* entropy contributions. Hence, quite generally, it may be deduced from the third law that:

> *By no finite series of processes is the absolute zero of temperature attainable.*

Such is the principle of unattainability of absolute zero.

References

Callen, H. B. (1985), *Thermodynamics and an Introduction to Thermostatics*, J. Wiley & Sons, New York.

de Fontaine, D. (1979), *Configurational Thermodynamics of Solid Solutions*, in *Solid State Physics*, Ehrenreich, H., Seitz, F. and Turnbull, D., **34**, pp. 73-274, Academic Press.

Landau, L. D. and Lifshitz, I. M. (1958), *Statistical Physics*, Addison-Wesley.

Kestin, J. (1966 and 1968), *A Course in Thermodynamics*, Vols. I and II, Blaisdell Publishing, Co.

Pippard, A. B. (1961), *The Elements of Classical Thermodynamics*, Cambridge University Press.

Part II

Materials Applications

Chapter 9

Temperature-Composition Phase Diagrams

In his classic textbook on phase diagrams, Rhines (1956) has this to say:

> Phase diagrams mean more to the metallurgist than mere graphical records of the physical states of matter. They provide a medium of expression and thought that simplifies and makes intelligible the otherwise bewildering pattern of change that takes place as elemental substances are mixed one with another and are heated or cooled, compressed or expanded. They illumine relationships that assist us in our endeavor to exercise control over the behavior of matter.

This text emphasizes very eloquently the importance of phase diagrams for metallurgists, although today we would refer more generally to "material scientists." The textbook cited has particularly appealing sections on ternary (three–components) and quaternary (four–components) phase diagrams illustrated by three-dimensional perspective figures of phase regions. However, modern readers may be surprised by the near–total absence of thermodynamic theory in this text; previously phase diagrams were constructed by determining experimentally which phases occupied which regions of temperature-composition space, drawing a boundary line around those regions, while respecting the Gibbs phase rule. Another text on phase diagrams is that of Prince (1966, unfortunately out of print) which also has excellent geometric representations of multicomponent phase diagrams, but with greater emphasis on the basic thermodynamic aspects of the subject.

9.1   Binary Systems

In the present section, only binary systems (two components only, say A and B) will be treated. Multicomponent systems, ternary and higher, which pose considerable graphic representational problems, will be treated in subsequent sections. As another basic simplification, phases, even solid ones, will be considered as subjected only to hydrostatic pressure. Solid phases thus will be assumed to be present in their equilibrium state, *i.e.* we shall allow implicitly structural defects (dislocations, incoherent interfaces, and so on) to relax all non-hydrostatic stresses in solid phases. This is an important simplification, as stressed solids are, for example, found *not* to obey the elementary phase rule. This rule, as Gibbs cautioned over a hundred years ago, is strictly valid only for fluid phases, or hydrostatically stressed solids.

Let us now show how, in principle, binary phase diagrams can be derived graphically from knowledge of free energy functions. Free energy curves for as many phases as desired may be combined at any given temperature. Stable equilibria are obtained by constructing the lowest possible common tangents, as explained in Sec. 7.9.

One Free Energy Curve

For this qualitative study, assume that the molar Gibbs free energy of mixing is adequately approximated by a regular solution model model (see Sec. 7.10). The g_M portion of Eq. (7.87), specialized to the binary case, may be written (with the unique interaction parameter W in place of Ω),

$$g_M(c_A, c_B, T) = 2W c_A c_B + RT(c_A \ln c_A + c_B \ln c_B) \qquad (9.1a)$$

or, with only the independent concentration variable $(c \equiv c_B)$:

$$g_M(c, T) = 2W c(1 - c) + RT[c \ln c + (1 - c) \ln(1 - c)]. \qquad (9.1b)$$

Indices A, B have been dropped from the single remaining effective interaction parameter:

$$W = W_{AB} - \frac{W_{AA} + W_{BB}}{2}. \qquad (9.2)$$

Note that Eqs. (9.1) and (9.2) are invariant in the exchange of A and B, or c and $(1 - c)$, hence all regular solution free energy of mixing curves are symmetric about the mid-point of the concentration axis, *i.e.* about the value $c = \frac{1}{2}$. First and second derivatives of g with respect to the independent variable c are, successively

$$g' = 2W(1 - 2c) + RT[\ln c - \ln(1 - c)] \qquad (9.3)$$

and

$$g'' = -4W + \frac{RT}{c(1-c)}. \tag{9.4}$$

Let us consider the case $W > 0$. The effective interaction parameter W is positive when the unlike bond energy W_{AB} is larger than the average of the like bond energies, W_{AA} and W_{BB}. Hence, enthalpy will be minimized when as many A atoms (molecules) and B atoms cluster in separate domains. When W is positive it is said that the solution has "clustering" tendency.

A "clustering" phase diagram can be readily constructed based on this model. Since only a single free energy curve is considered, the energy of mixing alone suffices to describe equilibrium properties. At high temperatures, the entropy dominates, and the corresponding free energy curve will resemble the ideal entropy curve of Fig. 7.6 multiplied by the negative of the absolute temperature. Then $g(c)$ is everywhere convex so that, at fixed c, the equilibrium state of the system is that of the disordered (completely mixed) solution of molar free energy $g(c)$.

As temperature decreases, and since W is positive, a temperature will be reached at which the second derivative g'' will vanish at $c = \frac{1}{2}$. That temperature, called the critical temperature T_c, is given by

$$RT_c = W. \tag{9.5}$$

This critical temperature is very similar to the one located at the top of the vapor-liquid 2-phase region found earlier in the case of one-component equilibrium in Chapter 6. It follows that, at T_c the system's heat capacity, or compressibility will exhibit a profile similar to the one shown at Fig. 6.3 (b), characteristic of second-order transitions rather than the delta function expected for a first-order transition.

Just below T_c, $g(c)$ ceases to be convex, but develops concave regions, as shown in Fig. 9.1, upper portion. For any $T < T_c$, the quadratic equation $g'' = 0$ has two real solutions c_S, symmetric about $c = \frac{1}{2}$. These are the so-called *spinodal points* (see Fig. 9.1, upper curve, constructed for $\frac{T}{T_c} = 0.8$). Because of the symmetrical properties of the regular solution free energy curve, the common tangency points c^α, c^β may be determined by solving the transcendental equation $g' = 0$, *i.e.*

$$\ln \frac{c}{1-c} = -\frac{2W}{RT}(1-2c). \tag{9.6}$$

Below T_c, loci of solutions of $g'' = 0$ (spinodal) and $g' = 0$ (equilibrium common tangency points, or binodals) can be plotted to yield the *spinodal curve* (dashed line, Fig. 9.1, lower portion) and *binodal curve* (full line) or *miscibility gap* (MG).[1]

---

[1] There is more on the spinodal concept in Chapter 18.

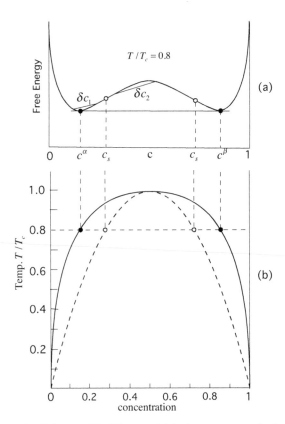

Figure 9.1: Regular Solution Model: panel (a), free energy calculated at reduced temperature $T/T_c = 0.8$; by the "chord rule", concentration fluctuation δc_1 increases the free energy locally, fluctuation δc_2 decreases the free energy. Panel (b), miscibility gap locus (MG, full curve), spinodal locus (dashed curve).

Figure 9.1 is the *phase diagram* for the system under study: regular solution model with $W > 0$. This diagram is not merely a schematic one: the free energy curve and the phase diagram were in fact calculated from the equations given above. The phase diagram should be interpreted as follows: for $T > T_c$, the (mixed) disordered solution of two types of constituents, A and B, say, is stable. Below T_c, and for any point lying inside the MG, the solution decomposes into a mixture of two phases, α and β, the concentrations of which are given by $g' = 0$. Above T_c, the distinction between "phases" α and β disappears. The spinodal line separates regions where the curvature of g is positive from the region where it is negative. In that latter region (*inside* the spinodal), any small fluctuation in composition away from the average concentration lowers the average free energy according to the *chord rule* of Sec. 7.9. For an average concentration in regions *outside* the spinodal, but inside the MG, any small fluctuation δc raises the average free energy. In that region, only a sufficiently *large* composition change can actually lower the average free energy. In that case, a *nucleation* event is required (nucleation theory will be discussed in Chapter 17). Systems with $W < 0$ cannot be handled in the present context; in that case, *unlike* pair bonds are favorable, and the constituents tend to order on crystal lattice sites. In that context, crystallography becomes an important aspect of the problem, requiring different sets of tools. This topic will not be pursued here, but has been treated extensively in the author's review papers [de Fontaine (1979 and 1994)] and in Chapter 18.

Dilute solutions

When a given constituent of a solution is present in very small proportions, that solution is said to be *dilute* in that constituent. If the dilution is high enough so that the minority constituent does not interact with itself, then the free energy of mixing (here simply noted g_M) is effectively ideal, and the relevant thermodynamics simplifies. If, furthermore, the material in question is crystalline, then the solute elements are considered as point defects, such as vacancies or interstitials, to be treated in more detail in Chapter 12.

Recall from Eqs. (7.43) and (7.87) that we can write quite generally the Gibbs energy as $g = g_L + g_M$: a linear term plus a mixing term. In the present case, it will suffice to take the regular solution model as the free energy g_M given by Eq. (9.1b) since a dilute solution is regarded as practically ideal. Indeed, ideality requires that the interaction term in that equation vanish, leaving only the ideal entropy of mixing plus the linear term which by the definitions Eqs. (7.40) and (7.42) will have the form $z_1 + (z_2 - z_1)c$ with c, as usual, being the concentration of the minority component $c \equiv c_2$ and $c_1 = 1 - c$. Since we are dealing with Gibbs free energies $(z \to g)$, let us write the coefficient of c in the linear term as $g° \equiv (g_1 - g_2)$, where $g°$ can be written as

$$g° = h° - T s° . \tag{9.7}$$

Then the resulting Gibbs energy g is just g_M (from Eq. (9.1b)) plus the linear term $g_L = w_o + g°c$ (the constant w_o is from the W_o of Eq. (7.87)) but with the interaction term left out because of ideality. We are left with the simple expression

$$g_\dagger = w_o + g°_\dagger c_\dagger + RT[c_\dagger \ln c_\dagger + (1 - c_\dagger)\ln(1 - c_\dagger)] , \tag{9.8}$$

the dagger symbol indicating that the subscripted variables pertain to dilute solutions only, so that $c_\dagger$ actually designates the concentration of *point defects*.

The equilibrium concentration is obtained by setting to zero the derivative of Eq. (9.8)

$$g'_\dagger = w_o + g^\circ_\dagger + RT[\ln c_\dagger - \ln(1 - c_\dagger)] = 0 \qquad (9.9)$$

which can be solved to yield

$$c_\dagger = e^{-g^\circ_\dagger/RT} . \qquad (9.10)$$

From Eq. (9.7) we may also write

$$c_\dagger = C_o\, e^{-h^\circ_\dagger/RT} \qquad \text{with} \qquad C_o = e^{s^\circ_\dagger/R} , \qquad (9.11)$$

showing that the primary temperature dependence of the solute concentration is given by the exponential of the reciprocal of the absolute temperature, in the ubiquitous form of the Boltzmann factor. Such equilibrium concentrations can also be arrived at by means of the equilibrium constant of chemical reactions, as shown later in Chapter 11.

It is seen from Eq. (9.11) that the only way that the equilibrium concentration $c_\dagger$ can vanish at non-zero temperature is for $h^\circ_\dagger$ to increase without limit. Theoretically then, the entropy of mixing always favors at least some solubility, some contamination of "pure" substances, as was already mentioned in Sec. 7.7. Often, the actual solubility near a pure substance or compound is so restricted as to be invisible on a phase diagram, leading to what is called "line compounds", an example of which is shown schematically on Fig. 9.6.

Two Free Energy Curves

Consider now two distinct phases, α and β, with their respective free energy curves g^α and g^β, for alloy system A, B. One phase (α) could represent the liquid, the other (β) some crystalline structure. For simplicity, imagine the shape of each curve to be represented, at least qualitatively, by a regular solution model. Since, in the present case, the curves and their relative positions need to be compared to one another, the linear portions g_L may no longer be ignored. In fact, for each pure element ($i = A, B$) and each phase ($\gamma = \alpha, \beta$), we have the Gibbs free energy expressions $g^\gamma_i = h^\gamma_i - Ts^\gamma_i$. For each pure component, transition temperatures T_i may be found whenever Δh_i and Δs_i have same signs, with Δz_i ($z = h, s, g$) being defined collectively by $\Delta z_i = z^\alpha_i - z^\beta_i$. Let us assume (without loss of generality)

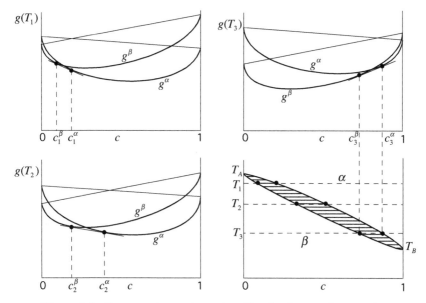

Figure 9.2: Common tangent construction for isomorphous system.

that α is the high-entropy phase (liquid, say) and β is the low-entropy phase (solid); then Δs_i is positive, and Δh_i as well, so that the enthalpy of the crystal is less than that of the liquid. The transition temperatures T_t for each pure element are determined, as in Chapter 6, by

$$\mu_i^\alpha = \mu_i^\beta \quad \to \quad g_i^\alpha = g_i^\beta \tag{9.12}$$

or

$$\Delta h_i - T_t \Delta s_i = 0. \tag{9.13}$$

Individual pure-i free energy curves should behave as in Fig. 6.1. At intermediate concentrations ($0 < c < 1$), $g^\alpha = g^\beta$ does *not* determine equilibrium. Instead, lowest common tangents must be constructed whenever possible so as to satisfy the equilibrium criteria $\mu_i^\alpha = \mu_i^\beta$ ($i = A, B$). Recall that the common tangent construction is always possible whenever two free energy curves intersect, as explained in connection with Figs. 7.8 and 7.10. At high temperature, the entropy term dominates in

$$\Delta g = \Delta h - T \Delta s \tag{9.14}$$

so that the curve for the high-entropy phase[2] (here α) will lie entirely below the low-entropy, low-enthalpy curve. As temperature decreases, several
things happen: (1) g^α moves "up" (in temperature) and g^β "down," relatively; (2) the difference in slopes of the linear portions g_L^α, g_L^β become
less pronounced, at least if the implied pure-states enthalpy and entropy
differences in $\Delta g_i = \Delta h_i - T\Delta s_i$ are much less pronounced than the explicit
T-dependence of Δg_L; (3) the shapes of the curves change as the "bulging
out" due to the ideal entropy contribution will decrease. The dominant
effect is the first one, causing the g^α and g^β curves to "slide past one another" as the temperature changes. Over the temperature interval (roughly
$T_A - T_B$) of curve intersection, the common tangent construction allows a
locus of tangency points to be plotted, as in Fig. 9.2, illustrating the case
of a so-called *isomorphous* system. The resulting phase diagram shown at
lower right of Fig. 9.2 indicates single-phase regions (α, above, β, below)
and in between a lenticular two-phase region ($\alpha + \beta$) made up of *tie lines*
joining conjugate points of common tangency or *co-nodes*. If α is the liquid
phase and β the solid, the corresponding upper and lower co-node lines are
called *liquidus* and *solidus*. The *variance* (at constant P), or *degrees of
freedom*, in the two-phase region is $f = 2 - 2 + 1 = 1$ according to the phase
rule (Sec. 4.6).

First or last (in temperature) contact between α and β free energy
curves need not occur at the pure states A and B. If these initial or final
contacts occur within the $0 < c_0 < 1$ interval (extremities excluded), then
at the point of contact (c_0, T_o), liquidus and solidus curves must present a
unique double extremum.

Three Free Energy Curves

New features may arise when free energy curves for three phases are com-
bined. As three curves, g^α, g^β, and g^γ (say) evolve with changing temperature, almost inevitably, a unique, fixed temperature will exist at which all
three curves tangentially touch a common straight line. At states of triple
tangency the variance ($f = 2 - 3 + 1$) must of course vanish according to
the phase rule. Such situations are depicted in Fig. 9.3, illustrating these
"invariant" states (because there we have $f = 0$ for fixed pressure).

Let us assume that λ (for "liquid," for example) is that high-temperature
phase, and α and β are two (crystalline) low-temperature phases. Only
two distinct cases can arise: the point of tangency of high-temperature
phase (λ) lies either inside or outside the (α, β)-tangency interval. The

---

[2]The term "high-entropy phase" must be taken merely as "the phase having the higher
entropy", not to be confused with its more recent use in the expression "high-entropy
alloy" (HEA), to be defined in Sec. 10.4.

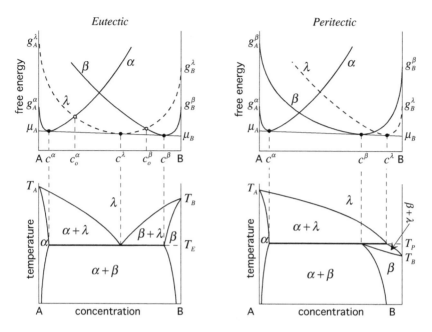

Figure 9.3: Eutectic (left) and peritectic (right) phase diagrams (lower figures) and corresponding free energy curves at the invariant temperature (upper figures). Chemical potential μ_A stands for $\mu_A^\alpha = \mu_A^\beta = \mu_A^\lambda$, and similarly for μ_B.

former is called the *eutectic* case, the latter, the *peritectic* case, occurring at, respectively, the eutectic (T_E) and the peritectic (T_P) temperatures. These words are of Greek origin: $\varepsilon \upsilon \tau \varepsilon \kappa \tau \varepsilon \iota \nu$ (to melt easily), $\pi \varepsilon \rho \iota \tau \varepsilon \kappa \tau \varepsilon \iota \nu$ (to melt on the periphery). Indeed, a two-phase ($\alpha + \beta$) eutectic mixture will melt at a temperature which is lower than that of the pure A and pure B constituents. In the peritectic case, a two-phase ($\alpha + \beta$) mixture will, upon heating, give way to a ($\alpha + \lambda$) two-phase mixture, melting originating typically at the β phase periphery.

The two possible three-phase equilibria can be described symbolically by chemical-like reactions or topological relations (see Fig. 9.4). There are no other possibilities (at given pressure). If λ represents some high-temperature *solid* phase, then the correct expressions to use are *eutectoid* and *peritectoid*.

Note that α and β tangency can occur on two separate free energy curves (dashed extensions) or on a single continuous curve (full curve), that of a sub-regular solution model with positive interaction parameter W, for example. In the latter case, the underlying lattice of both α and β must be the same. Note finally that phase boundary lines in the phase diagrams

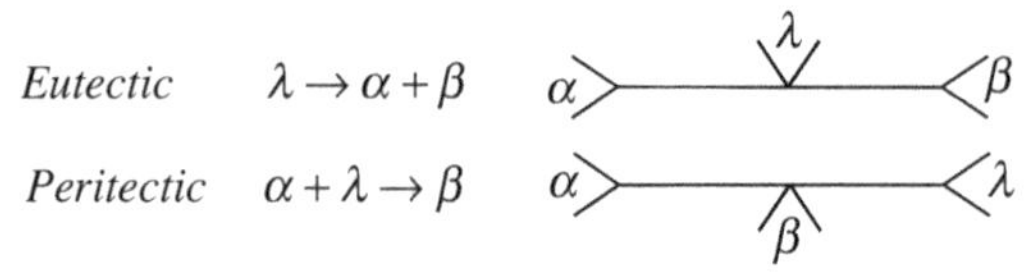

Eutectic $\lambda \rightarrow \alpha + \beta$

Peritectic $\alpha + \lambda \rightarrow \beta$

Figure 9.4: Eutectic and peritectic invariant reactions are mirror images of each other. Here the symbol λ is assumed to represent the liquid.

must extrapolate into two-phase fields as can be proved by considering metastable equilibria.

More Free Energy Curves

More than three free energy curves can be combined to produce a rich variety of phase diagrams. In all cases, however, the Gibbs phase rule must always be obeyed: in a binary system no more than three phases can coexist at equilibrium at a given temperature and at arbitrary pressure. There may exist a finite set of particular pressures at which four phases coexist: these pressures are those of special *quadruple points*, the generalization to binary systems of the *triple points* of "unary" (single component) systems studied in Chapter 6. The topological features of binary temperature-composition phase diagrams are always the same: single-phase and two-phase (locus of co-nodes) regions must be present; in most cases, three-phase equilibria will also appear and will be of two types only: eutectic(-oid) or peritectic(-oid). Phase diagrams will differ only in the manner in which one-, two-, and three-phase regions are connected. Let us consider four phases, one of

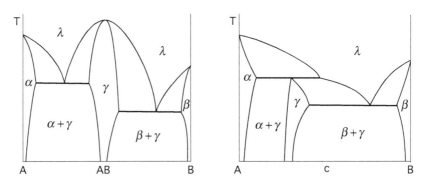

Figure 9.5: The two cases of appearance of compound γ in a binary phase diagram: by congruent reaction (left panel) and peritectic reaction (right panel). Labeling of the upper two-phase regions involving the liquid (λ) is not indicated for lack of space.

which (λ) is liquid: $\lambda, \alpha, \beta, \gamma$. The curve g^γ may be that of a compound, $A_m B_n$, say. If the addition of either A or B to the stoichiometric compound raises its energy drastically, the g^γ curve will be sharply peaked close to its stoichiometric composition $c = \frac{n}{m+n}$ (in atomic or molecular fraction). If this deep-minimum free energy curve makes first contact with the liquid free energy curve as the temperature is lowered, then a phase diagram like that of Fig. 9.5 (left figure) is obtained. It is then said that the compound γ (in the case of Fig. 9.5, $m = n = 1$) melts *congruently*, usually at close to its stoichiometric composition. If the g^γ curve first intersects an $[\alpha - \lambda]$ common tangent as the temperature is lowered, then γ will appear as a result of a peritectic reaction. The resulting phase diagram is depicted in Fig. 9.5 (right figure). That type of reaction may repeat in a sort of *cascade of peritectics*, for example in the succession of phases $\beta, \gamma, \delta, \epsilon, \eta$ in the familiar Cu-Zn phase diagram.[3]

It may also happen that one of both components (A, B) exist as two or more crystalline phases, or allotropes. Separate phases will possess separate free energy curves and the temperature at which two such curves intersect on the pure A (or B) axis determines the allotropic transition temperature for that component. An example is given in Fig. 9.6: component A is found in three allotropic varieties α, γ, and δ with melting temperature (of δ) at T_A, and allotropic transitions at T_1 ($\delta \to \gamma$) and T_2 ($\gamma \to \alpha$). On the "B" side of the diagram, phase β is taken to have no measurable solubility range and thus "collapses" on the B axis. The diagram of Fig. 9.6 resembles the carbon-rich region of the Fe-C phase diagram : δ is a body-centered cubic phase (bcc), γ is face-centered cubic (fcc) and α (ferrite) is again bcc. B is then not pure carbon itself but the compound Fe_3C (cementite) which tolerates no departure from stoichiometry. For that reason, it is called a "line compound."

9.2 Multicomponent Systems

For binary systems ($A - B$), the graphical representation is particularly useful: one sees at a glance which are the phases in presence ($\alpha, \beta, \gamma \ldots$) and their interrelationships. For multicomponent systems, *i.e.* with three or more chemical components, ternaries ($A - B - C$), quaternaries ($A - B - C - D$) and so on, the graphical representation becomes problematic. Still, it is worth presenting here some of the geometrical aspects of the problem, at least in the case of ternaries, and, very superficially, that of quaternaries, in the next two subsections.

---

[3] Binary and ternary experimentally-determined phase diagrams may be found on the Web, or in handbooks.

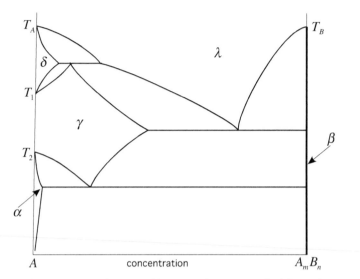

Figure 9.6: Phase diagram showing allotropic phases α, γ, δ. A hypothetical "line" compound β is shown at fixed stoichiometry $A_m B_n$; λ denotes the liquid.

A suitable geometrical figure for describing the composition space in temperature-composition phase diagrams is the regular simplex.[4] For binary diagrams, the composition simplex is that of a line segment of length 1 (though it was not called "simplex" in the trivial case of the previous section), for ternary systems it is the equilateral triangle, for quaternary systems it is the regular tetrahedron, and so on. It follows that, for n-component systems in general, a temperature-composition phase diagram can be regarded as a regular (n-1)-simplex with the temperature axis orthogonal to it. Such a geometrical figure is called an n-dimensional prism.

Ternary Systems

For ternary phase diagrams the equilateral triangle is often called the Gibbs triangle in honor of the one who first introduced it in thermodynamics (Gibbs, 1875). The Gibbs triangle is shown in Fig. 9.7, where the three concentrations c_1, c_2, c_3 are seen to obey the general sum rule $\sum_{i=1}^{n} c_i$, with $n = 2, 3, 4, \ldots$, here for the case $n = 3$, with indices $1, 2, 3$ written for A, B, C respectively. A ternary phase diagram is one composed of three binaries: $(A - B), (B - C), (C - A)$, corresponding to the three edges of

[4]In geometry, a k-simplex is a figure in k dimensions consisting of k+1 points, called vertices, linked to all other k vertices by lines called edges. A regular simplex is one in which all edges are of equal length, say unity.

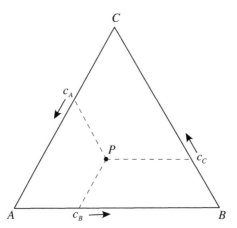

Figure 9.7: Composition simplex for a ternary system: the Gibbs triangle.

the triangular simplex. The simplest possible case is one consisting of three so-called isomorphous binaries, each one resembling the binary diagram of Fig. 9.2 of the previous section. A somewhat more complex case is the one involving only one isomorphous binary and two eutectic binary systems. A perspective drawing for that case is shown in Fig. 9.8 [from Rhines (1956)]. This diagram features one three-phase equilibrium $\lambda + \alpha + \beta$ (λ denoting the liquid) represented by isothermal triangles inside the ternary. What happens is that the highest binary eutectic line $A + B$ opens up, as it were, to form the triangle in question, slides on down (the expression "down" naturally means falling temperature) to collapse into a single line, the binary eutectic on the $C - B$ side. As the 3-phase isothermal triangles glide, their three sides sweep out three ruled surfaces[5] which are loci of 2-phase equilibria, just as the tie lines of binary systems encountered in the binary systems of the previous section. Here the tie lines need not be parallel to one another, however. In such a 2-eutectic system, there can be no 4-phase equilibrium involving the liquid, but in general, 3-component thermodynamic systems at constant (arbitrary) pressure can exhibit isolated 4-phase equilibria of type $\lambda + \alpha + \beta + \gamma$, where λ could also denote a solid phase. There are in fact three types of 4-phase invariants in any given ternary system (5-phase equilibria are disallowed by the phase rule in ternaries at arbitrary P). The three types are indicated in Fig. 9.9 with the associated transitions indicated. Indeed in a plane a fourth point can lie inside a triangle formed by the other three, or lie outside. Cases (a) and (c) (fourth point

---

[5]In geometry, a surface S is ruled, if through every point of S there is a straight line that lies on S.

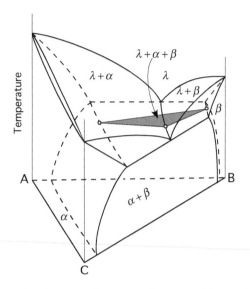

Figure 9.8: Perspective drawing of ternary diagram with one isomorph and two eutectic binaries [from Rhines (1956)]; one 3-phase equilibrium triangle shown shaded.

outside) are mirror images of one another, just as eutectics and peritectics are mere mirror images of each other in the binary case of Fig. 9.4. The interpretation is as follows: in case (a), three 3-phase triangles merge at the invariant temperature and split, just below the invariant temperature where the liquid λ disappears, leaving only the 3-phase equilibrium $\alpha + \beta + \gamma$. In case (c), its opposite, the triangle $\lambda + \alpha + \beta$ splits just below the invariant temperature (as indicated symbolically by the dashed lines in the figure). In case (b) the 4-phase equilibrium is represented by a quadrilateral which is made up of two triangles on the way "down" involving λ, then splitting along the other diagonal below the invariant 4-phase equilibrium.

Let us consider a simple ternary case consisting of three binary miscibility gaps systems combining to produce $\alpha - \beta - \gamma$ equilibria at low temperatures (see Fig. 9.10) *miscibility gap*. Isothermal sections are shown for six temperatures (reduced temperatures t, on arbitrary scale). Note how binary miscibility gaps merge to form ternary two-phase regions, and finally three-phase regions. The values of the pair interaction parameters used in the calculation are indicated in the figure legend. The analysis of phase relations can also be supplemented by vertical sections, parallel to the temperature axis.

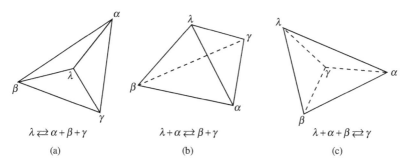

$$\lambda \rightleftarrows \alpha + \beta + \gamma \qquad \lambda + \alpha \rightleftarrows \beta + \gamma \qquad \lambda + \alpha + \beta \rightleftarrows \gamma$$

(a)   (b)   (c)

Figure 9.9: The three cases of invariant 4-phase equilibria in ternary systems.

Quaternary Systems

According to the general model described above, a quaternary diagram is a 4-dimensional prism consisting of a regular tetrahedron with the temperature axis orthogonal to it, a geometrical figure difficult to imagine. Still, such diagrams may be analyzed by means of projections to lower-dimensional spaces. This is what Rhines (1956) and Prince (1966) have done in their textbooks, by also providing drawings of phase regions in perspective. Perhaps the most systematic treatment of quaternaries, however, is that of Cayron in his doctoral dissertation [Cayron (1760)]. A quaternary phase diagram contains 4 ternary systems and 6 binary systems. A way of representing the associated 4-dimensional phase diagram has been proposed by Cayron, whose coordinate system is shown in Fig. 9.11. The top left figure is a projection onto the $A–B$ vertical plane containing the temperature axis; it is normal to the $A–B–C–D$ hyperplane, itself reduced to a simple line in this projection. A projection of binary system $C–D$ is on the right since it is collapsed to a single line on the left projection. The two lower figures are projections of the tetrahedral 3-simplex $A–B–C–D$.

At fixed pressure, at most five phases can coexist at (invariant) equilibrium. Just as in the previous cases of binaries and ternaries, the extra phase point can be inside or outside a 4-phase monovariant simplex (*not* regular, in general), and if inside (for example), the various additional cases correspond to the various ways for tetrahedra to collide or split up, in ways much similar to what exists for lower-dimensional phase diagrams, as illustrated in Fig. 9.12 whose caption explains the tetrahedron collapsing process.

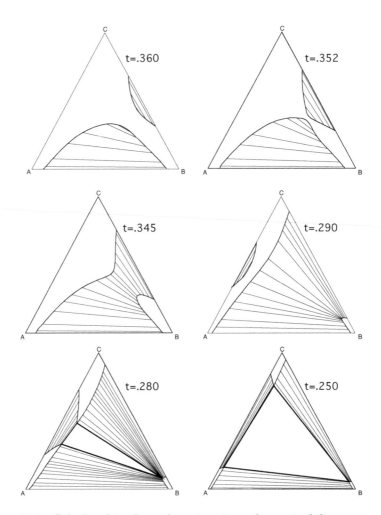

Figure 9.10: Calculated isothermal sections in a three-miscibility gap ternary RS model with interactions $w = 1.0, .75, .60$ for 6 temperatures t. Heavy-line triangle indicates three-phase equilibrium. Note the short temperature interval between $t = 0.352$ and $t = 0.345$. Densities of tie lines have no physical significance, depending only on requirements of numerical convergence of the computer algorithm.

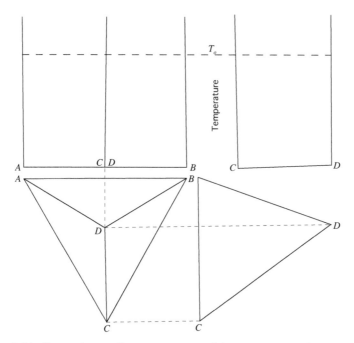

Figure 9.11: Cayron's coordinate system used for quaternary phase diagrams.

9.3 Experimental Determination

Experimental determination of phase diagrams in a challenging and time-consuming undertaking. That is probably why that important task is rarely accomplished nowadays. In principle, the problem is straightforward: determine the regions in temperature-concentration space where certain phases of a substance can be found at thermodynamic equilibrium. Then construct phase boundaries around those regions, taking care that the rules of Gibbsian thermodynamics are correctly obeyed.

Easier said than done; phase identification requires X-ray or neutron diffraction and electron microscopy and diffraction. Quantitative thermodynamics also requires calorimetry, and sample preparation needs to be particularly clean. Given such difficulties, it is not surprising that many published phase diagrams are not very reliable, or at best, incomplete. Experimentally determined phase diagrams are published online can be found, for example at ASM Handbook, volume 3 (2016).

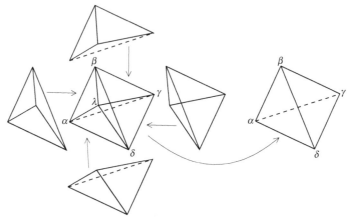

Figure 9.12: Illustration of invariant 5-phase equilibrium: $\alpha, \beta, \gamma, \delta$ tetrahedron with point λ inside. At a temperature slightly above this quaternary eutectic are four monovariant 4-phase equilibria involving λ; all four will collapse into the central figure at the eutectic temperature, as shown. Just below the invariant 5-phase equilibrium phase λ disappears, leaving only the 4-phase (solid) tetrahedron figure on the right. The arrows indicate the sense of decreasing temperature.

9.4   Computation of Phase Diagrams

The equations of thermodynamics are essentially *identities*, so cannot really provide quantitative descriptions for specific materials. To achieve such practical goals, it is necessary to complement the thermodynamical formalism with numerical data, which must be obtained by other means, the most obvious one being, of course, doing experiments. There exist, of course, a vast amount of thermodynamic data available in handbooks and in electronic databases. The problem is that too often databases appear to contain all possible data except what one is looking for.

For that reason, if for no other, workers in the field have been looking for ways to obtain thermodynamic data from various computational methods. With the advent and availability of powerful computers, such techniques have become increasingly attractive. In particular, and given the importance of phase diagrams, exemplified by the quote from Rhines given at the head of this chapter, significant efforts have been made to calculate temperature-composition phase diagrams based minimally (or not at all!) on experimental data. How feasible is the goal given the immense number of possible? For n elements, taken m at a time, the total number N_n^m of binary ($m = 2$), ternary ($m = 3$), ... combinations is[6]

$$N_n^m = \frac{n!}{m!(n-m)!} = \frac{n(n-1)\cdots(n-m+1)}{m!}, \qquad (9.15)$$

the second equality being less "symmetric" but more expedient to handle numerically, as it results from canceling $(n - m)!$, top and bottom from the first version of the formula.

Suppose we consider that about 60 elements of the periodic table are really useful in applications. Then with $n = 60$ total number of components there will be, according, to those formulas, 1,770 binaries, 34,220 ternaries, ... and so on, the numbers quickly becoming astronomical.[7] Hence, some form of numerical help is imperative. Basically, two types of computations have been proposed: one based on fitting analytical functions[8] to a restricted set of data (Calphad method, see below), the other based on first principles calculations, i.e. on resolution of the Schrödinger equation itself.

The following table summarizes the geometry of temperature-concentration phase diagrams for different numbers of components.

Table 9.1: The column headings are, from left to right, the number of components / the dimension of the phase diagram / the simplex order / and the maximum number of phases φ at equilibrium at fixed (arbitrary) pressure.

n^o comp.	n^o dim.	simplex	max. n^o φ
1	1	0	2
2	2	1	3
3	3	2	4
4	4	3	5
...	...	...	...

Calphad Method

The Calphad method (CALculation of PHAse Diagrams) proposes the calculation of Gibbs free energies from fitting free energy functions to available thermodynamic data and/or phase diagrams already determined or partly determined. The free energy functions are usually regular solution models

---

[6]the following formula is a special case of the general one, Eq.(8.20).

[7]As I once was explaining this point at a meeting devoted to phase diagram calculations, one person in the audience mentioned that after $m = 31$ the number of diagrams begins to decrease; indeed at $m = 60$, there is only *one* diagram to calculate in a space of 60 dimensions, all others being *mere* sections in lower-dimensional spaces!

[8]A function is analytic if it can be expressed as a convergent power series, *i.e.* a Taylor's expansion, being infinitely differentiable.

(see Sec. 7.10), or *i.e.* sub-regular models containing polynomials in concentration and temperature with adjustable parameters used for fitting. It is important to use polynomials in c_i of degree less than three because the higher terms interact with the logarithms of the configurational entropy possibly risking extraneous extrema to the Gibbs energy functions [see de Fontaine and Hilliard (1965)]. Also, the number of free parameters must be limited because, in the words of mathematician Marc Kac, "with five adjustable parameters you can fit an elephant". Another problem with the exclusively fitting method is that a model Gibbs free energy curve may be fitted adequately to experimental data but with no guarantee the extracted enthalpies and configurational entropy values may be correct individually (even though their algebraic sum may be fitted precisely).

The Calphad method has been extensively described in the book by Lukas *et al.* (2007), although the basic approach was first proposed in a much earlier book by Kaufman and Bernstein (1970). It is surely a useful technique as it provides directly free energy functions from which other thermodynamic quantities may be obtained by taking derivatives. Conversely, the experimental procedure generally starts with the measurement of heat capacities from which free energies may be obtained by integration. Figure 6.2 illustrates this for the case of a single substance: it is indeed preferable to differentiate than to integrate, to go "down" Fig. 6.2 rather than "up". Of course, to obtain free energy functions by curve fitting, one must have something to fit *to*. Thus, the Calphad method must rest at least partially on experimental data. But its strength lies in the possibility of interpolating between data points or indeed to extrapolate thermodynamic values beyond what is available, even into regions of instability of known phases. Much useful data may be obtained by this method, and industry has often made fruitful use of data thus obtained. So, Calphad practitioners do not usually perform thermodynamic measurements experimentally, but they collect data, make assessments of it, tabulate it, and derive from it other thermodynamic results. All of this work is carried out by a firm called Thermocalc (www.thermocalc.com) which also teaches classes on the handling of their database and their varied software. Of course all these data can be used not only for calculating phase diagrams but also for calculating general thermodynamic data for equilibrium or metastable phases of systems of interest.

First Principles Calculations

At the other extreme of computational procedures we find so-called First Principles Calculations which, at least ideally, rely on *no* experimental data whatsoever. Accomplishing such a task reliably and efficiently would be the holy grail of thermodynamics. Considerable progress has been made in that direction over recent years, but much remains to be done. The technique of first-principles calculations was made possible by advances in electronic structure computations through an approximate method of solving the Schrödinger equation in condensed matter, a technique generally referred to as *density functional theory* (Kohn and Sham, 1965), actually based on a theorem by Feynman. The first general computer codes based on these techniques were given by Marvin Cohen and collaborators (see for example Cohen, M. L. and Louie, S. G. 2016), later followed by such computer packages such as VASP *Vienna Ab-initio Simulation Package, Abinit* and more. It is important to note that more than just the energy of a given crystalline structure is calculated with atoms located at ideal lattice sites; in alloys with atoms of different "sizes", or in materials with imperfections, atoms may be displaced from their ideal sites. In general, structures must be *relaxed*, meaning that atomic displacements must be determined by minimizing energies with respect to internal (atomic) displacements and external displacements (lattice cell shape).

These computational techniques have proved to be extremely precise in computing energies of atomic structures at absolute zero of temperature, but are, by themselves, incapable of producing *free* energies, needed for thermodynamics: the all-important *entropy* is missing! With the Calphad method, the distinction between *energy* and *entropy* is to some extent unimportant: all that is required is that the Gibbs energy provide satisfactory results. In first principles calculations, energy and entropy must be calculated separately, and generally calculated by different techniques. The final result, of course, will be no better than the weakest step in the calculation.

A completely first principles calculation (*ab initio*) for solids would require, in addition, the prediction of crystal structures themselves. That is a tall order as for any system of chemical elements, there exist a non-denumerable infinity of possible crystal structures. Still, some progress has been made in this direction by such techniques as data mining [see for example Morgan *et al,* (2004)]. Thus, complete first principles calculations necessitate a variety of numerical techniques based on sound physics. Fortunately, some ab-initio packages of computer codes are now available, for example the one called ATAT (*Alloy Theoretic Automated Toolkit*) created by van de Walle (2009 and 2013), for which the cited paper gives a detailed description, along with extensive references to the theoretical methods used. The author tells us that ATAT has been used for ionic conductors, refractory materials, semiconductors, metallic alloys, minerals, interfacial and point defect thermodynamics.

The staring point for the ab-initio calculations is, as is fitting, the fundamental statistical thermodynamical definition of the Helmholtz free energy $F = -k_B \ln \mathcal{Z}$, where $\mathcal{Z}$ is the partition function (or "sum over states" and k_B is Boltzmann's constant, as defined in Sec. 8.2. Following van de Walle (2013), we have

$$\mathcal{Z} = \sum_{\sigma \in L} \sum_{v \in \sigma} \sum_{e \in v} e^{[-E(L,\sigma,v,e)/k_B T]} , \tag{9.16}$$

where $E(L, \sigma, v, e)$ is the energy of the system in state characterized by the variables L, σ, v and e. In this equation, the sums are performed successively over the atomic configurations σ (denoting types of crystal structure L, atomic occupation of lattice sites σ, atomic displacements v, and electronic states e), which means that the degrees of freedom of the thermodynamic system under study are broken down in separate levels of microscopic states, in other words, the partition function $\mathcal{Z}$ of Eq. (8.27) has been *factorized* [see Callen (1985) for a brief treatment]. Still, the number of atomic configurations is prohibitively large, but can be managed by describing these by a so-called *Cluster Expansion*, introduced first by Kikuchi (1951) in his *Cluster Variation Method*, and developed further by Sanchez *et al.* (1951) and by Zunger (1994) and collaborators.

Kikuchi used his CVM mostly to improve the treatment of the configuration entropy in second-order transitions and in solid solutions, thereby giving good accounts of short and long-range order. It is now apparent that Monte Carlo simulations are better suited for such tasks, but that the cluster expansion in fact provides an excellent parametrization of the state of order in alloys, and also of various thermodynamic properties. The resulting *Cluster Algebra* can be expressed as a complete set of orthonormal *Cluster Functions* [see Sanchez, J. M. and de Fontaine, D. (1978)] just as in the case of familiar orthonormal sets. The configurational entropy does not suffice, of course, vibrational and electronic entropy must also be included, the first one by standard lattice dynamics [see the textbook by Fultz for a detailed treatment (2014)], the second one by electron statistics.

Trial Calculation of the Cd-Mg Phase Diagram

This section presents a calculation of the solid state portion of the Cd-Mg phase diagram, at least partially from "first principle" (Asta *et al.*, 1993; see also de Fontaine, 1994). Details of the calculation are not given here, the objective being simply to indicate the direction being taken by thermodynamics at present. Calculations were performed before the ATAT system was operational, and were based instead on the LMTO (Linear Muffin–Tin Orbitals) the atomic sphere approximation of density functional theory in

(ASA). Configurational free energy calculations were performed by the cluster variation method (CVM) rather than by Monte Carlo simulation, plus some experimental data used for atomic relaxation and vibrational entropy.

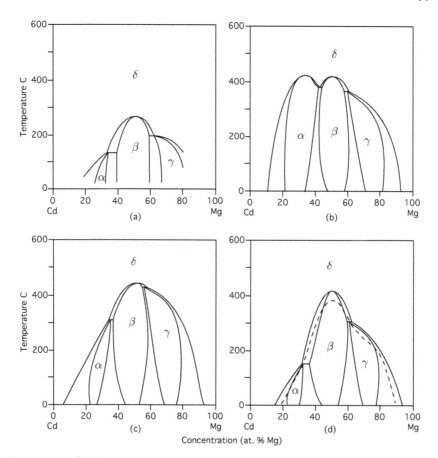

Figure 9.13: Solid-state portion of Cd-Mg phase diagram: (a) from experimental phase diagram, (b) diagram calculated from configurational free energy only, (c) calculated including also relaxation energy, (d) calculated including vibrational and electronic free energy contributions. The dashed line in this panel is the $\langle\frac{1}{2}00\rangle$ *ordering spinodal*. Phases α and γ have the $D0_{19}$ structure and phase β has the B_{19} structure. Phase δ is the hcp solid solution.

The Cd-Mg system of interest is the hcp (hexagonal close-packed) equivalent of the well studied fcc–based Cu-Au binary. For both of these alloys, liquid-phase equilibria are located at temperatures higher than those of the ordered solid phases, thereby isolating those phases which may then be calculated independently of liquid equilibria (for which there are no reliable free energy approximations).

Also, the two metals Cd and Mg are very different in that Mg is less anisotropic than Cd, in the sense that the former has c/a ratio much closer to the ideal value than the latter; it is thus instructive to see whether the computational method used can do justice to these differences, causing large asymmetries in the shape of the phase boundaries between the Cd and the Mg sides of the diagram.

LMTO calculations of seven competing structures indicate that, besides the hcp pure solid phases, only three structures are energetically stable at low temperatures: Cd_3Mg and $CdMg_3$, having crystal structure $D0_{19}$, and CdMg having structure B_{19} (see Asta, et al. 1993). The ground state energies of these structures were then used by so-called *structure inversion* method to obtain cluster expansion coefficients, leading to the required CVM free energies, hence to the phase diagram shown in Fig 9.13(b), which includes only configurational energy effects. Panel (a) of the same figure shows the solid state equilibria of the experimentally determined phase diagram (see ASM Handbook). Panel (c) shows the result of the same calculations as were performed for panel (b), but this time with relaxation included, *i.e.* taking into account the large deviation from ideality of the c/a ratio for Cd. On comparing panels (b) and (c) it is apparent that relaxing this ratio to its equilibrium value has a large influence on the phase diagram on the Cd side, as expected. The fourth panel, (d) additionally incorporates vibrational and electronic free energies, which tend to lower the stabilities of the ordered phases with respect to the disordered δ. The dashed line in this panel is the $\langle\frac{1}{2}00\rangle$ *ordering spinodal* (see Chapter 18).

It is encouraging to notice that incorporating effects known to be important in phase diagram calculations progressively improves the appearance of the phase diagram with respect to that of the experimental one, (a). The calculated equilibrium temperatures are still too high; at this time the reason for such discrepancy — which is also present with more sophisticated calculations — is not known. It is possible that the neglect of short wavelength fluctuations is the cause; in which case coarse-graining might be the required remedy.[9]

To summarize, the two main advances which have led to what may be called generally *First Principles Thermodynamics* are (1) Density Functional Theory for the energy contribution and (2) the Cluster Expansion or Monte Carlo simulations for configuration descriptions and the entropy contribution, and also for the parametrization of the energy and the de-

---

[9]Originally the calculations of panel (c) indicated phase boundaries located even higher on the temperature axis, so a slight compression of the temperature scale was applied in that panel for ease of comparing the "Gothic cathedral" *shapes* of the diagrams.

termination of crystalline ground states. Additionally, electronic structure calculations can be sped up in so called *high throughput calculations* which allow a large number of calculations to be performed in very short times [see for example Curtarolo *et al.* (2012)]. These automated processes are particularly useful in comparing properties of a large number of compounds of different structures in times extremely short compared to what would be required for experimental methods. The future when the computer will replace the calorimeter is not far away.

In comparing Calphad-based and ab-initio calculations, we note that ATAT is a freely distributed and open source, whereas Thermocalc is not. That is not quite a fair comparison as Thermocalc includes a great deal of thermodynamic data whereas ATAT does not. Both of course require learning, but the learning curve for ATAT and similar packages is necessarily steeper. More fundamentally though, the basic philosophies of the two approaches differ: one (Thermocalc) is top-down, the other is bottom-up. That is to say that the former is basically a curve-fitting procedure while the latter is based on fundamental physics. Also, the Calphad codes, being simpler, can handle large and more complex multicomponent systems whereas the first-principle codes are limited, as of now, to supercells of a few hundred atoms.

It has been emphasized above that the calculation of thermodynamic data requires the contributions of both energy and entropy. This seems obvious, but was not always recognized by some specialists.[10]. Actually we must deal with both together, so that thermodynamics, with its *free* energies, is now seen as a unifying field, which is still developing (and is no piece of cake).

9.5 Phase Diagrams and the Third Law

What is the influence of the third law on the structure of phase diagrams? From a practical standpoint the issue is not of great importance since atomic mobility is so slow at very low temperatures as to make any changes in thermodynamics unobservable. Nevertheless, theoretical answers to that

---

[10]I recall an early symposium attempting to bring together the two communities, that of the energy (electronic structure) specialists, and that of the entropy (statistical mechanics) specialists. One participant summarized proceedings of the conference thus: "The energy community views the quantum mechanics of the problem as really hard, while the entropy contributions (ideal entropy) to the free energy are a piece of cake; the entropy community views the statistical mechanical aspects of the problem as very difficult, while the energy contributions (mean field interactions) to the free energy are a piece of cake."

question in the past have been controversial, so that a discussion of the problem may be in order.[11]

For well-behaved systems, it is expected that the thermodynamic limit exists, *i.e.* that internal energy, entropy, free energies,..., for which we have given the generic symbol X in this book, be *extensive* quantities whose values are proportional to the number of atoms (or molecules) or volume. Fortunately, this is generally true (except strictly where elastic interactions are present in inhomogeneous systems). Thermodynamic equilibrium is then independent of the size and the shape of the system, if it is large enough. In such cases, one may then define *unit* quantities in the following convergent manner:

$$x = \lim_{N \to \infty} \frac{X(N)}{N}, \tag{9.17}$$

where N denotes "size", *i.e.* number of moles, of atoms, of molecules, of unite cells or volume, ... whatever the normalization may be. The normalized extensive quantity x is now an *intensive* quantity, but still retains its extensive quality in thermodynamic expressions. We may then denote the variables x as generalized *densities*. The *total* quantities X thus diverge with N, as expected. The third law states that it is the *total* entropy $S(N)$ which tends to zero as the absolute zero is reached. That law fixes the entropy zero, since without it, S being an extensive quantity, only the *difference* of entropies could be determined.

Section 8.3 gives the following formula for the *residual* entropy, *i.e.* that for T tending to absolute zero (a proof is given in Appendix F):

$$\lim_{T \to 0} S_{res}(N) = k_B \ln g_o(N), \tag{9.18}$$

k_B being Boltzmann's constant, and $g_o(N)$ being the degeneracy (or multiplicity) of the ground state.[12] If $g_o(N)$ is finite, then the total residual entropy is non-vanishing, and the third law is broken. But that does not really matter because, as soon as the temperature is non-zero, no matter how slightly, the total entropy will be proportional to N, and therefore much greater than the residual entropy S_{res} which increases more slowly than N. The third law is then saved (provisionally). If, however, the third law were formulated for the entropy density s, then *a fortiori* the third law would be obeyed. It therefore follows in either case of the definition of the entropy, that the third law is unbroken, but marginally so.

---

[11]A most enlightening treatment of the question is that of Alphonse Finel in a two page document in French [Finel, A. *Entropie résiduelle et principe de Nernst*, Nov. 2003, unpublished] of which the following text is practically a translation, with a few minor modifications here and there. Figure 9.14, however, is the present author's.

[12]The ground state is the state of lowest energy at absolute zero.

Consider now the case where the degeneracy increases much faster with N than linearly, exponentially as $g_o(N) = \lambda a^N$, (λ and a being some positive constants) for example. This time the residual entropy is not infinitely small with respect to the total entropy at non-zero temperature, and the third law is broken, but not just marginally so. It is therefore then *not* permissible to use the third law to lift the indeterminacy of the second law with respect to the entropy. In short, we can say that the Nernst principle is effectively broken when the residual entropy *density* is non-zero. Figure

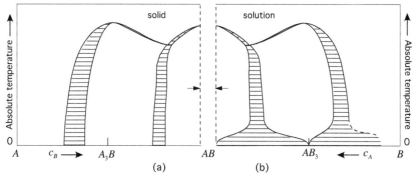

Figure 9.14: Low-temperature behavior of ordering phase diagram on the fcc lattice (exploded view); left panel (a) calculated by R. Kikuchi and C.M. van Baal by CVM with first-neighbor tetrahedron cluster interactions [unpublished, see also de Fontaine (1979)]; right panel (b) with weak long-range interactions, showing "elephant feet" two-phase regions (partly conjectured).

9.14 addresses the points raised by Finel in a somewhat schematic manner. Two different cases are shown side by side. The top portion of the figure (high temperature) is taken from an unpublished calculation by van Baal and Kikuchi with the CVM in the nearest-neighbor ordering interaction approximation. It is seen that for the A_3B and AB compounds a wide range of solubility is allowed due to the multiple degeneracy of the ground state. At low temperature, on the left hand side, the third law is effectively broken since disorder is present at zero K except in narrow two-phase regions (indicated by tie-lines). On the right- hand side, the lower portion is appropriate for cases where the atomic interactions are more complex and long-range. In that case, there is no solubility away from either AB or AB_3 stoichiometry. Tetrahedron-octahedron calculations were actually performed with longer-range interactions (Sanchez and de Fontaine, 1978 and 1980) and the degeneracy was lifted so that solubility was eliminated near absolute temperature, as expected. At high concentration, near pure B, there may be more ordered phases present and also 0^oK disorder, so only a dashed line is shown, out of ignorance. The lack of solubility at low temperature for AB or AB_3 ordered compounds is achieved by pinching of the ordered phase regions and a widening of the two-phase regions in a structure that some authors have dubbed "elephant feet" in the phase diagram.

Note that the phase diagram in question has been recalculated much more accurately (Finel and Ducastelle, 1986; Tétot, Finel and Ducastelle, 1990) by use of the CVM in both tetrahedron and tetrahedron-octahedron approximations and by Monte Carlo simulations over the whole temperature and composition range of solid solutions, down to almost absolute zero. The general shape of the diagram pictured in Fig. 9.14 (a) is confirmed, except that the triple point (where three phases A_3B, AB and the $[A \leftrightarrow B]$ solid solutions meet) is of peritectoid rather than eutectoid type. These calculations by Finel and co-workers also show that over almost all of the phase diagram, the agreement between CVM and Monte Carlo simulations is practically perfect. That is an interesting result; presently, the CVM tends to be replaced in thermodynamic calculations by Monte Carlo methods with the unfortunate result that analytic thermodynamic functions are lost. Note also, and that was the main point of the whole exercise, that in these calculated phase diagrams, and others calculated by other authors (see Tétot et al. 1990), the third law of thermodynamics is indeed violated, but not Eq. (9.18) for the entropy at low T. The fault lies not in the third law, as formulated on page 117, but in the fact that the present example of failure was somewhat artificial: an fcc-based phase diagram governed by first neighbor interactions only. In real alloys, the interactions are expected to be more long-range or more complex, and the degeneracy will be lifted.

It is clear, in closing this section, that phase diagram complexities described therein will generally not be observed in experimental practice, but they are of considerable theoretical interest.

References

Ab Initio, https://www.abinit.org/

ASM International Handbook, Volume 3: *Alloy Phase Diagrams* (2016), https://www.asminternational.org.

Asta, M. D., McCormack, R. and de Fontaine, D. (1993), *Phys. Rev. B* **48**, 748.

Cayron, R. (1960), *Etude Théorique des Diagrammes d'Equilibre dans les Systèmes Quaternaires*, Institut de Métallurgie, Louvain, Unpublished.

Ceder, G. (1994), *Modelling: the ab-initio computation of phase diagrams*, Encyclopedia of Advanced Materials X.

Cohen, M. L. and Louie, S. G. (1916), *Fundamentals of Condensed Matter Physics*, Cambridge University Press, Cambridge, U. K.

Curtarolo, S., Morgan, D., Persson, K. and Ceder, G (2003), *Phys. Rev. Letters*, **91**, 135503-1-135503-5.

Curtarolo, S., Morgan, D. and Ceder, G. (2005), *Computer Coupling of Phase Diagrams and Thermochemistry*, **29**, 163-211.

Curtarolo, S. *et al.* (2012), *Comput. Mater. Sci.*, **58**, 218.

de Fontaine, D. and Hilliard, J.E. (1965), *Acta Metall.* **13**, 1019.

de Fontaine, D., *Solid State Physics*, **34**, 73 (1979); **47** 33 (1994).

Ducastelle, F.(1991), *Order and Phase Stability in Alloys* (New York: Elsevier Science.

Finel, A. and Ducastelle, F. (1986), *Europhys. Lett.*, **1** 135.

Fultz, B. (2014), *Phase Transitions in Materials*, Cambridge University Press, Cambridge CB2 8BS UK.

Gibbs, J. W. (1875), *On the Equilibrium of Heterogeneous Substances. Trans. Connecticut Acad.*, pp. 108-248; (1877), *ibidem*, pp. 343-524. See for example (1993), *The Scientific Papers of J. Willard Gibbs*, Ox Bow Press.

Hohenberg, P. and Kohn, W. (1964), *Pys. Rev. B* **136**, B864.

Kaufman, L. and Bernstein, H. (1970), *Computer Calculations of Phase Diagrams.*

Kikuchi, R (1951) *Phys. Rev. B*, **81**, 988.

Kohn, W. and Sham, L.J. (1965), *Phys. Rev. A* **140**, A1133.

Lukas, H. L., Fries, S.G. and Sundman, B. (2007) *Computational Thermodynamics*
−The Calphad Method, Cambridge University Press.

Prince, A. (1966), *Alloy Phase Diagrams*, Elsevier Publishing Co.

Rand, M. H. and Kubaschewski, O. (1966), *The Thermochemical Properties of Uranium Compounds*, Oliver & Boyd (1963). Publishing Co.

Rhines, F. (1956). *Phase Diagrams in Metallurgy*, McGraw-Hill Book Co.

Sanchez, J. M. and de Fontaine, D. (1978) *Phys. Rev. B*, **17** 2926.

Sanchez, J. M. and de Fontaine, D. (1980) *Phys. Rev. B*, **21** 216.

Sanchez, J. M., Ducastelle, F. and Gratias, D. (1984), *Physica A* **128**, 334.

Tétot, R., Finel, A. and Ducastelle, F. (1990), *J. Stat. Phys.*, **61** 121.

van de Walle, A.(2009), *CALPHAD*, **33** 266.

van de Walle, A.(2013), *JOM*, **65** 1523.

Vienna Ab-initio Simulation Package (VASP), https://www.vasp.at/

Zunger, A. (1994) *NATO ASI on Statistics and Dynamics of Alloy Phase Transformation*, **319**, p. 361, ed. Turchi, P.E. and Gonis, A., (Plenum Press.

Chapter 10

Topological Disorder

This book is concerned mainly with equilibrium thermodynamics, but some aspects of the equilibrium theory apply as well to amorphous materials, also called glasses. Hence the decision to include a section on this subject since it is not usually found in thermodynamics textbooks.[1]

10.1   Glass Thermodynamics

The left side of Fig. 6.2 showed the behavior of the Gibbs energy, entropy and specific heat as a function of temperature at constant pressure in the vicinity of a first-order change of phase at T_o, from phase α (the low-temperature phase) to phase β (high temperature). At present, in Fig. 10.1, we are interested in the vicinity of the melting transition T_m where the entropy must, at phase equilibrium, indicate a discontinuity at the transition, according to the results of Chapter 6. Also, on this latter figure, we show the extrapolation of the entropy curve of the high-temperature phase (dot-dash curve), corresponding to the extrapolation of the Gibbs free energy curves of Fig. 6.1. In practice, the liquid can indeed be retained at temperatures lower than T_m in a metastable state (dot-dashed line in Fig. 10.1). How far the metastable liquid can be retained depends on how rapidly it was cooled from above T_m and under what conditions.

In some instances, the metastable liquid can be extended down to the so-called glass transition temperature T_g (upper open circle), at which tem-

---

[1]With a nod towards Dorothy Parker:

> Instructors seldom teach classes
> On facts relating to glasses.

This is hardly true today.

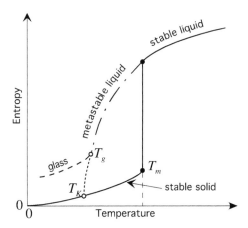

Figure 10.1: Entropy curves *vs.* temperature: equilibrium (full curves), meta-stable liquid (dot-dashed and dotted), and glass (dashed). T_m, T_g, T_K are respectively melting, glass and Kauzmann temperatures.

perature the liquid "freezes" in a disordered state, preventing the long-range motion of atoms and molecules, with a precipitous rise in the viscosity, so that the resulting (metastable) glass, or *amorphous state*, presents thermodynamic properties similar to those of the solid. The dotted curve is that of the entropy that the metastable liquid would have had if it had continued on down.

Somewhat facetiously, the glass transition is said to have been invented by Nature in order to avoid the so-called "Kauzmann catastrophe" (Kauzmann 1948) at which the extrapolated liquid entropy would fall below that of the solid (below the lower open circle, at T_K). Normally, the low-temperature phase is a crystalline solid, meaning that the structure of this phase has translational symmetry, though contrary to what was commonly believed in the past, exceptions have been found, such as quasicrystals, for instance. The Kauzmann catastrophe is not particularly severe as the specific volume curve of Fig. 6.2(e) is expected to behave in a way similar to that of the entropy curve at the discontinuity, and yet, for example, the specific volume of liquid water is less than that of the solid (ice). Moreover, (apparently) disordered phases of certain colloidal suspensions appear to be more stable as ground states (most stable at low temperatures) than the ordered ones; as already mentioned, it all depends on one's definition of the word "ordered" [see Onsager, L. (1949); Frenkel, D. (1999)].

The precise nature of the glass transition is not well-understood, and the question of whether the glass transition is of first or second-order type is not settled. Perhaps the simplest way to dispose of the question is to state that the transformation from metastable liquid to glass is not a *transition* at all; it is merely a *transformation*. This approach is not merely a semantic one: strictly speaking, a transition is a classic thermodynamic concept and must occur *at* equilibrium. But the glass *transformation* is history-dependent, that is, its exact location on a phase diagram depends on how much of the metastable liquid has been retained, thus on how fast the cooling has been carried out. Hence the glass transformation itself does not qualify as an equilibrium *transition*, it is non-equilibrium.

10.2 A Little Geometry

Let us first investigate the three familiar phases of mono-atomic matter: gas, liquid, solid, by means of a very simple model, that of hard spheres of equal size in a fixed-volume container, large with respect to the sizes of the spheres. Then the geometry of condensed phases can be considered as analogous to that of sphere packing, a model which may apply quite well to simple metals, though *not* to certain alloys or molecular solids.

Physicists are still pondering why the liquid phase exists at all. Why does the vapor not transform directly into the solid phase, or conversely as in the sublimation process? It would simplify matters considerably for theorists: gases and solids are fairly easy to model, not so for the liquid, which could be looked upon as some hybrid state, intermediate between gas and solid (crystal). Yet this intermediate exists in nature, and must be dealt with. The three main phases of matter are often modeled in a "thought experiment" consisting of a large volume containing many small objects, here taken as small hard spheres. The simplest distribution of such objects, *i.e.* the gas phase, is obtained by assigning random numbers to all three coordinates to the centroids of all such objects. The "physical" example of such a distribution is that of an ideal gas if the objects, for instance hard spheres of same size, are small with respect to the size of the container. If the objects in question are sufficiently large that their size cannot be neglected, because the rules of random motion of the spheres are now subjected to the interdiction of sphere overlap, the number of vibrational and translational microstates available will be somewhat less than it was in the previous case, the physical analog of our model is that of a non-ideal gas. Non-ideality will also be present if the objects in question are mutually interacting at a distance. Still, if the mean free path of the spheres is large with respect to their diameter, the physical analog is still that of a gas, ideal or not.

Now imagine that the spheres are actually touching, while still being distributed at random. Such is the model which is usually accepted for

Table 10.1: The Five Platonic Solids. The parameters Vertices, Edges and Faces are related by the Euler equation $V - E + F = 2$.

Polyhedron	Schläfli symbol	V	E	F
Tetrahedron	$\{3,3\}$	4	6	4
Cube	$\{4,3\}$	8	12	6
Octahedron	$\{3,4\}$	6	12	8
Icosahedron	$\{3,5\}$	12	30	20
Dodecahedron	$\{5,3\}$	20	30	12

certain liquids, despite its inadequacies, particularly severe if the objects do not have approximately spherical symmetry. Contact plus randomness can be maintained if there is strong enough attraction between spheres, for example, even for inert objects, if they are acted upon by gravity, as was mentioned in connection with the hard sphere example of Sec. 7.8.[2]

The third familiar state of matter, the crystalline solid, can be obtained by careful periodic arrangement of elements on a lattice. Symmetry properties of lattices and crystals are well-known, but it is the long-range periodicity of such arrays, known as *translational symmetry*, which is the hallmark of crystals in nature. It is an open question as to whether translational symmetry is necessarily the state of lowest energy in materials. There are exceptions, such as that of quasicrystals, the discovery of which earned Dan Shechtman a solo Nobel prize in Chemistry in 2011. In most cases though, crystallinity is the state of thermodynamically stable matter at low temperatures.

Geometrical properties of the hard sphere model can be further elucidated by appealing to regular polyhedra. By "regular polyhedron" it is meant a figure in 3-space, all faces of which are the same regular polygons (all sides of equal length), and all vertices are surrounded alike. Such is the definition of so-called Platonic solids, of which there are exactly five. They are listed in Table 10.1 along with their Schläfli symbol $\{p, q\}$, *i.e.* p−sided polygons, q of which meet at each vertex[3] [see Coxeter (1954)]. The other entries of the table, V, E and F, designate respectively the number of vertices, edges and faces. These three parameters are not independent, but

---

[2]For an exotic example, consider the packing of nucleons in a neutron star.

[3]Plato identified the first four listed with the four fundamental elements: fire, earth, air, and water, in that order. The dodecahedron was identified by him as the whole world, the fifth fundamental element, or the *quintessence* (in Latin).

are related by the famous topological equation of Euler

$$V - E + F = 2 .$$ (10.1)

The table contains two pairs of solids which are each other's *duals*, *i.e.* their numbers for V and F are inverted; the tetrahedron is its own dual. It can be proved rigorously that there are no other regular polyhedra in Euclidean 3-space [see for example the elegant derivation of Coxeter (1954)].

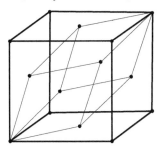

It is evident that *local* close packing of hard spheres is optimal for spheres located at the vertices of a regular tetrahedron {3,3}. However the regular tetrahedron does not pave space translationally with no empty interstices or overlap. Actually it can be shown that the only Platonic solid which does pave space is the {4,3} cube. The dodecahedron and icosahedron are included in Table 10.1 for completeness, and also because their 5-fold symmetry is seen in quasicrystals. Optimal packing of hard spheres and long-range translational symmetry can be achieved by arranging spheres at the lattice points of a face-centered

Figure 10.2: Conventional face-centered cubic unit cell (heavy lines) and primitive unit cell (light lines).

cube (fcc) and its variant, the hexagonal close-packed structure (hcp) obtained by appropriately shuffling the close-packed diagonal planes of the cube (themselves forming planar arrays having hexagonal symmetry) in such a way that each sphere of one hexagonal plane fits exactly over the interstices of the planar array just below or above. Periodic three-dimensional structures can be constructed by translating a *unit cell* of the structure, such as that shown in Fig. 10.2, along the three dimensions of space, the heavy lines in the figure denoting the conventional cell, and the lighter lines denoting the primitive unit cell, containing just one sphere. The cubic unit cell of the fcc lattice is thus made up of a central octahedron with tetrahedra at the corners; the primitive unit cell of all 14 (Bravais) lattices always contains only one lattice point point.

Another important element pertaining to sphere packing is that of Dirichlet regions or Voronoi polyhedra. Coxeter has stated the problem elegantly (1959):

> We consider two kinds of arrangements of equal spheres: *packings*, where no point is inside more that one sphere, and *coverings* where no point is outside every sphere. Each arrangement determines a "honeycomb" whose cells are Dirichlet regions (DR). The Dirichlet region for any particular sphere is

a polytope (polygon in 2-space, polyhedron in 3-space) whose interior consists of all points that are nearer to the center of the sphere than to the center of any other. [...] We are interested in cases where the density is as near to one as possible: close packings and thin coverings. In such cases the spheres are respectively inscribed in and circumscribed about the Dirichlet regions.

An image of Dirichlet regions in 2 dimensions is shown in Fig. 10.3 for an array of points distributed randomly in a plane. The DR about a given point are constructed by drawing straight (dashed) lines from that point to points in its vicinity, and constructing lines perpendicular to those "bonds" at their mid-points and tracing the convex hull of the resulting polygon. In three dimensions, the DR are polyhedra constructed by drawing bisecting planes to the "bonds" and taking the convex hull of resulting polyhedra.

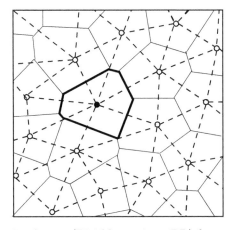

Figure 10.3: Voronoi polygons (Dirichlet regions, DR) for a quasi-random distribution of points in 2-dimensional space (light lines). The DR around the black circle is outlined in heavy lines. From Ziman (1979).

It is interesting to note that the fcc structure has closest packing (as already mentioned), and the bcc structure thinnest covering, an observation that is to be related to the fact that, in most pure metals, closest-packing structures are found at low temperatures, and thinnest-covering structures are found at high temperature. This is particularly apparent in rare earth metals. A heuristic argument for that tendency may be that, because the bcc structure is more "open" than the close packed structures, it offers more vibrational microstates in the former structure that in the latter, thus favoring high vibrational-entropy structures at high temperature.

10.3 Random Close Packing

The notion of Dirichlet regions (Voronoi polyhedra) can also be used in attempts to elucidate the structure of simple liquids or glasses, assuming that their structure can be approximated by that of random arrays of equal spheres. Such structure results from *random close packing* (RCP) where each sphere touches a (local) maximum number of its neighbors. As already mentioned, the RCP is seen as being intermediate between that of a gas, and that of a crystalline solid. If we accept this analogy, then the liquid is regarded as similar to a RCP structure but with spheres being able to escape their immediate neighborhood by random exchanges or hopping. In a glass, the hopping of spheres is severely restricted and motion is generally limited to vibrations inside the cage formed by near neighbors of the sphere. The transition between these two types of motions — long-range motion and local vibrations — as the temperature is lowered defines the so-called *glass transition* T_g. It is not clear why this transition is so sharp. Because of the disordered nature of the RCP structure, it is difficult to describe it conveniently. Diffraction patterns produce diffuse rings which give some information about distances between pairs of particles, be they atoms or molecules or hard spheres, but detailed information about structure itself is hard to come by.

Attempts have been made to construct models of RCP structures. The simplest one is obtained by simply pouring hard spheres of equal diameter into a large container, but ways must then be found to determine the statistics of sphere positions. Here is one model: many plasticine spheres are fashioned so as to have as near equal sizes as possible. The spheres are then rolled in flour so that they will not stick to one another before dropping them in random fashion inside the container. Pressure is then applied so as not to leave interstices between the now deformed spheres in the container. The objects are now the shape of irregular polyhedra, which pave space completely. By construction they are Voronoi polyhedra which can be examined one by one. The location of their centroids are recorded, including their number of vertices, edges and faces.[4]

Coxeter (1954) has proposed a very interesting mathematical model for RCP structures. First the Schläfli symbol $\{p, q\}$ must be extended to the case of a regular array of polyhedra of a three-dimensional array of *cells*. The extension is $\{p, q, r\}$ for polyhedra of faces which are regular polygons of p sides, q of which meet at a vertex, and r of which surround a common edge in the array. It is shown that for an infinite "honeycomb", or "tessellation" (possessing translational symmetry) the following relation

---

[4]Actually only two of the parameters V, E and F need to be recorded because these parameters must obey the Euler formula (10.1).

must hold (Coxeter 1954):

$$\cos \frac{\pi}{q} = \sin \frac{\pi}{p} \sin \frac{\pi}{r} . \tag{10.2}$$

The only solution for which this equation holds in integers is for the tessellation $\{4,3,4\}$, a regular array of cubes, as we saw earlier. Also as was mentioned earlier, the tetrahedron $\{3,3\}$ is the geometrical figure of equal spheres with closest *local* packing. To extend close packing *mathematically* we seek an artificial array of tetrahedra of the type $\{p,3,3\}$. The parameter p cannot be an integer; indeed the solution of (10.2) for $q = r = 3$ is

$$\sin \frac{\pi}{p} = \cot \frac{\pi}{3} = \sqrt{\frac{1}{3}} .$$

Thus

$$p = \frac{180}{35.264\ldots} = 5.1044\ldots$$

in agreement with the observation that pentagons are prevalent as faces of the polyhedral cells. The corresponding number of faces F and vertices V are calculated from the knowledge of the parameters (p, q, r), giving values

$$F = 13.398\ldots \quad \text{and} \quad V = 22.796\ldots .$$

Note that a polyhedron which does pave space exactly is the Kelvin tetrakaidecahedron, or truncated octahedron, with $F = 14$; *not* a Platonic solid. The image that emerges from such calculations is thus that of a complete covering of *average* Voronoi polyhedra with fractional numbers of faces, edges and vertices, which appear to be in good agreement with averages obtained by taking averages from actual physical models. What we are tacitly claiming here is that the average of a set of objects (the real; DR polyhedra) is the same as the set of average objects (the average polyhedron with fractional values of $[V, E, F]$), which in general is not correct.

Entropy-driven Transformations

Phase transformations are driven by matter seeking lower free energy, either changing discontinuously to a new phase (first-order phase transformation), with different free energy, or by changing continuously, usually through a singularity (second-order phase transition). A reduction of free energy may be achieved either through a decrease in value of the energy (or enthalpy) term, or by an increase in the entropy term. In the latter case, we fully expect changes of phase from condensed matter to gas phases upon a temperature increase. Conversely, a change from a disordered to an ordered phase can take place if the loss of entropy is compensated by a decrease in internal energy. But, as emphasized quite pertinently by Frenkel (1999), intuitive notions of order and disorder can be misleading. Moreover, it is not correct to state naively that order necessarily corresponds to low entropy, and disorder to high entropy. There is no such law of thermodynamics.

What is correct is the famous Boltzmann equation Eq. (8.13) which relates entropy S to the number of available microstates Ω, but that number is not always easily related to *disorder*. Frenkel (1999) states the following example as follows:

> The earliest example of an entropy-driven ordering transition is described in a classic paper of Onsager (1954) on the isotropic-nematic transition in a (three-dimentional) system of thin hard rods. Onsager showed that, on compression, a fluid of thin hard rods [...] *must* undergo a transition from the isotropic fluid phase, where the molecules are translationally and orientationally disordered, to the nematic phase. In the latter phase, the molecules are translationally disordered, but their orientations are, on average, aligned. [...] At first sight it may seem strange that the hard rod system can *increase* its entropy by going from a disordered fluid phase to an orientationally ordered phase. Indeed, due to the orientational ordering of the system, the orientational entropy of the system decreases. However, this loss in entropy is more than offset by the increase in translational entropy of the system: the available space for any one rod increases as the rods become more aligned.

It must be emphasized that the situation here is quite different from that described in Sec. 7.8: in that section, the objects, hard spheres, were implicitly subjected to gravity, which indirectly provided mutual attraction. In the present section, gravity has been eliminated, which is easy to do theoretically (in computer simulation, for example), but hard to do experimentally. Also, this section concerns itself mainly with objects, or packing units, whose shapes depart strongly from the approximately spherical. Such units are for example molecules which are inherently anisotropic. In a particularly interesting study, Sharon Glotzer and co-workers (2009) have investigated the packing properties of truncated regular polyhedra as a function of the truncation parameter t, with $t = 0$ for the regular tetrahedron (Schläfli symbol {3,3}) to $t = 1$ for the regular octahedron (Schläfli symbol {3,4}). Computer simulation at constant volume with 20 to 100 million Monte-Carlo steps produced various self-assembled crystal structures including diamond, β-tin, high-pressure lithium and others. The fact that the crystallization polyhedra were inert and the absence of gravity meant that the polyhedra neither attracted nor repeled each other, thus reducing the relevant free energy to only entropy terms; the resulting phase transitions were thus entirely entropy-driven.

10.4   Amorphization Criteria

Many types of materials can be obtained in amorphous (*i.e.* non-crystalline) form. Indeed, for many ceramic or polymeric substances, the amorphous structure, albeit metastable, is the normal solid state, primarily because the crystallizing unit is a complex molecule which is not compatible with translational symmetry and packing. For metallic systems, particularly for pure metals rather than alloys, the crystallization unit is the atom itself, and crystallization can take place on cooling just below the thermodynamic melting temperature, T_m. In order to obtain a metallic glass it is then necessary to prevent crystallization, and this section is devoted to finding criteria for easy amorphisation. The practical reason for obtaining metallic glasses is that such substances can have interesting mechanical and electro-magnetic properties, so this section will be concerned with metallic glass formation, but only with its thermodynamic aspects.

To form a glass from the liquid, what must be done is to bring the system by cooling (rapidly) past the melting temperature T_m down and past the glass temperature into the region of the phase diagram where the glass is at least metastable, thereby avoiding crystallization. Referring to the entropy plot of Fig. 10.1, that means following the metastable liquid curve (dot-dash curve) to T_g, beyond to the dotted curve, and on down to room temperature. In most cases, that is a long path, so it is best to shorten it. Also, since crystallization from the melt is a first-order transition, it requires a nucleation event (see Chapter 17) which is normally catalyzed by impurities or defects (including the walls of the container), hence it is important to work under very clean, defect-free condition, if possible. It is also necessary to work with small samples so that heat can be evacuated readily form the sample. Importantly, rapid cooling is necessary in order to avoid nucleation of crystals, for instance at rates of order 10^5 or $10^6 \, \mathrm{K \, s^{-1}}$.

Clustering and Ordering

Assuming that such outside conditions are met, as far as is possible, we must find out what internal conditions might be favorable for glass formation, *i.e.* what are the favorable *thermodynamic* conditions. It had been noted for some time (Greer 1995) that, in binary systems to take a simple example, association together of the components A and separately of components B

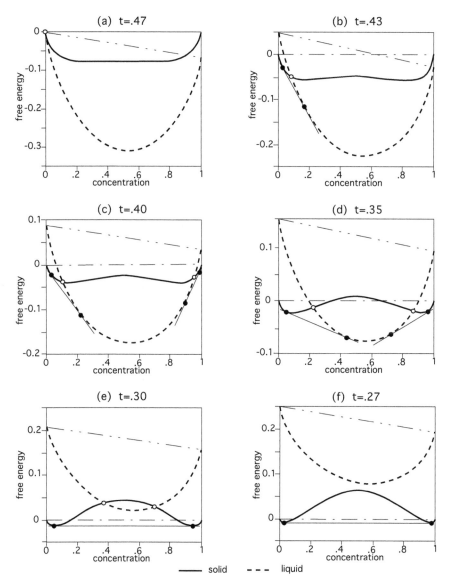

Figure 10.4: Free energy curves calculated from a regular solution model at 6 different "reduced temperatures, t". Closed circles indicate pairs of tangency points. Open circles indicate intersections of free energy curves, generating the T_o lines of Fig. 10.5. The invariant eutectic temperature is at about t=0.32.

in the liquid phase tends to promote amorphization, and that formation of AA and BB pairs, rather than AB pairs in the solid is favorable as well. How can we have both apparently contradictory tendencies? The key is to have atoms (molecules) with very different "sizes" so that a preponderance of AB bonds will tend to create unfavorable stresses in the solid, though such size disparity can be readily accommodated in the liquid. A classic example is the system $Pd_{40}Ni_{40}P_{20}$. In this alloy, the situation is even more favorable because the small ion Pd tends to stabilize metallic ion tetrahedra with Pd at the center. These centered regular geometric figures are the furthest from spherical symmetry of all Platonic solids, thus hindering crystallization.

To make these heuristic rules a bit more theoretically plausible, we appeal to the simplest possible model: that of a binary alloy (AB) regular solution (RS, Sec. 9.1). Furthermore, instead of plotting separate free energy curves for the α and β crystalline phases, for simplicity and for minimizing the number of physical parameters, we consider a single free energy curve with two minima, as shown in Fig. 9.1. This choice does not really restrict generality since all we need is to illustrate the handling of the common tangent rule; it does not matter that this approach is strictly confined to cases where the two crystalline phases have a common lattice structure. However, the RS equation (7.87), sufficient for describing equilibrium with a single free energy function, is not sufficient if more than one free energy function is present: the linear term of Eq. (7.42) must be included so as to treat the relative position of these curves as a function of temperature. It is known that one may always add a linear function of concentration to a given free energy without changing the equilibrium properties of that phase by itself. So, we may include the linear terms in only one free energy function, here in the liquid RS formula. The difference in linear terms thus appears in the RS formula for the liquid, as given below in Eq. (10.3b). Physically, this term represents the difference in enthalpies of melting Δh, and the difference of entropies of melting Δs for solid and liquid in their pure states A and B. For simplicity, Δs is considered to be the same for both A and B. The RS formulas for solid and liquid are thus given by:

$$g_s = v\,c(1-c) + RT[c\log c + (1-c)\log(1-c)] \qquad (10.3a)$$

for the solid and

$$g_l = \Delta h - T\Delta s + w\,c(1-c) + RT[c\log c + (1-c)\log(1-c)] \quad (10.3b)$$

for the liquid, with

$$\Delta h = a + b\,c, \qquad (10.3c)$$

where v and w are respectively solid and liquid interaction parameters, positive for like pairs (AA, BB) favored and negative for unlike pairs (AB) favored. Constants a and b in Eq. (10.3c) are fitting parameters for the enthalpy of melting. The reference state for the free energies (10.3) is the zero of the crystalline state at the pure element extremities, joined by the horizontal dot-dash line. However, the liquid free energy curve is not symmetric about the $c = 0.5$ concentration, thereby resulting in different melting points T_A and T_B. A double-dot-dash line emphasizes this non-symmetry. Free energy curves for the case $v = 1$ and $w = 0.2$ are calculated at six different temperatures indicated over each plot by the parameter $t \equiv k_B T / v$. Note the different labelings of the free energy axis, required as the solid and liquid curves "slide past" one another as the temperature is decreased.

Resulting phase diagrams for various sets of values of v and w are shown in Fig. 10.5, with sample free energy curves shown in Fig. 10.4 (de Fontaine 2002). The resulting phase diagrams are all of simple eutectic type with phase boundaries which are very nearly straight lines. A constant glass transition T_g has been chosen for all four pairs of interaction parameters, for the sake of illustration. Two different values of the interaction v have been investigated, both positive, because of the requirement of favoring like pairs, *i.e.* of phase separation. The top two diagrams are calculated with $v = 1$, the lower two with $v = 1.8$, the top left diagram being that corresponding to the free energy curves of Fig. 10.4. For the liquid free energy, the parameter w is taken to be equal to $+0.2$ (leftmost diagrams) and to the negative of that value (rightmost). In that way, the importance of, respectively, phase separation and ordering in the liquid is exhibited. The calculated so-called "T_o" line is plotted in the diagrams as a dashed line, where the symbol T_o itself is shown at the lowest point of the dashed curves. These lines are the loci of the intersection of liquid and solid free energy curves. The physical interpretation of the T_o line is as follows: it is the locus of the *equality of the free energies* of the liquid and the solid if no redistribution of the constituents in the liquid state were to take place on cooling, obviously a non-equilibrium situation. For simplicity, a single line for the solid free energy has been considered instead of separate curves as in Fig. 7.10 for phases α and β. In the latter case, the T_o lines for α and β do not connect smoothly and a unique minimum cannot be defined. No matter, the general conclusions, to be outlined below will still be valid, particularly so because we are investigating here properties of metastable states which strictly do not belong to equilibrium thermodynamics.

As was mentioned previously, glass formation which takes place with difficulty, as with metallic glasses, is facilitated by very rapid cooling which

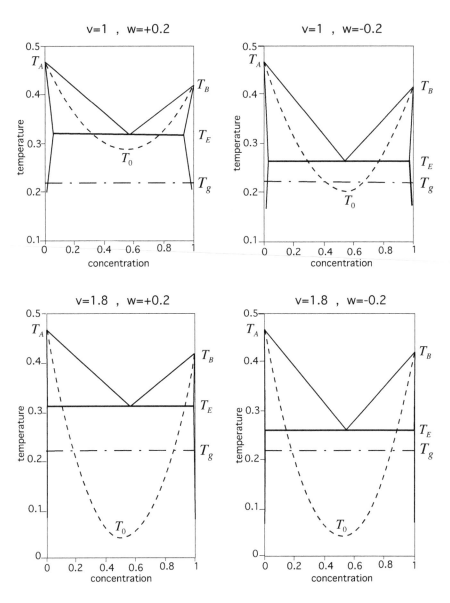

Figure 10.5: Temperature-concentration phase diagrams constructed from four different regular solution models, the top left one corresponding to the free energy curves of Fig. 10.4. By construction, all pairs of melting temperatures for A and for B are the same for all four binary systems. T_g is an arbitrary, but constant glass transformation temperature. T_o is the lowest temperature of the locus of free energy equality between solid and liquid phases.

preferably must start from the equilibrium liquid at a temperature as near as possible to the glass transformation temperature, *i.e.* at low temperature. Hence we are looking for a starting composition near the equilibrium eutectic one, and with a low T_o so that liquid and compositionally unrelaxed solid may remain in (metastable) equilibrium as long as possible during rapid cooling. A glance at Fig. 10.5 indicates that these conditions are met preferentially in the lower right panel of the figure, or possibly at the lower left one. In both cases, the key is to choose the larger value of v, the solid state interaction parameter, which means, in the hypotheses adopted, high size mismatch for the crystallizing unit, which for metals and alloys, means the atoms themselves. This conclusion matches that discovered in many experimental studies (see for example the review paper by Greer 1995). For example, Greer notes that "No metallic glass is known to be based on components that have difference in atomic diameters less than 10%". It is also observed that glass formation is promoted by $A-B$ ordering, rather than by $A-A$ and $B-B$ clustering: in all cases calculated here, $w < 0$ lowers the eutectic temperature T_E.

Number of Components

In practice it is also known that the number n, of components has an important effect of glass formation: the more components, the merrier, *i.e.* the more that crystallization can be retarded, all other things being equal. That idea is known as the anthropomorphic sounding "confusion principle", which is difficult to characterize. In an attempt at quantitative

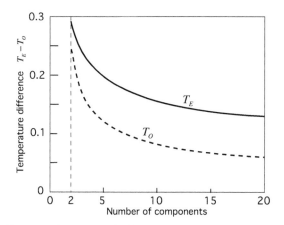

Figure 10.6: Calculated eutectic (T_E) and T_o lines for a completely symmetric multicomponent eutectic system, for number of components $n = 2$ to 20.

illustration, consider an artificial system of n components for which all binaries are regular solution models model with identical parameters. It then also follows that all melting temperatures for all components are the same. This may sound like a very unrealistic way of proceeding, but it really is a way of taking the caveat "all other things being equal" at full value. The postulated completely symmetric system has the additional advantage that the temperatures of its multicomponent eutectic (always at the very center of the n-dimensional regular simplex) and of the minimum T_o can be calculated quite simply.[5] It is seen in Fig. 10.6 that both the values of multicomponent T_E and T_o drop precipitously in temperature as n is increased from the binary case, $n = 2$, to about $n = 5$, then rapidly levels off after about $n=10$ or 15.[6] The drop in eutectic is therefore found to be highly beneficial at first, then levels off for the higher multicomponent systems.

High-entropy alloys

The topic of high-component alloys of near equiatomic concentration, mentioned in the previous section, naturally leads to the topic of so-called *high-entropy alloys*, a recently discovered class of mostly single-phase materials featuring many components at nearly equiatomic concentrations.[7] Some of these materials, in favorable cases, were indeed found to possess mechanical and corrosion properties that were quite exceptional; actually avoiding, at least partially, the mutual contradiction between high strength and high ductility [see for example Gludovatz (1153)]. The key element of high-entropy alloys is the presence, near the center of the regular composition simplex, of generally five or more components of a wide range of solubility, or at least of metastable solid solutions.

It is out of the question to look up, or even calculate a phase diagram for such high-multicomponent systems, but heuristically one expects to find the required solid solutions in systems for which the lower-dimensionality component systems exhibit high-temperature solid solutions, which in turn require component elements to be fairly similar to one another in crystal structure, melting points and atomic sizes. Favorable systems would thus include Fe, Co, Cr, and Ni, for example. A mixture of such elements formed at high temperature can then be tested for stability by general methods, as described in Sec. 16.2. Ideally, a multivariable free energy should then be perturbed at various temperatures and compositions to see how the solid

---

[5]D. de Fontaine and N. Speed, unpublished work at UC Berkeley (2002).

[6]Actually the calculation was continued to the unrealistic $n = 40$, and the trend continued smoothly, as expected.

[7]I am grateful to Prof. R. Ramesh for alerting me to work on high-entropy alloys.

solution responds. Given the large number of components involved, nothing more sophisticated than a regular solution model, as given by Eq. (7.87), can be envisaged for the free energy.

According to that approximation, the enthalpy is assumed to be given by a sum of binary interaction terms, themselves defined by the effective pair interaction parameters Ω_{ij} of Eq. (7.87), the configurational entropy being the ideal one $-Ts_{id}$. The problem of testing for the stability of the solution then reduces to finding reasonable values for the Ω_{ij} parameters, for example by fitting to binary enthalpies calculated from so-called *first principles*, as mentioned in the previous chapter. Fortunately, such density-functional computations have already been performed for a large number of binary compounds (Curtarolo, S. *et al.*, 2003). That calculated data

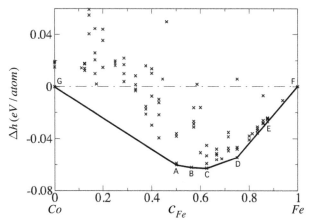

Figure 10.7: Calculated heats of formation from a large number of binary compounds of Fe and Co (crosses) and associated convex hull (heavy broken line). Ground states are indicated by letters A through G (adapted from Troparevsky, M. C. *et al.*, 2015).

base has been used to construct the plot of Fig. 10.7 (Troparevsky, M. *et al.*, 2015), where the crosses indicate the formation enthalpies of various binary compounds of the Co-Fe alloy. The heavy broken line is the so-called *convex hull* relative to the crosses (see page 160). Letters A to G indicate locations on the plot of ground state (*i.e.* lowest formation enthalpy) Co-Fe compounds, including pure elements. Thus the point at A corresponds to the stoichiometric compound CoFe having lowest formation enthalpy. Plots similar to that of Fig. 10.7 were constructed for other binary systems, particularly those relative to the CoCrFeMnNi system, and the minimum value of the convex hull in each case was used to obtain the effective binary enthalpy required to investigate the stability of various multicomponent solid solutions in the search for optimum high-entropy alloys. The search proved to be quite predictive for suitable multicomponent systems. Although it appears to be quite different, these authors' method is really similar to the multicomponent regular solution model proposed in this section, which, however has not been tested. The present model is based on approximations of the free energy, so that the ideal entropy term is "automatically" included and the limitation to sums of binary compounds for the total enthalpy is introduced naturally from the regular solution approximation. The general conclusions of the two studies, that of Troparevsky *et al.* (2015) and those presented here in Fig. 10.6 are similar: that the benefit of including more components increases very rapidly at first with the numbers of components, becomes significant after about 5 components, then more slowly thereafter, the reason for this similar behavior being found naturally in the common use of the ideal entropy of mixing.

The very same plot as that of Fig. 10.7 has been reproduced in an article by Jain *et al.* (2016) to illustrate the power of accurate, high-throughput density-functional calculations, allowing zero-Kelvin energy computations to be performed very rapidly.

References

de Fontaine, D. (2002) "Calculation of Phase Diagrams", Chapter 10 in *Intermetallic Compounds*, Vol. 3, J.H. Westbrook and R.L. Fleischer, Editors, J. Wiley & Sons, pp. 185-208.

Coxeter, H. S. M. (1954) *Acta Math. Acad. Sci. Hungaricae*, **5**, 263.

Curtarolo, M., Morgan, D., Persson, K., J. Rodgers, J. and Ceder, G. (2003) *Phys. Rev. Lett.* 91, 135503.

Frenkel, D. (1999). *Physica A.* **263**, 26.

Gludovatz, B., Hohenwarter, A., Cantoor, D., Chang, E. H., George, E. P., and Ritchie, R. O. (2014) *Science* **345**, 1153.

Greer, A. L. (1995), *Science* **267**, 1947.

Haji-Akbari, A., Engel, M., Keys, A. S., Zheng, X., Petschek, R. G., Palffy-Muhoray, P. and Glotzer, S.C. (2009) *Nature*, **462**, 7274.

Jain, A., Persson, K. A., and Ceder, G. (2016) *APL Mater.*, **4**, 053102.

Kautzmann, W. (1948) *Chem. Rev*, **43**, 219.

Onsager, L. (1949), *Proc, NY. Acad. Sc.*, **51**, 627.

Troparevsky, M. C., Morris, J. R., Daene, M., Wang, Y., Lupini, A. R., and Stocks, G. M. (2015) *JOM*, **67**, 2350.

Ziman, J. M. (1979) *Models of Disorder*, Cambridge University Press.

Chapter 11

Chemical Reactions

In this chapter we study *chemical reactions*, simple cases first, more general ones later. The subject is important for both scientific and technological reasons; for example, it is an excellent application of the general principles of classical thermodynamics. For heavy industry — chemical, metallurgical and energy — make extensive use of the formalism of chemical reactions. At the "light" extreme of "micro-application", lies that pertaining to defects in crystalline substances, such as in semiconductors. In this application, point defects in crystals "react" with one another to produce new types of defects by processes which may be described symbolically by means of chemical equations.

Up to now, we have assumed that constituents, be they atoms or molecules, cannot be created or destroyed. We now lift part of this restriction by allowing *molecules* to react with one another, *i.e.* to appear or disappear from a reacting system. Only *atoms* must be conserved. If that constraint is lifted, one enters the realm of *nuclear* reactions. Regardless of the nature of constituents, these thermodynamic equilibria can be described by "chemical equations" between reactants.

11.1 One Chemical Reaction

Familiar reactions are, for example,

$$H_2 + \frac{1}{2}O_2 = H_2O \tag{11.1}$$

or

$$H_2O = H^+ + OH^- , \tag{11.2}$$

the first equation referring to the synthesis of water from the elements, the second the dissociation of water into ionic constituents. *A priori*, it is not known how a molecule will break up; however, with experience we deem

$$NH_4Cl = NH_3 + HCl$$

plausible, but not

$$NH_4Cl = NH + H_3Cl.$$

Chemical reactions can be combined:

$$Ti + O_2 = TiO_2 \qquad\qquad (11.3a)$$
$$SiO_2 = Si + O_2 \qquad\qquad (11.3b)$$

the "sum" of which is

$$Ti + SiO_2 = TiO_2 + Si. \qquad\qquad (11.3c)$$

Clearly these three equations are not mutually independent, any one of the three results from the other two, and atomic species must balance on either side of each chemical equation, as the following text will show.

Mass Action Law

By convention, components on the left hand side of chemical equations are called *reactants*, those on the right hand side, *products*. In general, a chemical reaction can be written as

$$0 = \sum \nu_i A_i \qquad\qquad (11.4)$$

where the A_i designate molecular symbols and the ν_i (Greek *nu*) are *stoichiometric coefficients*, negative for reactants, positive for products. The corresponding *mass balance equation* has the same form:

$$0 = \sum \nu_i M_i \qquad\qquad (11.5)$$

where the M_i designate "masses", which could be taken as the number of protons in the molecule, or constituent atoms, or the mole numbers. Equation (11.5) can be regarded as a *chemical constraint*.

As the reaction (11.4) proceeds, small molar quantities δN_i are created or destroyed in such a way as to obey the *law of definite proportions*:

$$\frac{\delta N_1}{\nu_1} = \frac{\delta N_2}{\nu_2} = \ldots = \frac{\delta N_n}{\nu_n} = \delta\xi \qquad\qquad (11.6)$$

where $\delta\xi$ measures, in some sense, the degree of advancement of the reaction, just like the *affinity function* defined in Section 4.5, with $\delta\xi > 0$ for a reaction proceeding from left to right, and conversely. Equations (11.6) can be written more succinctly as

$$\delta N_i = \nu_i \delta\xi \qquad (i = 1, 2, \ldots n) \,. \tag{11.7}$$

Thermodynamic equilibrium at constant T, P requires that the Gibbs free energy be stationary for any small variation in the δN_i away from equilibrium:

$$\delta G = \sum_i \mu_i \delta N_i = 0 \,. \tag{11.8}$$

Direct substitution of (11.7) into (11.8) gives

$$\delta G = \sum_i \mu_i \nu_i \delta\xi = 0 \tag{11.9}$$

for arbitrary (non-zero) $\delta\xi$. Hence we have, at equilibrium

$$\sum_i \nu_i \mu_i = 0 \,. \tag{11.10}$$

This equation, which is formally similar to Eqs. (11.4) or (11.5) is the fundamental one for chemical equilibrium.

The same result may be arrived at by combining Eq. (11.4) and the constraints (11.7) multiplied by Lagrange multipliers (see Appendix D):

$$\lambda_i(\delta N_i - \nu_i \delta\xi) = 0 \,.$$

Then we have

$$\sum_i \mu_i \delta N_i - \sum_i \lambda_i \delta N_i + \sum_i \lambda_i \nu_i \delta\xi = 0 \,,$$

which leads to

$$\mu_i = \lambda_i \quad (\text{all } i) \quad \text{and} \quad \sum_i \nu_i \lambda_i = 0 \,. \tag{11.11}$$

These Eqs. (11.11) are equivalent to Eq. (11.10).

Spontaneous evolution, at constant T and P, is governed by the inequality

$$\delta G = \Delta g \, \delta\xi \leq 0 \tag{11.12}$$

where the unit free energy change ΔG is defined as

$$\Delta g = \frac{\delta G}{\delta\xi} = \sum_i \nu_i \mu_i \,. \tag{11.13}$$

It is then clear that the reaction will proceed towards the products when $\Delta G < 0$, towards the reactants when $\Delta G > 0$. Of course, $\Delta G = 0$ is the equilibrium condition (11.10).

Equation (11.13) may also be written, by use of

$$\mu_i = \mu_i^\circ + RT \ln a_i \tag{11.14}$$

(see derivation of Eq. (7.12)) as

$$\Delta g = \Delta g^\circ + RT \ln \prod_i a_i^{\nu_i} \tag{11.15}$$

in which

$$\Delta g^\circ = \sum_i \nu_i \mu_i^\circ \tag{11.16}$$

is the *standard free energy change* of the reaction, *i.e.*, the change in the Gibbs free energy when $\{\nu_r\}$ moles of reactants combine at constant T, P to produce $\{\nu_p\}$ moles of products, all reactants and products being in their *standard states*, meaning: in their *pure* states, at their arbitrarily chosen reference pressures (which need not be the same for all constituents).

At equilibrium, we have from Eqs. (11.15) and (11.16)

$$\Delta g^\circ = -RT \ln K \tag{11.17}$$

where K, the so-called *equilibrium constant*, is the equilibrium value of the product of activities:

$$K = \left(\prod_i a_i^{\nu_i} \right)_{eq}. \tag{11.18}$$

Equations (11.17) and (11.18) constitute the *Law of Mass Action*.

Equilibrium Constant

From Eq. (11.17) it also follows that

$$K = e^{-\frac{\Delta g^\circ}{RT}} = K_o e^{-\frac{\Delta h^\circ}{RT}} \tag{11.19}$$

with

$$K_o = e^{\frac{\Delta s^\circ}{R}} \tag{11.20}$$

where

$$\Delta g^\circ = \Delta h^\circ - T\Delta s^\circ \tag{11.21}$$

has been used. It is seen that the equilibrium constant K (and also K_o) depends on temperature alone, since Δg° does not depend on the concentration, *i.e.* and on the degree of mixing of the components. In Eqs. (11.16),

(11.19) and (11.20) it is understood that the operator Δ acts in the following way, on arbitrary thermodynamic function Z° (in pure state):

$$\Delta Z^\circ = \sum_i \nu_i z_i^\circ \qquad (11.22)$$

where the z_i° are the molar quantities of the substances i. ΔH° and ΔS° are called standard heat and entropy of reaction, respectively.

At the reference pressure(s), Δg°, and thereby the value of the equilibrium constant K, depends on temperature alone. This temperature dependence can be calculated, in principle, by integrating the standard chemical potentials as follows: from Eq. (7.8) we have, for any substance i,

$$\mu^\circ(T) = g(p^\circ, T) = -\int_{T_\circ}^T s(p^\circ, t)\, dt\,. \qquad (11.23)$$

The entropy in Eq. (11.23) may be evaluated by considering the specific heat at constant reference pressure p°:

$$c_p^\circ \equiv c_p(p^\circ, T) = T \left(\frac{\partial s}{\partial T} \right)_{p^\circ} = T \frac{\partial s^\circ}{\partial T}$$

so that

$$s(p^\circ, t) = \int_{t_\circ}^t \frac{c_p^\circ(t')}{t'}\, dt'$$

and

$$\mu^\circ(T) = -\int_{T_\circ}^T dt \int_{t_\circ}^t \frac{c_p^\circ(t')}{t'}\, dt'$$

where $t_\circ$ and $T_\circ$ (which may be identical) denote temperatures at some known initial state, for example room temperature, at which entropy and Gibbs energy are known. By using the same definition of the operator Δ as given in Eq. (11.22) we finally arrive at

$$\Delta g^\circ(T) = -\int_{T_\circ}^T dt \int_{t_\circ}^t \frac{\Delta c_p^\circ(t')}{t'}\, dt'\,. \qquad (11.24)$$

In a slightly different form, this is known as the *Nernst equation*.

Of course, one still needs to know the temperature dependence of the specific heats for example, for substances of interest. Often, experimentally determined values can be found in tables. It is often assumed that, over restricted temperature ranges, c_p can be approximated by the empirical formula

$$c_p(T) = a + bT + cT^{-2} \qquad (11.25)$$

for example. Then parameters, a, b, c can be tabulated in reference handbooks, and thermodynamic functions can be approximated by successive
integrations, according to the formulas given above.

One more relation bears mentioning, that between the temperature
derivative of the equilibrium constant and the standard heat of reaction:

$$\frac{\mathrm{d}\ln K}{\mathrm{d}T} = \frac{\Delta h^\circ}{RT^2}. \qquad (11.26)$$

This relation is known as the van't Hoff equation, which can be obtained
immediately from the Gibbs–Helmholtz equation (7.37).

Note the convenient way of presenting results pertaining to chemical
reactions: it is to plot $\ln K$ as a function of $\left(\frac{1}{T}\right)$ according to the relation

$$-\ln K = \frac{1}{R}\left[\Delta h^\circ\left(\frac{1}{T}\right) - \Delta h^\circ\right], \qquad (11.27)$$

easily derived from Eqs. (11.17) to (11.21). If the thermodynamic variables
Δh° and Δh° are practically temperature independent, then the logarithm
of the equilibrium constant as a function of the reciprocal of the absolute
temperature plots as a straight line. Such a construction is known as an
Arrhenius plot, an example of which is given in the next chapter, If the
experimental data points do not fall on a straight line *systematically*, then
the temperature independence is not warranted for the system in question.

11.2   Several Chemical Reactions

Now let there be, in a single phase, n reacting components whose masses
are related by q equations

$$0 = \sum_{i}^{n} \nu_{ji} M_i \quad (j = 1, 2, \ldots, q). \qquad (11.28)$$

Equations (11.3) are examples of systems of chemical reactions (11.28)
where, in each of the equations of the set, several of the ν_{ji} stoichiometric
coefficients may have value zero. For each possible reaction the following
hold:

$$\delta N_{ji} = \nu_{ji}\delta\xi_j \quad (i = 1, 2, \ldots n) \quad (j = 1, 2, \ldots q) \qquad (11.29)$$

where δN_{ji} denotes incremental changes in the number of moles of component i in reaction j. Overall, the change δN_i in i is given by

$$\sum_{j=1}^{q} \delta N_{ji} = \delta N_i = \sum_{j}^{q} \nu_{ji}\,\delta\xi_j. \qquad (11.30)$$

Direct substitution into the condition of thermodynamic equilibrium (at constant T, P) gives, successively,

$$\delta G = \sum_i \mu_i \, \delta N_i = 0 \tag{11.31a}$$

$$= \sum_i^n \mu_i \sum_j^q \nu_{ji} \, \delta \xi_j = 0 \tag{11.31b}$$

$$= \sum_j^q \left(\sum_i^n \nu_{ji} \mu_i \right) \delta \xi_j = 0 \,. \tag{11.31c}$$

At this point it would be tempting to set the coefficients of all $\delta \xi_j$'s equal to zero to obtain q equations analogous to the equilibrium condition (11.10). Strictly, this cannot be done as there is no guarantee that the q increments $\delta \xi_j$ are independent. Recall, for example, that only two of the three Eqs. (11.3) are independent. Usually, a set of r independent chemical reactions can be extracted from a larger ($q \geq r$) system by inspection. In complicated cases, however, it may be necessary to proceed more systematically. For that purpose, Eqs. (11.28) may be considered as a set of q equations in the n masses M_i. First note that this rectangular $n \times q$ system always has a "solution" for the M_i, even for $q > n$, since the equations were derived from valid atomic conservation conditions. Hence the equations are mutually compatible. From linear algebra it is known that the number of independent equations in the system is equal to the *rank r* of the rectangular $n \times q$ matrix, r being the order of the largest non-vanishing determinant which may be extracted from this matrix by elimination of rows and columns.

If $q < n$, then clearly $r < n$ since $r \leq q$. If $q \geq n$, it may be concluded immediately that $r \leq n$. Actually, a stronger statement can be made by noting that any $n \times n$ square system extracted from system (11.28) is a *homogeneous* one which, by physical argument, must have non-zero solution for the M_i. Hence the $n \times n$ determinant of the system must vanish, hence the rank r of the n-matrix must be *strictly less* than n: $r < n$.

By the compatibility condition, it does not matter which $r \times r$ system of independent equations is extracted from (11.28); Eq. (11.31) must result with the summation over j extended to r equations only. Since only independent increments $\delta \xi_j$ remain, the coefficients (in brackets) may now be set equal to zero, giving the equilibrium conditions

$$\sum_i^n \nu_{ji} \mu_i = 0 \,, \qquad (j = 1, 2, \ldots r) \,. \tag{11.32}$$

Note the restriction on the value of the index j.

11.3 Multi-Phase Equilibrium

Assume, at first, that all substances are present, at least in minute amounts, in all phases at equilibrium. Conservation of atoms will still be expressed by equations of the type (11.28) applied to the multiphase system as a whole. Whether or not reactions actually take place, and in which phases, is information which is not provided by these equations. For that, general conditions of equilibrium must be derived. For complete generality, consider both reacting ($i = 1, 2, \ldots m$) and inert ($k = m+1, m+2, \ldots n$) components. For the former, Eqs. (11.29) are obeyed for all phases together, for the m reacting species related by r independent chemical reactions.

As in previous chapters, denote by N_i^α the number of moles of substance i in phase α, *etc...* Then we have, by summation:

$$\sum_\alpha \delta N_i^\alpha = \delta N_i = \sum_j^r \nu_{ji}\,\delta\xi_j = \sum_j^r \delta N_{ji} \quad (i = 1, 2, \ldots m) \qquad (11.33a)$$

and

$$\sum_\alpha \delta N_k^\alpha = \delta N_k = 0 \qquad (k = m+1, m+2, \ldots n). \qquad (11.33b)$$

Equilibrium conditions will be derived through the use of Lagrange multipliers. Equations of constraints are, corresponding to Eqs. (11.33):

$$\lambda_i \left(\sum_\alpha \delta N_i^\alpha - \sum_j^r \nu_{ji}\,\delta\xi_j \right) = 0 \qquad (i = 1, 2, \ldots m) \qquad (11.34a)$$

and

$$\lambda_k \sum_\alpha \delta N_k^\alpha = 0 \quad (k = m+1, m+2, \ldots n). \qquad (11.34b)$$

These equations, combined with the variation of Gibbs energy

$$\delta G = \sum_\alpha \sum_{i=1}^n \mu_i^\alpha\,\delta N_i^\alpha$$

produce

$$\sum_\alpha \sum_i^m (\mu_i^\alpha - \lambda_i)\delta N_i^\alpha + \sum_\alpha \sum_k^n (\mu_k^\alpha - \lambda_k)\,\delta N_k^\alpha + \sum_i^m \lambda_i \sum_j^r \nu_{ji}\,\delta\xi_j = 0\,.$$

$$(11.35)$$

Equating to zero the coefficients of the mole number variations in the first two summations yields

$$\mu_i^\alpha = \mu_i^\beta = \ldots = \mu_i^\varphi = \lambda_i. \quad (\text{all } i) \tag{11.36}$$

After exchanging the order of summations in Eq. (11.35) we obtain, by setting to zero the coefficient of $\delta\xi_j$,

$$\sum_{i=1}^m \lambda_i \nu_{ji} = 0$$

or

$$\sum_{i=1}^m \nu_{ji}\mu_i = 0, \quad (j = 1, 2, \ldots r) \tag{11.37}$$

since, by Eq. (11.36) the phase index on the chemical potentials may be discarded. Equations (11.36) and (11.37) constitute the equilibrium conditions for an arbitrary multicomponent, multiphase, multi-reaction system. "Mass action laws" may be written down for the various reactions as was done for a single reaction in Sec. 11.1.

11.4 Phase Rule

The Gibbs phase rule, derived earlier for "inert" components, may now be generalized to reacting ones. Once again, the number of degrees of freedom (or variance) f is obtained by taking the difference between the number of variables and the number of equations. Here, "new" equations of equilibrium must be considered: those given by Eq. (11.37). In addition, certain other constraints may be imposed. These are generally of two basic types:

Constraints on Initial Amounts

If, for example, all hydrogen and oxygen present in a gas phase arise from the dissociation of liquid water, we must have at all times, and for all degrees of advancement of the dissociation Eq. (11.1), for example

$$N_{H_2} = 2N_{O_2}. \tag{11.38}$$

If arbitrary amounts of H_2 and O_2 were present initially in the gas phase, Eq.(11.1) would not hold.

Electroneutrality

A mixed substance, in gaseous or condensed phase, must remain electrically neutral. Hence, in the ionic dissociation reaction Eq. (11.2), we must always have

$$N_{OH^-} = N_{H^+} . \tag{11.39}$$

Equations such as (11.38) or (11.39) are additional constraints which, when required, must be included into the total count of the number of equations. It is not necessary here to repeat the arguments of Sec. 4.6 leading to the Gibbs phase rule, it is merely required to include r *independent* chemical reactions and, say p *independent* "initial amounts" or "electroneutrality" conditions to obtain the modified phase rule

$$f = n^* - \varphi + 2 \tag{11.40a}$$

with

$$n^* = n - r - p. \tag{11.40b}$$

The quantity n^* thus appears as the *effective number of independent components* in the system.

Note that the validity of Eqs. (11.40) does not depend on each substance's being present in all phases. If k (say) is absent from β, then there is one less variable (c_k^β) to take into account, but also one less equation since $\mu_k^\beta = \lambda_k$ is absent. Thus, the net count remains the same.

11.5   Ellingham Diagrams

Consider reactions occurring between a pure metal and its pure oxide, both in condensed form (solid or liquid). Assume that the gas phase contains molecular oxygen and a mixture of arbitrary non-reacting gases. The relevant chemical reactions are written:

$$xM(c) + O_2(g) = M_xO_2(c) \tag{11.41}$$

when the symbol c designates the condensed phase, g the gas phase. For the purpose of constructing Ellingham diagrams, the stoichiometric coefficient of the metal M $(x \equiv \nu_M)$ must be such as to make that of molecular oxygen (O_2) equal to unity. Even in multiphase systems the condition for equilibrium is given by Eq. (11.37), leading to the mass action law Eq. (11.15).

In the present case we can invoke the rule whereby the activity of a pure condensed phase is approximately unity. Hence, from Eq. (11.18), the equilibrium constant is given by

$$K = \frac{a_{M_xO_2}}{a_M^x a_{O_2}} \cong p_{O_2}^{-1} \qquad (11.42)$$

in which we have set the reference (standard state) pressure equal to 1 atm. From Eqs. (11.17), with $\Delta G = 0$ and (11.42) it follows that

$$\Delta G^\circ = \Delta H^\circ - T\Delta S^\circ = RT \ln p_{O_2} . \qquad (11.43)$$

In typical oxidation reactions, and over wide temperature ranges, the standard enthalpy and entropy changes are observed to be roughly temperature independent. Hence the standard free energy plots as a straight line against temperature for a given reaction, as shown schematically as heavy lines in Fig. 11.1.[1] Symbols A and B stand for pure metal reactants.

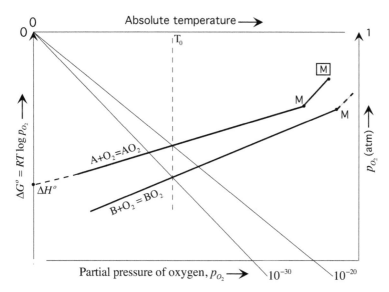

Figure 11.1: Schematic Ellingham diagram. Isolated chemical symbol M indicates melting of that element. Symbol enclosed in a square indicates melting of the oxide.

The intercepts of the straight lines on the $0^\circ K$ axis give ΔH°, and the slopes give ΔS°. Note that representative lines for chemical reactions have discontinuous slopes at first-order phase transitions (isolated symbol for the

---

[1]Experimentally determined Ellingham diagrams for real substances may be found on the Web.

element indicates melting point, symbol inside square indicates vaporization
of the element).

Note also that the straight lines drawn from ($\Delta G^\circ = 0$, $T = 0^\circ K$) represent lines of constant oxygen partial pressure. At the point of intersection of
such lines with the representative line of a given chemical reaction, chemical
equilibrium is achieved at the temperature of the intersection point (such
as T_o) for example). Hence, it is possible to determine oxygen partial pressures by constructing such straight lines from the (0,0) origin (upper left
corner) all the way to the p_{O_2} scale found at bottom and at right of the
Ellingham diagram.

Combined reactions, such as Eq. (11.3c) can also be studied by means
of this diagram. If the partial pressure appearing in Eq. (11.43) is denoted
more explicitly as $p_{O_2}^{eq}$, p_{O_2} then denoting an arbitrary partial pressure, we
have, by Eqs. (11.15) and (11.43)

$$\Delta G = RT \ln \frac{p_{O_2}^{eq}}{p_{O_2}} . \tag{11.44}$$

It follows that, for $p_{O_2}^{eq} > p_{O_2}$, $\Delta G > 0$ and a *reduction* reaction is expected.
Conversely, for $p_{O_2}^{eq} < p_{O_2}$, *oxidation* should proceed. It may then be
deduced that an oxidation process for a reaction whose representative line
is located lower on the ΔG scale will be more favorable than one for which
the relevant line is located higher. For example then, reaction (11.3c) should
normally proceed towards the right at any temperature. As a consequence,
it would be disastrous to attempt to melt Ti in silica crucibles, the affinity
of Ti for oxygen being significantly greater than that for Si.

Chapter 12

Point Defect Equilibrium

Thus far, little explicit mention has been made of the crystalline nature of solids. Real crystals may differ from ideal ones however, in the sense that localized defects may break the perfect translational symmetry of the ideal crystal. In fact, considerations in Sec. 7.7 leading to Figs. 7.7 and 7.8 guarantee that at least some point defects (vacancies, interstitial atoms) should be present even in "pure" substances. In this section we shall be concerned only with *point defects* as these may exist in equilibrium in actual crystalline solids. By contrast, more extended defects such as dislocations (one-dimensional) and interfaces (two-dimensional) are not equilibrium defects although they are almost always present.

12.1 General Definitions

Recall that in Chapter 7 constituents present in a system could not be created or destroyed, only exchanged between phases. In the previous chapter (Chapter 11), such restrictions were partially lifted by allowing some constituents, the *molecules*, to be created or annihilated: conservation of constituents was replaced by the less restrictive one of conservation of atoms. As a result, new types of conservation relations were introduced, which took the form of *chemical reactions*. In this section, the notion of atomic conservation is modified by recognizing that atoms in solids may occupy or leave different crystallographic sites, thereby creating local imperfections: point defects. Not surprisingly, new constraints will be introduced, and these will be expressed formally, as before, as chemical reactions. For illustration, it is convenient to consider a definite example, such as the one depicted in Fig. 12.1 which represents a hypothetical "two-dimensional oxide" M_2O, M (metal) being monovalent, O (oxygen) being divalent. Vacancies ($\square$) are

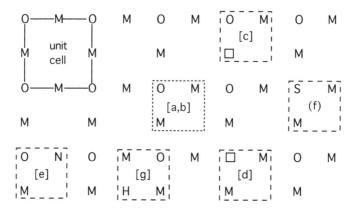

Figure 12.1: Two-dimentional M_2O crystal. The structure consists of three simple square sublattices (two of M and one of O). The conventional unit cell is outlined by full lines. Examples of building units are shown by squares: normal building unit [a,b] enclosed by dotted lines, others enclosed by dashed lines representing defective units [c to f], described in the text.

shown on both oxygen and metal sites.[1] Substitutional imperfections could also be present, such as molecules of type "N" substituting for "M", say. Interstitials may also exist at equilibrium. In the general case Kröger and co-workers (1956, 1964) propose the following definitions:

Constituents

Constituents are defined as atoms or group of atoms, or molecules in a crystal, *i.e.* atomic nuclei and their core electrons (or group of same). Vacancies, for example, are not constituents according to this definition.

Unit Cell

The primitive crystallographic unit cell is the smallest repeating unit of a perfect crystal, while the conventional unit cell may contain several such cells. The unit cell, primitive or conventional, contains an integral number of *formula units* (such as M_2O in our example). It is important to distinguish in a crystal those sites which are crystallographically equivalent and those which are not. A set of equivalent sites with translational symmetry forms a lattice. The crystal structure as a whole thus consists, usu-

---

[1] Kröger and co-workers (1956) adopt the letter V as a symbol for vacancies. In this text, the square symbol will be used to avoid confusion; besides, the symbol □ is rather cute.

ally, of several interpenetrating sublattices, each sublattice being normally occupied predominantly by a particular constituent. In the Kröger–Vink notation, the sublattice is designated by the atomic symbol of its major constituent. A given atomic constituent, say "A", may occupy preferentially the sites of several sublattices; together, then these sublattices contain the set of A-type sites. For example, in Fig. 12.1, four square sublattices may be recognized: the O sublattice, two distinct M sublattices, and one empty sublattice which is that of the interstitial sites.

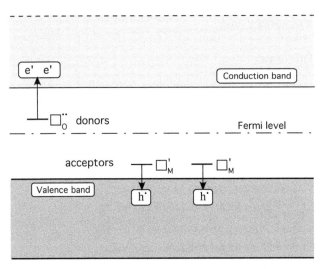

Figure 12.2: Energy level diagram illustrating the effect of imperfections on the electronic structure of valence and conduction bands.

Building Unit

Material may be added or removed from a crystal but, if the particular crystal structure under consideration is to be preserved, addition or removal of material must be performed in such a way that *the crystal structure is conserved.* Preserving structural integrity is accomplished by adding or removing crystallographic sites one primitive unit cell at a time. The unit cell need not be added physically, only geometrically, *i.e.* the sites of the unit cell are added (or removed) but not necessarily occupied by the "correct" atoms, or not necessarily all occupied. Actually, what is being added (removed) is a *building unit*, defined as a unit consisting of the crystallographic sites of the primitive unit cell, occupied by structural elements. The latter, to be defined more fully in the next paragraph, consist of atoms and vacancies. Building units consisting of the "molecule" M_2O

are shown in Fig. 12.1. Relative building units may also be defined as the difference between an imperfection and the perfect crystalline state.

Thus, sites may be created or annihilated in a crystalline phase, but only in integral amounts of building units. Physically, such units are added (removed) to the surface of the crystal. Formally, however, the surface need not be considered: each unit added to the surface creates a new surface unit but converts a previous surface site to a bulk site. The net effect is thus a change of sites in the *bulk* since, for a sufficiently large crystallite, addition (removal) of small numbers of building units does not sensibly alter the total number of surface sites.

Structure Elements

Structure elements are defined as any atom (or small cluster), or the absence thereof occurring at a crystalline site(s). According to the Kröger–Vink notation (slightly modified), we have:

[a] M_M : atom M at normal site M (or "M" sublattice)

[b] O_O : same for oxygen on site of oxygen sublattice

[c] $\Box_M$: vacant site on M sublattice

[d] $\Box_O$: vacant site on O sublattice

[e] N_M : N atom substituting on M site

[f] S_O : S atom substituting on O site

[g] $H_\bullet$: H atom in interstitial position.

Electronic Defects

Crystalline solids are assumed to be made up of neutral atoms, while core electrons remain localized on the host atom. In metals and alloys, outer-shell electrons may spread out, quasi free electron fashion, in partly empty bands, the conduction bands. In insulators and semiconductors, outer-shell electrons occupy filled bands, the *valence bands*, being separated by a gap from the normally empty conduction band. In typically covalently bonded solids the valence electrons are localized in saturated covalent bonds between ions; in typical ionic solids, the valence electrons are located on the more electronegative element (anion).

Regardless of electronic redistribution, or local charge transfer, the unit cell must remain electrically neutral in the perfect crystal. That neutral cell will be considered as the reference state for local charges. Any excess local negative charge relative to this neutral state will be designated by a dash (*/*) appended to the symbol for the structure element located at the site in question. Any excess local positive charge will be designated by a dot ($\bullet$). Likewise, if in a semiconductor a valence electron is excited to the conduction band, it will impart excess negative charge locally, designated by the symbol e'. The excited electron leaves behind an electron vacancy or hole (h) in the valence band, designated by $h^\bullet$. It is convenient (though not essential, and often misleading) to consider the "Lewis Octet" picture: in the present example, each M gives up one electron which becomes tightly bound to the O atom in the perfect crystal, thereby completing the octet of electrons around O, and leaving a complete electronic shell configuration for M. The oxide formula should then actually be $M_2^+ O^{--}$. The *relative charges* on both M and O are zero, however, since a structure element is considered neutral if it has the normal charge appropriate for that site. This site neutrality can be emphasized by writing $M^\times$ and $O^\times$ for an electronically neutral site.

Let us assume that a neutral oxygen atom is removed from the crystal. A vacancy is left behind accompanied by two electrons. As long as these electrons remain bound to the vacancy, the structure element is neutral (it has the charge appropriate to the crystal): $\square_O^\bullet$ or $\square_O^{\bullet\bullet}$. Likewise, if $\square_M$ captures an itinerant electron, then that structure element in turn becomes ionized: $\square_M'$.

It is customary to depict electronic defects in crystals by a schematic energy level diagram of the type shown in Fig. 12.2. Energy is plotted vertically, and position along the crystal is plotted horizontally schematically. Imperfections which readily give up electrons to the conductive band are called *donors*, and the corresponding donor levels are situated close to the conduction band. Conversely, imperfections which readily trap electrons are called *acceptors*, with energy levels close to the valence band. Electron trapping produces holes in the valence band. Electrons locally trapped at an imperfection behave very much like electrons in a free atom: they possess discrete energy levels, ground and excited ones, and "orbit" around the point defect with characteristic Bohr radii. The difference is that, particularly in the case of so-called shallow level impurities, the binding of the electron is very weak, the levels spaced very close together and the Bohr radii very large, of the order of tens of Angstroms. Thermal vibrations, or other perturbing influences (applied electric field, incident radiation) can easily excite the trapped electrons into the conduction band. Such is the origin of a great many electronic devices, radiation detectors in particular.

Point Imperfections

Point imperfections (or defects) are thus structure elements other than neutral atoms on their normal sites. Neutral imperfections have the normal charge associated with that site. Ionized imperfections carry excess relative charge, positive or negative. Point defects can be of four types: vacant sites (*i.e.* "vacancies") ($\square$), interstitial atoms ($H_{\bullet}$), substitutional defects (N_M) and the intrinsic electronic defects e' and $h^{\bullet}$ (electrons and holes).

Imperfections play an essential role in the physics of electronic devices. In particular, charged defects are carriers of electric charge and provide electrical conduction, with conductivity proportional to mobility and concentration of carriers. The concentration of defects, in turn, is determined by the thermodynamics of the crystal in equilibrium with its environment characterized by such intensive variables as temperature and pressure. The general thermodynamic formulation used is that of chemical reactions. Here, only basic principles will be given along with a few examples. For more details, the reader is referred to the textbooks by Kröger (1964) or Kofstad (1972). The first of these texts is also an excellent thermodynamic treatise and contains a wealth of quantitative information about the thermodynamics of point defects in crystals.

12.2   Electrochemical Reactions

It was seen in the previous chapter that chemical reactions, of the general form Eq. (11.4) must be constructed in such a way that masses are conserved, as expressed by Eq. (11.5). In the present case of "chemical reactions" between structural elements in crystals, additional constraints must be taken into account:

Mass Balance Constraint

The sum of the masses of the constituents must balance on the right and left of the equation. Vacancies and electronic defects ($\square, e, h$) have no mass.

Site Ratio Constraint

Regardless of site occupancy, the ratio of anion to cation sites must be a constant, *i.e.* the number of sites on sublattices $\alpha, \beta, \gamma \ldots$ must always remain in constant proportion. This means that sites may be created or destroyed but only in such a way that integral numbers of building units are added or removed.

Electroneutrality Constraint

The system must remain electrically neutral. Therefore, the total effective charge must remain the same before and after the formation of defects. These constraints can be handled by means of Lagrange multipliers. It was shown in Chapter 11 that the introduction of Lagrange multipliers resulted in the formulation of the conditions for thermodynamic equilibrium expressed as a chemical reaction in which the molecular symbols are replaced by corresponding chemical potentials. Constraints relative to imperfections in crystals can be handled in exactly the same way, as will be shown by simple examples.

In the past, the use of chemical potentials for constituents or structure elements, the amounts of which were not in fact independent variables, was criticized by some specialists in the field. Thus, in a definite compound AB, the chemical potential of A cannot strictly be defined as

$$\mu_A = \left(\frac{\partial G}{\partial N_A} \right)_{P,T,N_B} \tag{12.1}$$

since, by the constraint of site ratio conservation, A_A cannot be added to AB without simultaneously adding B_B and $\square_B$. Hence, only *differences* of chemical potentials have meaning in the sense of being measurable quantities; this was discussed in Sec.7.6 on independent variables. However, the use of Eq. (12.1) has no adverse consequences in actual calculations. Kröger *et al.* (1959) actually show that the use of such "virtual potentials" is consistent and convenient. These authors reserve a special symbol for these virtual potentials η (actually, it seems hardly necessary to do so, and we shall continue to use the symbol μ). It will turn out that these virtual potentials always occur in linear combinations which correspond to measurable quantities.

12.3 Examples

Examples will now be given of the application of the mass action law to point defect equilibria.

Intrinsic Electronic Defects

If the band gap is not too large, electron-hole pairs can be created by thermal excitation. The appropriate "reaction" is

$$\varnothing = e' + h^\bullet \tag{12.2}$$

where $\varnothing$ symbolizes the recombined state (or ordinary lattice site). By the mass action law, (11.19), we have

$$K = \frac{a_e a_h}{a_\varnothing} \cong [e'] [h^\bullet] \tag{12.3}$$

with

$$[e'] = n\,\gamma_e^\circ, \qquad [h^\bullet] = p\,\gamma_h^\circ \tag{12.4}$$

where the γ° are the Henry's law constants and n and p are the number of electrons (negative) and holes (positive) per unit volume, respectively, and the square brackets are the chemist's way of indicating concentration. The γ° can be incorporated conveniently into the equilibrium constant, thus

$$np = K' \quad \text{with} \quad K' = \frac{K}{\gamma_e^\circ \gamma_n^\circ}. \tag{12.5}$$

Henceforth, this "incorporation" will be tacitly assumed. Electroneutrality imposes

$$n = p = \sqrt{K'}. \tag{12.6}$$

Hence, electrical current carrier density can be determined if, at temperatures of interest, the equilibrium constant is known. This knowledge in turn requires that of ΔG° for the reaction. In the present case, ΔG° is estimated from band structure theory and Fermi–Dirac statistics (see Kröger 1064), for example.

Schottky Defects

Formally, this defect arises when the "building unit" of normal bulk sites is transferred to the surface, leaving behind corresponding vacant sites. In our "M_2O" oxide, for example, the reaction is

$$2\,M_M + O_O = 2\,\square_M + \square_O + (2\,M_M + O_O)_{\text{surf}}. \tag{12.7}$$

Actually, because of surface "self-similarity," reaction (12.7) does not actually increase the number of surface sites so the $2M_M + O_O$ cancels from both sides of the equation leaving the "net" Schottky reaction.

$$\varnothing = 2\,\square_M + \square_O \tag{12.8}$$

with equilibrium constant

$$K = [\square_M]^2 [\square_O]. \tag{12.9}$$

The vacancies may become charged:

$$\square_M = \square_M' + h^{\bullet}, \tag{12.10a}$$

$$\square_O = \square_O^{\bullet} + e', \tag{12.10b}$$

$$\square_O^{\bullet} = \square_O^{\bullet\bullet} + e', \tag{12.10c}$$

for example, each reaction introducing new equilibrium constants.

Vacancy complexes (associates) may also be created, for example

$$\square_M + \square_O = (\square\square)_{MO}. \tag{12.11}$$

More vacancies can be added to form small vacancy clusters, or mini-voids in crystalline material.

Frenkel Defects

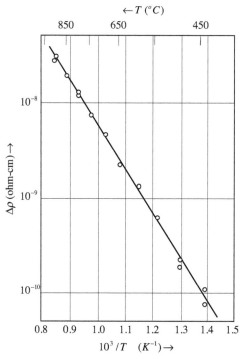

Figure 12.3: Resistivity change of quenched gold wires as a function of quench temperature. From J. E. Bauerle and J. Koehler, *Phys. Rev.*, **107**, 1493 (1957).

Another frequently occurring defect is the so-called "Frenkel pair," which may be created spontaneously by thermal excitation but which is generated in abundance under irradiation conditions (electron, or neutron bombardment), even in elemental crystals. The Frenkel reaction is

$$M_M + \square_\bullet = M_\bullet + \square_M \tag{12.12}$$

whereby a metal atom on a normal (substitutional) site is displaced into an interstitial site. The interstitial lattice is normally empty so that the concentration of $\square_\bullet$ is nearly unity. The Frenkel equilibrium constant is then expressed as

$$K = [M_\bullet][\square_M]. \tag{12.13}$$

In the case of metals, the vacancy concentration does not have to be balanced electronically by interstitials, so that the equilibrium constant for the creation of vacancies is simply given by $c_\square$ itself. The concentration of point defects was already given in Eqs. (9.10) or (9.11) but by means of quite a different method. Not surprisingly, the results are the same, but in different notation. The experiment leading to Fig. 12.3 is the following: it is assumed that the electrical conductivity in a metal is proportional to the concentration of vacancies, which can thereby be determined by the measuring the resistivity, in the present case a gold wire heated to temperatures ranging from about 450 C to 900 C. The actual measurements were performed at low temperature on samples rapidly quenched from the high temperatures stated. The logarithm of the measured resistivity, actually, of the increase in resistivity, $\Delta\rho$, is plotted vertically in an Arrhenius plot, Fig. 12.3 , as a function of $(1/T)$. It is seen that the experimental points lie very close to a straight line, indicating that enthalpies and entropies are sensibly temperature-independent over the range of temperatures considered, *i.e.* almost 450 C.

Off-Stoichiometric Compounds

As an illustration, let us consider here only the case of oxygen-deficient oxides with oxygen vacancies predominant; there may exist other possibilities. Loss of oxygen to the gas phase may be written as

$$O_O = \square_O + \frac{1}{2}O_2 \tag{12.14}$$

with equilibrium constant

$$K_1 = [\square_O]\sqrt{p_{O_2}}. \tag{12.15}$$

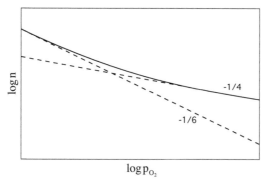

Figure 12.4: Electron concentration in off-stoichiometric oxide.

The oxygen vacancies may become ionized, according to Eqs. (12.10b) and (12.10c), with corresponding equilibrium constants

$$K_2 = \frac{n\,[\square_O^\bullet]}{[\square_O]}, \qquad K_3 = \frac{n\,[\square_O^{\bullet\bullet}]}{[\square_O^\bullet]}. \tag{12.16}$$

Electroneutrality imposes

$$n = \square_O^\bullet + 2\square_O^{\bullet\bullet}. \tag{12.17}$$

The vacancy concentrations may be eliminated by using Eqs. (12.15), to (12.17), so that the electron concentration may be expressed as

$$n^3 = K_1 K_2 \, p_{O_2}^{-\frac{1}{2}} [2K_3 + n]. \tag{12.18}$$

Equation (12.18) is awkward to solve explicitly for n as a function of the partial pressure of oxygen, say. However, a graphical method may be used: first one considers two limiting cases for Eq. (12.18) (dashed lines), then the combined case (full curve).

$$n \gg 2K_3 \tag{12.19a}$$
$$n \ll 2K_3 \tag{12.19b}$$

which, by Eq. (12.10c), and by the ratio

$$\frac{n}{K_3} = \frac{[\square_O^\bullet]}{[\square_O^{\bullet\bullet}]} \tag{12.20}$$

give corresponding inequalities

$$[\square_O^\bullet] \gg [\square_O^{\bullet\bullet}] \tag{12.21a}$$
$$[\square_O^\bullet] \ll [\square_O^{\bullet\bullet}]. \tag{12.21b}$$

Solutions for the electron concentration n for the two limiting cases are then, respectively:

$$n = p_{O_2}^{-\frac{1}{4}} (K_1 K_2)^{\frac{1}{2}} \qquad (12.22a)$$

and

$$n = p_{O_2}^{-\frac{1}{6}} (2 K_1 K_2 K_3)^{\frac{1}{3}} . \qquad (12.22b)$$

These two solutions are plotted as straight dashed lines on a schematic $\log n / \log p_{O_2}$ plot (Fig. 12.4). On the scale of several decades of partial pressures, it turns out that the deviations of the plot of the full equation (12.18) from the limiting straight lines is minimal; the full curve thus follows closely the upper branches of the two limiting straight dashed lines which are logarithmic plots of Eqs. (12.22a) and (12.22b).

References

Kröger, F. A. and Vink, H. J. (1956), *Solid State Physics*, **3**, 237.

Kröger, F. A., Stieltjes, J. W. and Vink, H. J. (1959), *Philips. Re. Reports*, **14**, 557.

Kröger, F.A. (1964), *The Chemistry of Imperfect Crystals*, North Holland.

Kofstad, Per (1972), *Non-stoichiometry, Diffusion and Electrical Conductivity in Binary Metal Oxides*, J. Wiley, NY.

Chapter 13

Interfaces

Gibbs (1875) introduced the notion of *heterogeneous substances* into thermodynamics. By that he meant a macroscopic substance consisting of *phases* separated by relatively thin interfacial regions, each phase being regarded as having uniform properties such as composition or energy, or other extensive quantities. The neglect of non-uniform contributions from the interfacial regions is acceptable as long as the interface material (which in reality has non-zero thickness) has total volume vanishingly small with respect to the total volume of the system investigated. In today's world, emphasizing nano-materials, that approximation may no longer be reliable in many cases. Gibbs was already aware of this difficulty, and so devoted a long chapter of his famous treatise to the study of interfaces. The following is a simplified presentation of the Gibbs's model.

13.1 The Gibbs Stratagem

The concentration profile (or some other extensive quantity per unit volume) in the vicinity of a small portion of the interface between phases α and β may look like the curve drawn in the upper panel of Fig. 13.1. A quantitative treatment of the thermodynamics of the interface region would thus require the introduction of position-dependent parameters which Gibbsian thermodynamics does not allow. Indeed the word "Heterogeneous" in Gibbs's treatise implies that heterogeneities may be present in the system in the form of different phases, but the properties of each phase must be uniform throughout.[1] The problem of non-uniformity in fact transcends that of interfaces: for instance, how does one define continuous and smooth

---

[1] The case of non-uniformity will be examined in Sec. 14.1.

position-dependent functions, $f(\mathbf{x})$ in materials which are always discrete in nature at the atomic scale?

The answer lies in the procedure called *coarse-graining* which consists of taking averages of the property f in a small volume around the space point $\mathbf{x}$, as will be described in the next chapter. For now, let us simply define the function $f(\mathbf{x})$ by taking averages over very thin slices of material locally parallel to what we deem to be the surface of separation between α and β. As stated by Gibbs, that surface is chosen *such that its points are similarly situated with respect to the conditions of adjacent matter*. Gibbs then discusses the actual placement of the dividing surface through arguments which are hardly transparent, but which amount to placing the interface at that position which minimizes the intrinsic effect of curvature on the thermodynamic properties of the interface. Moreover, since the thermodynamics of interfaces between elastic solids is very complicated, we shall treat the thermodynamics of the bulk phases as if they were fluids, or solids hydrostatically stressed.

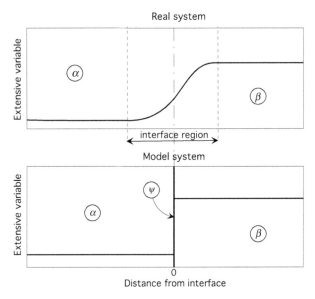

Figure 13.1: Schematic profile of extensive quantity z (per unit volume) at the interface between phases α and β; actual and model.

To treat the heterogeneities introduced by the interface between α and β, say, Gibbs introduces a very clever stratagem: in addition to the *real* system (upper panel of Fig. 13.1), he defines a model system of total volume identical to that (V) of the original system. In place of the original interface, of finite width, he places a "mathematical" interface of zero width, such as that indicated by the vertical line marked ψ at the lower panel of Fig. 13.1. In this *model* system, the properties of the two bulk phases extend unchanged right up to the dividing surface. Extensive properties of the interface (call them Z^ψ) are then defined in an operational way as follows: Z^ψ is simply the value of Z for a system containing the interface minus that for the system *not* containing that interface. We then have, by construction,

$$Z_{actual} = Z_{model} \tag{13.1a}$$

that is

$$Z_{actual} = Z^\alpha_{model} + Z^\beta_{model} + Z^\psi \tag{13.1b}$$

so that

$$Z^\psi = Z_{actual} - \left(z^\alpha_v V^\alpha + z^\beta_v V^\beta \right). \tag{13.1c}$$

In this equation, the notation z_v designates corresponding Z per unit volume, and the volumes V^α and V^β are those of the portions indicated in the lower panel of Fig. 13.1 in the model system (there are also natural cutoffs right and left at some distance away from the interfaced; not shown). Those volumes can, in principle, be measured once the position of the model dividing surface ψ is fixed; their sum of course equals the total volume V, actual or model. Equation (13.1c) gives a straightforward method for calculating extensive quantities Z^ψ in the interface, but the actual implementation of the method is difficult since these quantities are defined as very small differences of very large numbers.[2] The densities z_v themselves can be obtained by measuring these bulk values in regions far away from the actual interface, *i.e.* in regions where properties are uniform. By this procedure, Gibbs is able to treat a non-uniform system as a uniform one, while retaining quantitatively all the thermodynamic properties of the real interface. Actually, the Gibbs method is equivalent to a change of scale: the concentration profile shown schematically in the upper panel of Fig. 13.1 might be that observed with a very powerful electron microscope, while the one seen in the lower panel might be that of the very same profile observed with an ordinary light microscope. There is really no change in the system itself. In another viewpoint: a continuous, differentiable but non-uniform profile has been replaced by a uniform one but with a singularity (a discontinuity or delta function) at the location of the infinitely thin ψ "phase".

---

[2] It's a bit like determining the weight of the captain of a ship by first weighing the ship with the captain, then weighing the ship without the captain, and taking the difference.

13.2   Interfacial Equilibrium

The latter considerations suggest treating extensive quantities pertaining to surfaces as ordinary thermodynamic functions of extensive variables, as was done previously for bulk quantities,

$$U^\psi = U^\psi(S^\psi, A^\psi, N_1^\psi, N_2^\psi, \ldots N_n^\psi) \tag{13.2}$$

leading to the differential form

$$dU^\psi = T^\psi dS^\psi + \sigma^\psi dA^\psi + \sum_{i=1}^{n} \mu_i^\psi dN_i^\psi . \tag{13.3}$$

This differential expression requires explanations: the extensive variables are defined by the procedure explained by Eq. (13.1c). Since the volume of the interface is nil, it is replaced by the interface area A, which is a measurable geometrical quantity. The intensive variables are, for now, simply regarded as the coefficients of the corresponding extensive variables. Despite the discontinuity mentioned, it is seen that all elements of Eq. (13.3) are well-defined. As in earlier chapters, we shall henceforth use the abbreviated expression $\mu^\psi dN^\psi$ for the sum over all components (i) of the solution.

The list of extensive variables in expressions (13.2) or (13.3) is incomplete, as Gibbs knew full well: missing are the influence on the energy of the curvature of the interface, i.e. on the principal curvatures $C_1 = 1/R_1$ and $C_2 = 1/R_2$, where R_1 and R_2 are the principal radii of curvature. However, as mentioned earlier, Gibbs then shows that the *intrinsic* dependence of the energy of curvature can generally be neglected: by "intrinsic" is meant the change in energy due to the variations of the curvatures C_1 and C_2 of the interface but not of its area or composition. Of course, the energy of the interface depends significantly on its geometry, but the shape of an interface bounded by a closed, generally non-planar curve (minimal surface with a given boundary), at mechanical equilibrium, is a non-local problem, a very difficult variational one, known as "Plateau's problem".[3] Below, we shall treat only the case of a phase β enclosed by a spherical surface of radius R.

The condition of equilibrium is that the total energy of the system, closed, adiabatic, consisting of phases α and β and the interface ψ, with fixed overall volume, shall be stationary for reversible variations of the

---

[3]The general problem was solved only in 1930 by mathematician J. Douglas, who was awarded for the solution by the Field's medal in 1936. Small wonder that the solution will not be treated here.

relevant extensive variables. The equations of conditions are

$$S = S^\alpha + S^\beta + S^\psi, \tag{13.4a}$$

$$N_i = N_i^\alpha + N_i^\beta + N_i^\psi, \ (i = 1, 2, \ldots n) \quad \text{and} \tag{13.4b}$$

$$V = V^\alpha + V^\beta. \tag{13.4c}$$

The differential of the total energy is given by

$$dU = dU^\alpha + dU^\beta + dU^\psi \tag{13.5}$$

with the differential of the component parts given by

$$dU^\alpha = T^\alpha dS^\alpha + \mu^\alpha dN^\alpha - P^\alpha dV^\alpha, \tag{13.6a}$$

$$dU^\beta = T^\beta dS^\beta + \mu^\beta dN^\beta - P^\beta dV^\beta \quad \text{and} \tag{13.6b}$$

$$dU^\psi = T^\psi dS^\psi + \mu^\psi dN^\psi + \sigma^\psi dA^\psi. \tag{13.6c}$$

Equation (13.6c) is just Eq. (13.3) but with the shortcut convention on the sum over components.

The present equilibrium problem is thus that of minimizing the function U under conditions (13.4). That problem calls for Lagrange multipliers (see Appendix D). For the sake of clarity, let us break up the problem into a thermal, a chemical and a mechanical part. Lagrange multipliers (λ) will be used in the first two parts, but for the mechanical part the problem will be reduced to one with a single independent variable, by simply considering, for the β phase, a spherical inclusion of radius R. We have successively, following the rules of Eq. (D.10),

$$d\Phi_{\text{therm}} = (T^\alpha - \lambda_o)dS^\alpha + (T^\beta - \lambda_o)dS^\beta + (T^\psi - \lambda_o)dS^\psi \tag{13.7a}$$

$$d\Phi_{\text{chem}} = \sum_i \left[(\mu_i^\alpha - \lambda_i)dN_i^\alpha + (\mu_i^\beta - \lambda_i)dN_i^\beta + (\mu_i^\psi - \lambda_i)dN_i^\psi \right]. \tag{13.7b}$$

For the mechanical part we have

$$dU_{\text{mech}} = -P^\alpha dV^\alpha - P^\beta dV^\beta + \sigma dA, \tag{13.8}$$

where the superscripts ψ on the interface variables are no longer needed. Now, since $dV^\alpha = -dV^\beta$, we can write

$$dU_{\text{mech}} = -\Delta P \, dV^\beta + \sigma dA \tag{13.9}$$

with

$$\Delta P = P^\beta - P^\alpha. \tag{13.10}$$

For a sphere of radius R, the differentials of volume and area are respectively

$$dV^\beta = 4\pi R^2 dR, \qquad dA = 8\pi R \, dR, \tag{13.11}$$

so that, by combining Eqs. (13.9), (13.10) and (13.11), we obtain

$$dU_{\text{mech}} = 4\pi(-R\Delta P + 2\sigma)R \, dR. \tag{13.12}$$

Now that the dependent "mechanical" variables V and A have been replaced by the sole independent one R, and that the other extensive variables have been taken care of by Lagrange multipliers (as in Sec. 4.6), the condition of equilibrium can be obtained by setting to zero all coefficients of these variables in the expressions for the energy differential, producing the desired result

$$\begin{cases} \text{therm} & T^\alpha = T^\beta = T^\psi = \lambda_o = T \\ \text{chem}_1 & \mu_1^\alpha = \mu_1^\beta = \mu_1^\psi = \lambda_1 = \mu_1 \\ \cdots\cdots\cdots\cdots\cdots\cdots\cdots\cdots\cdots \\ \text{chem}_n & \mu_n^\alpha = \mu_n^\beta = \mu_n^\psi = \lambda_n = \mu_n \end{cases} \tag{13.13}$$

plus the "mechanical" condition

$$R\,\Delta P = 2\sigma. \tag{13.14}$$

Equation (13.14) is known as the Gibbs–Thomson equation. Equations (13.13) show that the intensive variables for the interface can indeed be regarded as, respectively, the absolute temperature and chemical potentials of the interface, furthermore, the thermodynamic equilibrium conditions are precisely those of the bulk as derived previously: across the sample, the temperature and the chemical potentials have constant value, regardless of which phase is traversed.

Equation (13.6c) provides definitions for the interfacial intensive variables which may now be written without the superscripts ψ:

$$T = \left(\frac{\partial U^\psi}{\partial S^\psi}\right)_{N^\psi, A}, \quad \mu_i = \left(\frac{\partial U^\psi}{\partial N^\psi}\right)_{S^\psi, N^\psi_{j\neq i}, A}, \quad \sigma = \left(\frac{\partial U^\psi}{\partial A^\psi}\right)_{S^\psi, N^\psi}. \tag{13.15}$$

Hence the variable σ may be interpreted as the change in energy due to a variation of the surface area at constant entropy and amounts. For fluids it is simply called the surface tension which causes the surface area to be as small as possible in order to reduce the surface energy. The other partials are very similar to the ones encountered for bulk thermodynamics, but with Area substituted for Volume.

The condition for mechanical equilibrium, as written here for a phase (β) enclosed by a sphere of radius R, is correct so far as it recovers the

Gibbs–Thomson equation,[4] a simplified form of which is

$$P^\beta - P^\alpha = 2\sigma/R.$$ (13.16)

One should not extrapolate from this the more general cases, however. Indeed, if we were to consider the case of a closed, non-planar curve with a thin elastic sheet stretched over a landscape of many hills and valleys, the Gibbs–Thomson equation(13.14) would tend to state that the pressures should alter continuously following the local curvature. That of course cannot be the case: there can only be two pressures, one for each side of the surface of separation. As mentioned previously, for the general case, the difficult, non-local Plateau problem must be solved along with the thermodynamic conditions. Of course, if the interface is flat ($R \to \infty$), then the pressures, P^α and P^β must be equal.

13.3 Gibbs Adsorption Isotherm

Since the surface energy, like its bulk counterpart is a homogeneous function of degree one in its (extensive) variables, the differential equation (13.2) can be Euler-integrated to obtain the finite form (in shorthand notation)

$$U^\psi = T S^\psi + \mu N^\psi + \sigma A.$$ (13.17)

After dividing by A and rearranging we arrive at an equation for *surface densities* or *surface excesses* (the sum over constituents has been made explicit again)

$$\sigma = u^\psi - T s^\psi - \sum_i \mu_i \Gamma_i$$ (13.18)

where the variables Γ_i are surface excess concentrations analogous to the bulk concentrations c_i: $\Gamma_i = N_i^\psi/A$. The Γ_i are thus surface densities (per unit area). Note that these surface concentrations can exceed those in the adjoining bulk phases: an interface, particularly in solids, can reject or adsorb solute to a considerable degree.

The sign of the Γ_i can be discussed within the framework of classical bulk thermodynamics as follows: first take the total differential of the interfacial (internal) energy (13.17) to obtain

$$dU^\psi = T\,dS^\psi + S^\psi dT + \mu\,dN^\psi + N^\psi d\mu + \sigma\,dA + A\,d\sigma.$$ (13.19)

---

[4]actually due to Laplace, as mentioned by Lupis (1993) in his very complete chemical thermodynamics textbook.

On comparison of this differential with the one given by Eq. (13.3), it follows, as expected from the derivation of the Gibbs–Duhem equation (4.4), that the following must hold:

$$A\,d\sigma + S^{\psi}dT + N^{\psi}d\mu = 0\,, \tag{13.20}$$

which is the Gibbs–Duhem equation for surfaces, an identity. By dividing through by A and rearranging we obtain the surface tension differential

$$d\sigma = -s^{\psi}dT - \Gamma d\mu\,, \tag{13.21}$$

indicating that temperature and the chemical potentials are the natural variables for surfaces: $\sigma = \sigma(T, \mu_1, \mu_2, \ldots, \mu_n)$, with partials

$$s^{\psi} = -\left(\frac{\partial\sigma}{\partial T}\right)_{\mu_i}, \quad \Gamma_i = -\left(\frac{\partial\sigma}{\partial\mu_i}\right)_{T,j\neq i} \quad (i = 1, \ldots n)\,. \tag{13.22}$$

Hence, at fixed temperature, and with the familiar relation for chemical potentials in terms of activities, $\mu_i = \mu_i^{\circ} + RT\ln a_i$ [Eqs. (7.20) or (7.25)] we find

$$\Gamma_i = -\frac{a_i}{RT}\left(\frac{\partial\sigma}{da_i}\right)_{T,j\neq i}. \tag{13.23}$$

If the solution is ideal for component i, then $a_i \approx c_i$ and we obtain simpler versions of the Gibbs adsorption isotherms:

$$\Gamma_i = -\frac{c_i}{RT}\left(\frac{\partial\sigma}{dc_i}\right)_{T,j\neq i} \quad \text{or} \quad \Gamma_i = -\frac{1}{RT}\left(\frac{\partial\sigma}{d\ln c_i}\right)_{T,j\neq i}. \tag{13.24}$$

Hence we obtain a simple rule for the adsorption or rejection of solute i at constant temperature:

$$(\partial\sigma/\partial c_i)_T \begin{cases} < 0\,, & \text{then } \Gamma_i > 0 \text{ and the interface absorbs solute} \\ > 0\,, & \text{then } \Gamma_i < 0 \text{ and the interface rejects solute}\,. \end{cases} \tag{13.25}$$

In crystalline substances, interfaces can also absorb and eject dislocations, but that is another matter.

13.4   Intersection of Interfaces

"All good things must come to an end," including phases, which must be bounded, finite. Even in single-phase solids, interfaces must be found in polycrystalline materials. In those cases, neighboring *grains* with identical *thermodynamic* properties (mostly, same composition), will be separated

by surfaces if their *crystalline* properties differ, *i.e.* if their relative orienta-
tions differ, for example. In what follows, the expression "different phases"
will refer to thermodynamic or crystallographic differences. The interfaces
in crystalline solids themselves will generally exhibit highly diverse and
complex structures, as described extensively, for example, in the 800-plus
page treatise by Sutton and Balluffi (1995) covering thermodynamics, crystallography, bonding and mechanical properties of interfaces in crystalline
materials.

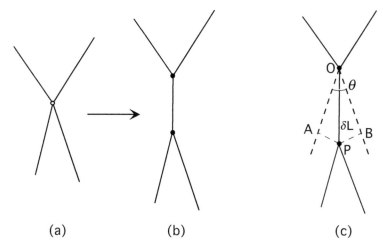

(a)　　　　　　　(b)　　　　　　　(c)

Figure 13.2: Replacing one 4-junction (a, open circle) by two 3-junctions (b,
closed circles). Proof (c) of instability of 4-junctions.

As stated previously, the present work considers mainly the thermodynamics of fluids, or of solids if these are subjected only to hydrostatic
stresses. With this caveat, we may then proceed to examine what happens
when several "phases" (grains) come together at one point. For instance, 4-
phase junctions could be formed, as seen in Fig. 13.2(a), but such junctions
are unstable and must break up at equilibrium into a pair of 3-phase junctions, as shown schematically at Fig. 13.2(b). To prove this, first note that
the four angles at a 4-junction cannot all be greater than 90°, so at least
one of them must be acute; so let it be the one depicted at the lower part
of Fig. 13.2(a). Then the proof follows with the help of Fig. 13.2(c): the
key to equilibrium is the reduction of total surface energy, hence reduction
of total surface area. In going from configuration (a) to (b) in Fig. 13.2,
the two segments $\overline{OA}$ and $\overline{OB}$ are replaced by the segment δL (bisecting
the angle θ), and corresponding surface areas. To show that a net reduction in surface area results, it suffices to show that the difference in lengths

$\delta L - (\overline{OA} + \overline{OB})$ is negative. This must be so since, because $\frac{1}{2}\theta \leq \frac{1}{4}\pi$ by construction, we must have

$$\overline{OA} + \overline{OB} \geq \overline{OA} + \overline{AP} > \overline{OP} = \delta L\,.$$

Therefore 4-junctions are unstable against dissociation into two triple junctions. *A fortiori*, n-junctions with $n > 4$ are also unstable.

Now consider three phases or grains (α, β, γ) that meet at a common junction such that the dihedral angles are $\theta_\alpha, \theta_\beta, \theta_\gamma$. Surface tensions can be represented by tangent vectors of lengths a, b, c, respectively proportional to $\sigma_{\beta\gamma}, \sigma_{\gamma\alpha}, \sigma_{\alpha\beta}$ (see Fig. 13.3 (a)). The equilibrium of forces requires the vanishing of the vector sum of tensions $\vec{a} + \vec{b} + \vec{c} = 0$, *i.e.* hence the triangle $\overline{ABC}$ closes (Fig. 13.3(b)). Angles and lengths of sides for any triangles we have the relations (proved in Appendix G):

$$\frac{a}{\sin A} = \frac{b}{\sin B} = \frac{c}{\sin C}\,. \tag{13.26}$$

Since A, B, C are the complements to 180° of $\theta_\alpha, \theta_\beta, \theta_\gamma$ respectively, Eq. (G.1) can be rewritten

$$\frac{\sigma_{\beta\gamma}}{\sin\theta_\alpha} = \frac{\sigma_{\gamma\alpha}}{\sin\theta_\beta} = \frac{\sigma_{\alpha\beta}}{\sin\theta_\gamma}\,, \tag{13.27}$$

which is the relation sought between surface tensions and wetting angles. The special case of 2-phase equilibrium (α, β) is of particular interest, as it

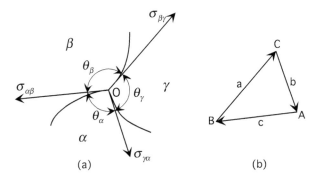

Figure 13.3: (a) Surface tension σ equilibrium at junction of 3 phases α, β, γ. (b) Vector sum of 3 tensions equals zero.

examines surface equilibrium for a phase (β) situated on a grain boundary, the two grains assumed here to be thermodynamically the same crystalline phase (α) but oriented differently. So, in Eq. (13.27) let γ be replaced by α. These equations then reduce to

$$\sigma_{\alpha\alpha} = 2\,\sigma_{\alpha\beta}\cos\frac{1}{2}\theta \tag{13.28}$$

where $\theta \equiv \theta_\alpha$ is the only independent angle to be considered (see Fig. 13.4). By the properties of the cosine, we must have at equilibrium

$$0 \le \frac{\sigma_{\alpha\alpha}}{2\sigma_{\alpha\beta}} \le 1. \tag{13.29}$$

Also, since the surface tensions are positive, the angle θ must be less than 180°. For true two-phase junction to exist, the following must then hold $\sigma_{\alpha\beta} \ge \frac{1}{2}\sigma_{\alpha\alpha}$. If however $\sigma_{\alpha\beta} \le \frac{1}{2}\sigma_{\alpha\alpha}$, then $\theta \to 0$ and the β phase will show affinity for one another, *i.e.* the β phase will completely wet the α phase, resulting in the practical elimination of $\alpha - \alpha$ contacts, as shown in Fig. 13.4(b), upper sketch, thereby creating as much $\alpha - \beta$ interface area as possible. If $\sigma_{\alpha\beta} \gg \sigma_{\alpha\alpha}$, then $\theta \to 180°$ and the β phase will tend to break away from $\alpha - \alpha$ junctions (see lower part of Fig. 13.4(b)).

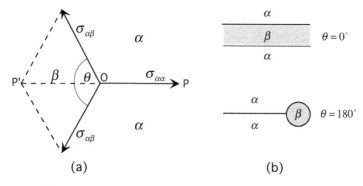

(a) (b)

Figure 13.4: (a) 3-junction equilibrium when two phases are thermodynamically identical but crystallographically misoriented, say. The three σ vectors form a closed triangle. (b) two extreme cases, $\theta = 0°$, full wetting; $\theta = 180°$, no wetting.

A hypothetical (and schematic) micrograph of a 2-phase microstructure is shown in Fig. 13.5, illustrating β precipitates, on grain boundaries and at a triple junction. In a crystalline single-phase system (but necessarily polycrystalline) we should have $\theta_\alpha = \theta_\beta = \theta_\gamma = 120°$. In two dimensions, a regular hexagonal structure will satisfy those requirements, but in three dimensions, no fully regular cell structures can fulfill those requirements everywhere (see Chapter 10 on Topological Disorder).

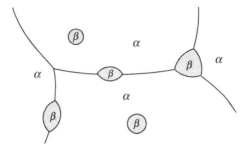

Figure 13.5: Schematic micrograph of parent α phase with β precipitates (shown shaded) in the bulk, at phase boundaries and at a triple junction. Wetting angle: $\theta \approx 120°$.

References

Gibbs, J. W. (1875). *On the Equilibrium of Heterogeneous Substances*, *Trans. Connecticut Acad.*, pp. 108-248; (1877). *ibidem*, pp. 343-524 . See also for example (1993). *The Scientific Papers of J. Willard Gibbs*, Ox Bow Press.

Lupis, C. H. P. (1993). *Chemical Thermodynamics of Materials*, New York, North-Holland.

Sutton, A. P. and Balluffi, R. W. (1995). *Interfaces in Crystalline Materials*, Oxford University Press, New York.

Chapter 14

Non-Uniform Systems

The topic of phase stability is a major one in what used to be called Physical Metallurgy, and is now one subject among many in Materials Science. Materials Science itself tends to bring together many individual disciplines pertaining to the properties of metals, ceramics, semiconductors, even biological substances. Such a unification is possible by placing less emphasis on different types of materials, and more emphasis on fundamentals, thus requiring a bit more in the way of physics and mathematics. By doing so, it has become possible to bring together pairs of subjects which used to be treated separately, but which can now be regarded as different facets of a common topic, that of *phase stability*. Examples are as follows, to be treated in this chapter rather briefly, as the field is very broad and can become rather involved. The following dichotomies arise:

$$[1] \text{ Type of space} \begin{cases} \text{Continuum medium} \\ \text{Discrete medium} \end{cases}$$

$$[2] \text{ Type of instability} \begin{cases} \text{Clustering reaction} \\ \text{Ordering reaction} \end{cases}$$

$$[3] \text{ Type of transformation} \begin{cases} \text{Nucleation and growth} \\ \text{Spinodal clustering and ordering} \end{cases}$$

All of these subtopics are extensions of Gibbsian thermodynamics whose medium was, originally, a fluid, *uniform* in its properties over large volumes. Hence, the first extension will be to *non-uniformity*, allowing local positional changes in such variables as concentration or electrical, magnetic, elastic properties, though here elastic effects will not be covered. A rational handling of non-uniformity must appeal to a method of describing local properties by continuous functions $f(\mathbf{x})$ of position $\mathbf{x}$, well-defined at every point. But solids, even fluids, are inherently discrete at the atomic scale, so care must be taken in approximating a discrete system by a continuous one. As mentioned several times in this book, the mechanism for performing that passage from the microscopic (atomic) to the macroscopic, plus the intermediate mesoscopic scale, is the so-called coarse-graining process, mentioned previously.

Duality number [1] will be discussed in Sec. 14.1, with an application to diffuse interfaces given in Sec. 14.2, where the idea of local bulk energy supplemented by a non-local "gradient" energy contribution is advanced. This idea appears to have first been proposed by the Russian school of L. D. Landau and collaborators (1958, 1965), and in the US by Cahn and Hilliard (1958). A very good introduction to the work of John Cahn (with or without his co-worker John Hilliard) may be found in the commemorative collection: *The Selected Works of John W. Cahn* (1998). In the opening article, entitled *Reflections on Diffuse Interfaces and Spinodal Decomposition*, Cahn relates the role that he played in shaping the subject in the early 1950's to the late 1900's. The author reminds us that, starting from about 1950, the main topic of studies in phase transformations was that of nucleation and growth, which was to be enlarged considerably starting in the early 1960's. Because of the traditional importance of the nucleation process, it will be covered in a separate chapter (Ch. 17).

In traditional Physical Metallurgy, changes of phase were mostly concerned with what is known as *clustering*, that is separation of a mixture of different atoms or molecules into regions rich in one dominant component. But it could be that like-components (identical) instead repel one another, preferring *unlike* near neighbors. That latter tendency tends to promote order on a small scale, hence that type of phase separation is known as an *order-disorder process*. Those two tendencies belong to what was listed above as dichotomy number [2]; it will be treated in Sec. 15.3. Dichotomy number [3], nucleation and spinodal decomposition, traditionally reflected two separate mechanisms; actually, as will be discussed later (in Chapter 18), there is no dichotomy, as those two apparently different mechanisms are actually one and the same if treated correctly in a rather more complex joint theory.

At another, broader level, the topic of "phase transformations" means different things to different disciplines. First of all, it is important to distinguish "transformations" from "transitions". The former is the province mainly of materials science, the latter of condensed matter physics or physical chemistry. When mentioning "materials science" it is important to realize that that field evolved rather recently from physical metallurgy. Why metallurgy? Probably because metals, being simpler materials compared to ceramics or biological substances, could lead the way towards more advanced theoretical treatments, now adopted by most subfields of materials science. But then, why the difference between transformations (transitions) in physics and in materials science? It is here suggested (and mentioned earlier) that the word "transition" should be reserved to phase changes occurring at or very close to thermodynamic equilibrium, as opposed to general phase changes occurring far from (or near to) thermodynamic equilibrium, thus including non-equilibrium processes. This dichotomy has tended to create a division in the field, reserving transformations mostly to Materials Science and transitions mostly to Physics and Physical Chemistry.[1] For example, a physicist will hardly ever venture into the realm of martensitic transformations, but a metallurgist will hardly delve into the intricacies of renormalisation group theory. The former will rely mostly on classical thermodynamics, the latter mostly on statistical mechanics.

For the majority of practicing engineers, the vast majority of phase transformations encountered are of first-order type, that of melting being the most familiar type. For scientists, only second-order transitions pose conceptual problems which require advanced mathematical analysis. Indeed, first-order reactions occur whenever two (or more) free energy curves (surfaces) intersect, one free energy function remaining practically unaware of the other. In second-order transitions, one single free energy function loses mathematical analycity at a critical point, and the mathematics plunges into the non-analytic (non-algebraic) domain. In what follows, mostly "coherent" instabilities will be addressed, coherent processes being those, in solids, which leave the lattice framework (topologically) unaltered.

14.1 Coarse Graining

The task of describing non-uniformity is thus to average the atomic occupancy, say $\sigma = 0, 1$ (or $= +1, -1$), *at* a crystallographic point $\mathbf{x}$ occupied — or not — in a small volume about $\mathbf{x}$, large enough to perform a meaningful average, but small enough to describe the concentration (or other) changes which occur in the system studied; such is indeed the procedure denoted as "coarse-graining" earlier. For the topic of *phase separation*, this procedure is reasonable enough, but for order-disorder transformations, one may wonder how this averaging prescription would work out. Indeed, since antiphasing usually occurs in order-disorder transformations, averages taken over large distances necessarily implies averages taken over antiphase domains (see Sec. 15.3), which means that the average concentration at a point $\mathbf{x}$ will vanish. To avoid that difficulty, one must appeal to "sublattice

---

[1]Of course I'm over-generalizing.

averaging" as was described elsewhere (de Fontaine, 1979) and which will
be taken up later in this chapter. This procedure allows the definition of
a continuous concentration *at* a crystallographic point, a method which is
not really *microscopic*, but actually *mesoscopic*.

Usually, this averaging procedure is assumed to have been carried out
implicitly. That is all right, but two important points must be kept in mind
for quantitative studies: (1) it must be ascertained that the final result
of a calculation does not depend too strongly on the size of the domain
over which the averaging is performed and (2) that calculated energies
(or other extensive quantities) must include the influence of fluctuations
occurring inside the averaging regions. At present, we merely assume that
this reduction of what should really be a quantum mechanical system to
a classical one has been performed, and allow mesoscopic thermodynamic
functions $f(c)$, along with the concentration $c = c(\mathbf{x})$, to be functions of
position $\mathbf{x}$, which means that we now allow non-uniformity, *i.e.* space-
dependent thermodynamic quantities: $f[c(\mathbf{x})]$, where f is a Helmholtz or
Gibbs free energy density (or other extensive quantity). For simplicity,
a single independent concentration c is considered here, thus only binary
systems.

It is also clear that, in non-uniform systems, local energy will depend not
only on the average concentration (c) in the small averaging volume considered, but also on the manner in which c (or other local variables) change
within that volume, *i.e.* the energy will depend also on local c but also on
the derivatives of the concentration with respect to position. Thus there
will be not only a *local*, but also a *non-local* contribution to the mesoscopic
function $f(c)$, in the form of concentration gradients, labeled collectively as
∇c, so that the required free energy function (internal, Helmholtz, Gibbs,...)
in non-uniform systems will be of the form $f(\mathbf{x}, c, \nabla c)$. The total energy
F will be obtained by integrating over the whole volume V to obtain the
functional

$$F[f] = \int_V f(\mathbf{x}, c, \nabla c) \, dV \,. \tag{14.1}$$

The equilibrium configuration $c(\mathbf{x})$ will be the one that minimizes the
total energy $F[f]$, hence the one that makes the functional (14.1) stationary.
This is the well-known *variational problem* in classical mechanics, summarized in Appendix H, where it is shown, for the one-dimensional case $c(x)$,
that the solution is obtained by setting to zero the variation δF of the
integral in Eq. (14.1), or at least in this case, its one-dimensional version
between the fixed points x_1 and x_2 and with the simplest approximation

for the gradient term:

$$\delta F = \int_{x_1}^{x_2} \left[\frac{\partial f}{\partial c} \delta c + \frac{\mathrm{d}}{\mathrm{d}x} \left(\frac{\partial f}{\partial c'} \right) \delta c \right] \mathrm{d}x, \quad \text{with} \quad c' \equiv \frac{\mathrm{d}c}{\mathrm{d}x}. \tag{14.2}$$

After various manipulations, it is shown that the equilibrium condition for arbitrary δc is given by the famous Euler–Lagrange equation (see Appendix A), in present notation (and one space dimension):

$$\frac{\partial f}{\partial c} - \frac{\mathrm{d}}{\mathrm{d}x} \frac{\partial f}{\partial c'} = 0. \tag{14.3}$$

14.2 Square-Gradient Approximation

Before any actual calculations can be made, however, it is necessary to make the function f explicit, in particular it is necessary to choose a formal expression for the gradient term. The form adopted by Cahn and Hilliard (1958, CH for short) is simply a coefficient, κ (kappa), times the squared gradient $(\nabla c)^2$; or in one space dimension, $(\frac{\mathrm{d}c}{\mathrm{d}x})^2$. That choice was dictated by the absence of clear evidence that higher gradients were necessary beyond the second power and by the fact that only even powers should appear in the formula for symmetry reasons. Also, for simplicity, the gradient energy coefficient κ was initially regarded as independent of concentration. Such appears also to be the rationale for the very same choice made by Landau and his school (1958, 1965) who proposed the same square-gradient approximation, in fact prior to the CH publications. Cahn and Hilliard (1958) state that the gradient energy coefficient κ should be positive, but add that they were unable to prove that this should necessarily be a fixed requirement. Indeed, it later appeared from the discrete-space work of the Russian school and that of Cook, de Fontaine and Hilliard (1969) that the coefficient κ must indeed be negative for the order-disorder case (see Chapter 15).

For the local term, it is perfectly reasonable to take the usual form $g(c)$ of energy (Gibbs or Helmholtz) generally used in thermodynamics studies, in this book in particular (see for example Fig. 14.1). Therefore we shall adopt the functional form

$$f(\mathbf{x}, c, \nabla c) = g(c) + \kappa (\nabla c)^2 \quad \text{or} \tag{14.4a}$$

$$f(x, c, c') = g(c) + \kappa (\mathrm{d}c/\mathrm{d}x)^2 \quad \text{for the one-dimensional case.} \tag{14.4b}$$

In these equations the functions f and g do not necessarily have the usual meanings of Helmholtz and Gibbs free energies; they are mere mathematical symbols for the free energies that the context specifies. To illustrate non-uniformity, Cahn and Hilliard (1958) chose the case of a planar interface, subject to various restrictions: the two phases, α and β must be fluid phases, or *coherent* crystalline phases of cubic symmetry, by which is meant that there is no lattice discontinuity across the interface (otherwise the interface might absorb of reject solute) and that the κ coefficient be a scalar. By Eqs. (14.1) and (14.4b) the total energy of the system is then

$$F = A N_V \int_{x_1}^{x_2} \left[g(c) + \kappa \left(\frac{\mathrm{d}c}{\mathrm{d}x} \right)^2 \right] \mathrm{d}x \qquad (14.5)$$

where N_V and A are number of atoms (or molecules) per unit volume and A is the area of the interface. The limits x_1 and x_2 are located arbitrarily provided that they fall sensibly outside the region of inhomogeneity of the interface. We now calculate the energy (σ) per area of the interface specifically by use of the "Gibbs stratagem" in the manner of Eq. (13.1c), where care must be taken to maintain the volumes constant $v^\alpha + v^\beta$, the volumes in question being defined by the positions of the integral limits x_1 and x_2. As shown by CH, and by arguments presented previously, the local free energy, Δg, is given by the length of the vertical lines in Fig. 14.1, in the direction from common the tangent to the curve g. The result is, following Eq. (14.5),

$$\sigma = N_V \int_{x_1}^{x_2} \left[\Delta g(c) + \kappa \left(\frac{\mathrm{d}c}{\mathrm{d}x} \right)^2 \right] \mathrm{d}x . \qquad (14.6)$$

The fact that the common tangent has been invoked means that phases α and β must be in thermodynamic equilibrium and, as noted in CH, a condition that is not required for calculating the energy of an infinitely thin interface, one that behaves as a singularity in the $g(c)$ function. Applying the Euler equation to the integrand (14.4b) gives the equilibrium condition

$$\frac{\partial g}{\partial c} - 2\kappa \frac{\mathrm{d}^2 c}{\mathrm{d}x^2} = \lambda \qquad (14.7)$$

where λ is an appropriate Lagrange multiplier (see Appendix D) taking care of constant integrated concentration. This multiplier acts as a chemical potential, actually a *difference* of chemical potential $\mu_B - \mu_A$, constant throughout the bulk phases and the interface region. In three space dimensions, Eq. (14.7) becomes

$$\frac{\partial g}{\partial c} - 2\kappa \nabla^2 c = \lambda . \qquad (14.8)$$

This unassuming little λ is actually the potential that appears in the expression "divergence of the gradient of the potential" which drives the kinetic equation that introduced the spinodal transformation that made John Cahn

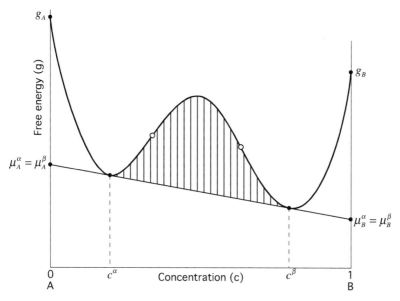

Figure 14.1: Free energy curve $g(c)$ used for the local part of the energy of a flat interface; the vertical lines inside the non-equilibrium portion of the free energy represent the quantity Δg, from the double tangent line to the curve. Open circles are spinodal concentrations.

(1961) justly famous, which also is at the heart of the kinetic equation which lies at the origin of the *phase field method*.[2]

Since the integrand in Eq. (14.5) is not explicitly a function of x, we may use the alternate Euler equation (H.9), derived in Appendix H, in present notation

$$\Delta g - c' \frac{\partial g}{\partial c'} = \text{constant}, \quad \text{with} \quad c' = \frac{dc}{dx}. \tag{14.9}$$

The constant must be zero since both Δg and $\frac{dc}{dx}$ must vanish at the limits x_1 and x_2, assuming these limits are sufficient far away from the interface.

---

[2]That equation first appeared as an unnumbered equation in J. W. Cahn's first spinodal paper (in 1961), somewhat incorrectly named "the Cahn–Hilliard equation" (incorrectly, since John Hilliard was not a co-author of the paper in question). It now appears that I was the first one to attempt to solve that non-linear equation by numerical techniques (de Fontaine, 1967), thereby becoming unwittingly the originator of the phase-field method. At the time, I also made a computer-generated movie of (non-linear) spinodal decomposition at the Argonne National Lab; a second, better one, was planned, which John Hilliard wanted to call "Son of Spinodal", but it was never produced.

Hence at equilibrium we must have

$$\Delta g(c) = \kappa \left(\frac{\mathrm{d}c}{\mathrm{d}x} \right)^2 . \tag{14.10}$$

Introducing this relation into Eq. (14.6) we find

$$\sigma = 2N_V \int_{x_1}^{x_2} \Delta g(c) \, \mathrm{d}x , \tag{14.11}$$

and, by a change of variables, the desired result

$$\sigma = 2N_V \int_{c^\alpha}^{c^\beta} \sqrt{\kappa \, \Delta g(c)} \, \mathrm{d}c , \tag{14.12}$$

a compact and elegant expression for the energy of a diffuse interface. The concentration profile in the vicinity of the interface can be calculated from the Euler differential equation [(14.9), here, first-order], but only after a definite mathematical model has been selected for the bulk free energy $g(c)$. Cahn and Hilliard (1958) chose a regular solution model model, as in Chapter 7, and obtained, as expected, a sigmoid curve such as the one seen at the upper panel of Fig. 13.1. The Landau school, and many present authors, chose a quartic function;[3] in general, it is required that the chosen free energy function should represent a double-well potential, which can be obtained by a quartic or regular solution, for example.

These important results can be analyzed qualitatively as follows: the gradient contribution can be decreased by broadening the diffuse interface, but that is done at the expense of having more non-equilibrium material (vertical lines in Fig. 14.1) inserted. Conversely, one can elect to sharpen the interface so as to include less non-equilibrium material, but then at the expense of a sharper gradient. Mathematically, the system choses correctly the best combination of these two contributions to the total energy by solving the Euler differential equation, *i.e.* the variational problem.

This seminal paper by Cahn and Hilliard (1958) introduced the mathematical treatment of non-uniform systems in the case of a diffuse interface. It was followed by another paper, using the same mathematical formalism, for the case of a critical nucleus, but will not be presented here.

These techniques suffer one major flaw: they make use of empirical analytic free energy models which contain portions of curves pertaining to non-equilibrium regions (the region in between the open circles, the spinodal points, on the free energy curve $g(c)$ in Fig. 14.1). Therefore much of the shaded region in this figure is somewhat fictitious, and different approximations (regular, sub-regular, polynomial, ...) can yield very different

---

[3]A quartic function is a polynomial of the fourth degree.

results, nor can experimental evidence help much because those portions of the curve are thermodynamically unstable. In other words, those portions of the curve are theoretically model-dependent and experimentally history-dependent, *i.e.* dependent on the way the system was prepared.

References

Cahn, J. W. and Hilliard, J. E. (1958), *J. Chem. Phys.*, **28**, 258.

Cahn, J. W. (1961), *Acta Metall.*, **9**, 795.

Cahn, J. W. (1998), *The Selected Works of John W. Cahn*, (eds. Carter, W. C. and Johnson, W. C.). TMS Publications, Warrendale, PA.

Cook, H. E., de Fontaine, D. and Hilliard, J. E. (1969), *Acta Metall.*, **17**, 765.

de Fontaine, D. (1967), *Doctoral Dissertation: A Computer Simulation of the Evolution of Coherent Composition of Variations in Solid Solutions*, Department of Material Science, Northwestern University, Evanston, IL.

Landau, L. D. and Lifshitz, I. M. (1958), *Statistical Physics*. Addison-Wesley, Reading, MA. Also: Landau, L. D., Lifshitz, I. M. and Pitaevskii, L. P. (1980), *Statistical Physics*, Third Edition, Pergamon Press (English versions).

Landau, L. D. (1965), *Collected papers*. Oxford, Pergamon Press, p. 546.

Chapter 15

Landau Theory

It is convenient to characterize compositional inhomogeneities by means of the variable $\xi = c - \bar{c}$, with the understanding that the variable c results from the local averaging process, or coarse-graining , as mentioned earlier, $\bar{c}$ being the average overall concentration (in this chapter, mostly binary solutions will be considered). This concentration difference will be regarded as an *order parameter*. But the order parameter can mean much more: its originators, L. D. Landau and collaborators (1937, 1958, 1965) defined the concept as a *variable which takes the value zero above a phase transition and non-zero below*. If the order parameter decreases continuously to the value zero as the transition temperature is reached from below, *i.e.* from lower temperatures, the transition is called a *second-order* or *continuous* transition; it is called *first-order* if the value zero is attained discontinuously. The Landau order parameter need not refer just to the concentration c; it can actually refer to many different physical quantities, such as magnetic spin or electronic dipole. In reality, these two are vector quantities, but if we consider only the simple case of dipoles parallel to a fixed crystallographic axis, then there are only two cases of dipole variables: $\{+1$ and $-1\}$, just as for the composition $[1$ and $0]$. Often the occupation variables are given the symbol σ, for "crystallographic site occupation variables". We call transformations in which all σ of the same kind bunch together "clustering reactions", or "phase separation transformations", or again, in the electromagnetism realm, "ferromagnetism" or "ferroelectricity" and those for which σ variables of different kinds alternate on lattice sites in one or other modes of "ordering reactions" or "order-disorder transformations", or antiferro-magnetism (-electricity).

Figure 15.1 illustrates these concepts schematically on a 2-dimensional square lattice. In this figure, white and black symbols can represent A

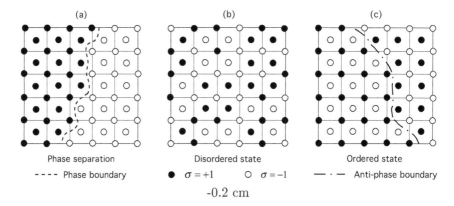

Figure 15.1: Three configurations: (a) Phase separation, two phases α and β; (b) Disordered state, random site occupation; (c) Ordered state, two sublattices α and β.

and B atoms (alloy problem), or up and down spins (magnetic or electronic Ising model), occupied and empty sites (lattice gas problem) or even simple cases of atomic displacements.

15.1   Ordering

It is clear from the schematic figure 15.1 that ordering processes can be quite varied — there are many categories of "order", but only one disorder and one "clustering" — and thus are more challenging. We can present here only a very brief account of the topic of ordering, which normally involves such mathematical techniques as Fourier transforms and crystallographic group theory [see for example books or review articles by Landau and Lifshitz (1958), Khachaturyan (1983), Ducastelle (1991) and the present author (1979, 1984) for treatments most relevant to materials science]. In addition to the processes of clustering and of many types of ordering, it must be mentioned that the order parameter itself can be multidimensional: this happens for multicomponent systems (more than two components in solution), or in the case of σ representing a vector or tensor quantity. But even for the simplest cases imaginable, such as the dipoles $\sigma = \{+1, -1\}$ interacting with their immediate neighbors on a simple square or simple cubic lattice, a model known as the *Ising model*, the difficulties can be overwhelming.[1] The Ising model was solved in one space dimension (linear chain) in 1925

---

[1]Note that we are mentioning here the correct, discrete Ising model with no averaging, no coarse-graining. That is what makes it so difficult.

by Ernst Ising who found, surely to his chagrin, that the one-dimensional model led to no transition at all, ferro- or antiferro-. At first, it was not known whether the two-dimensional Ising model with nearest-neighbor interaction on a simple square lattice gave rise or not to a transition until Lars Onsager (1944), in an extremely complex derivation, found an exact solution to the problem: the 2-dimensional Ising model did indeed possess a phase transition. Despite heroic mathematica efforts, an exact solution of the 3-dimensional problem is still not known, notwithstanding advances coming from *Renomalization Group Theory* (Wilson, 1975).

It follows that approximate solutions must be sought. In his early treatment Landau (1937) made use of purely phenomenological free energies (coarse-grained) curves, which could have been hand-drawn. That sufficed for qualitative treatments of the mechanisms of phase reactions, but not for quantitative studies; recall that Cahn and Hilliard had to resort to the regular solution model (Sec. 7.10) in order to calculate, for example, the shape of the diffuse concentration profile near a coherent phase boundary (Sec. 14.1). That model is a mean field approximation which regards atoms (molecules) as interacting with a homogeneous average medium, an approximation which, among other things, incorrectly predicts a phase transition in 1-dimensional systems. For order-disorder reactions, the Bragg–Williams (BW) approximation is also of this class; in fact, the BW method is to ordering what the regular solution model is to clustering. Ameliorated mean field models include the *Cluster Variation Method* (CVM) of Kikuchi (1951) (see also Sanchez *et al.*, 1984), according to which average *clusters* of lattice points, σ-occupied, interact with the surrounding medium. That method was mentioned previously on pages 146 and 151; it has often given good results where the BW method was not even qualitatively correct. The CVM is still a mean field theory, however.

The problem one faces is that the true free energy, according to statistical mechanics (Sec. 8.2), is given by the logarithm of the relevant partition function, and that function, the *sum over states*, is extremely difficult to calculate exactly except in the simplest of cases. Moreover, the sum in the partition function is generally replaced by its maximum term, thus introducing additional errors (see Sec. 15.7, below). Still, despite deficiencies, in the next two sections we shall be using mean field approximations consistently, including phenomenological expansions in direct and Fourier space.

15.2   Concentration Waves

The model of solid solutions — of atoms, of dipoles, ... — which is introduced in this chapter originated, to a great extent, with work performed in the Soviet Union [Landau and Lifshitz (1958), Ginzburg (1945), Krivoglaz (1958), Khachaturyan (1973)] and features a seemingly contradictory description of order and disorder in crystalline materials, in the sense that the model treats site occupation in a continuous manner $c(\mathbf{x})$, while at the same time treating space in a realistic discrete, crystallographic manner.

In dichotomy number [1], as we just saw on page 211 and following, discreteness and continuity were combined by means of coarse-graining, mentioned several times, in the case of variations of densities of microscopic objects over distances greater than atomic spacings, *i.e.* of *clustering* type. In the present section, we examine the process pertaining to *order-disorder*, represented schematically in Fig. 15.1. In this example, the point sites form a simple square lattice at 45° to the horizontal and vertical axes in the figure. The lattice sites themselves are shown "decorated" by "atoms" of two kinds, white or black. The Bravais lattice of the "decorated" structure (right-most panel) has now a larger unit cell than that of the original one: it has lost elements of symmetry. In crystallographic parlance, the ordering process has *broken the symmetry* in each *ordered* region. But in this panel, the right half has white objects located where the black objects are located on the left portion; in other words there has been a (geometrical) *phase change* across a so-called *anti-phase boundary*, shown as a dot-dash curve which separates the two *anti-phase domains*. In each domain, the white and the black objects now occupy the sites of two distinct *sublattices*, call them α and β.

In the early years of the subject "Phase Transformations", topics related to "phase separation" (or clustering) and "ordering" were usually considered as quite distinct, particularly in what was then called Physical Metallurgy (later to be transformed into Material Science). Now, the two topics are treated together, or should be, in order to resolve dichotomy number [2]. The analogy between *clustering* and *ordering* is formally the following: in clustering (phase separation), atoms (molecules, ...) separate, one type mostly in one phase (call it α), the other type mostly in another region of space (β). In an ordering reaction, different atom types segregate to different sublattices, α or β; in other words, *sublattices* in ordering play the role of *phases* in clustering.

How then does one perform the averaging process required to introduce continuity in the ordering case? One way, described elsewhere [de Fontaine (1979)], is to first define the sublattices appropriate to the ordering processes considered. Consider then a fully ordered structure (all black on one sublattice, all white on the other) and allow some disorder on each sublattice, perhaps caused by some diffusion (or hopping) of the two species, and *then* take averages over each sublattice in a single ordered domain, *not* across an antiphase boundary. This will generally result in some "gray" objects. Of course, it will not be necessary to consider actual atomic hopping processes, only fictitious ones.

In applying this technique, it is actually better to perform *lattice Fourier transforms* which introduce the idea of *concentration wave* (CW). Appendix I ("eye", not "one") describes the fundamentals of Fourier transforms in discrete (actually crystallographic) space, at least a simplified version thereof in one dimension. In particular, the notion of a *lattice delta function* is introduced: a continuous function which takes the value zero at all lattice points except at those for which the coordinates are exact multiples of a number N, related to the size of the sample considered, in which case its value is also N (see Fig. I.1). For $N \to \infty$, the lattice delta function becomes the familiar δ-function of continuum mechanics (when unnormalized by the factor $\frac{1}{N}$). Formally, the Fourier operations are accomplished through the Fourier operator $\mathcal{F}$ and its inverse $\mathcal{F}^{-1}$, respectively:

$$\mathcal{F} \stackrel{\text{def}}{=} \frac{1}{N} \sum_{\vec{p}} e^{-\mathbf{k}(\vec{h})\cdot\mathbf{x}(\vec{p})} \tag{15.1a}$$

$$\mathcal{F}^{-1} \stackrel{\text{def}}{=} \sum_{\vec{h}} e^{+\mathbf{k}(\vec{h})\cdot\mathbf{x}(\vec{p})} \tag{15.1b}$$

so that the symbolic expressions $\mathcal{F}^{-1}[\xi(\vec{p})]$ and $\mathcal{F}[X(\vec{h})]$ reproduce, for one dimension, the Fourier transform (I.5) and the Fourier expansion (I.1), respectively, in Appendix I. The three-dimensional notation used here [and in (de Fontaine, 1979)] requires a bit of explaining: the most rational manner of describing a 3-D lattice is by using a coordinate system based of the primitive unit cell, containing only one repeating motive (or atom, molecule, object) per cell. In that case, the symbol $\vec{p}$ must actually represent a triplet of integers, and the position vector is written $\mathbf{x}(\vec{p})$ for short, with

$$\mathbf{x}(\vec{p}) \equiv p_1\mathbf{a_1} + p_2\mathbf{a_2} + p_3\mathbf{a_3} \tag{15.2}$$

where the $\mathbf{a_i}(i = 1, 2, 3)$ are the (primitive) lattice vectors. The (wave) vector is

$$\mathbf{k}(\vec{h}) \equiv 2\pi(h_1\mathbf{a_1^*} + h_2\mathbf{a_2^*} + h_3\mathbf{a_3^*}) \tag{15.3}$$

where the a_i^* are the corresponding reciprocal lattice vectors defined by $\mathbf{a_i} \cdot \mathbf{a_j^*} = \delta_{ij}$, the δ symbol being equal to 1 if the two indices are identical, and 0 otherwise. The vectors $\mathbf{k}(\vec{h})$ are reciprocal lattice vectors, also called wave vectors, or k-space vectors, according to the physics terminology. The parameters h_i are not integers, but integers divided by N_i, the latter quantities being the number of primitive unit cells along the axes of the "computational unit cell", thus containing $N = N_1 \times N_2 \times N_3$ primitive unit cells. This seems complicated, but applications are quite simple in practice, as explained for instance in texts of X-ray, or neutron, or electron diffraction, where conventional (non-primitive) unit cells are generally used. In the present text, only one-dimensional transforms are used, as explained in Appendix I, and only simple examples will be treated, below. However, the 3-dimensional description was mentioned here for completeness.

In applications to diffraction theory, the reciprocal lattice points are the Bragg peaks in the diffraction patterns (black points on the cube of Fig. 15.2), and the open circles indicate the so-called superstructure peaks. These latter positions are located at the centers of the square faces of the tetrakaidecahedron shown in heavier lines in Fig. 15.2, centered at the origin 000. Such a polyhedron is called a Brillouin zone, which is the Voronoi polyhedron for the reciprocal lattice of fcc, *i.e.* bcc in the present case. Such polyhedra were depicted in 2 dimensions in Fig. 10.3, but in the present case the Voronoi polyhedron is unique because of the translational symmetry of the Bravais lattice; there is a Voronoi centered at each point of the reciprocal lattice (filled circles in Fig. 15.2), all identically oriented.[2] Diffracted amplitudes allow, in principle, the determination of the crystal structure when it is unknown; unfortunately, the amplitudes of the diffracted waves — which are complex numbers — are unknown, and only the intensities (amplitude-squared) are available, so that structure determination requires supplementary information and much ingenuity.

That, associated with the discretization of space, as we have seen, allows both types of inhomogeneities — clustering and ordering — to be viewed in the same framework: phase separation in crystalline solutions (long wavelength CWs, as in spinodal decomposition, see below), and ordering (short wavelength CWs, commensurate with atomic distances). In what follows, without elaborating on the mathematics, it must be kept in mind that the crystalline matter that is being investigated is non-uniform [in terms of $c(\mathbf{x})$], but uniform (or "coherent") in terms of the lattice structure, *i.e* the lattice planes must be continuous, in other words such singularities as dislocations are not treated in classical thermodynamics.[3] It is perfectly possible to perform Fourier transforms in discrete space, as was done long ago by Born and Huang in their classic textbook on lattice dynamics (1962), for example.

---

[2] For reminders, a Voronoi polyhedron, or Dirichlet region about a given point is that region of space closest to that point than to any other point.

[3] In fact, unlike the case of vacancies, dislocations are non-equilibrium defects.

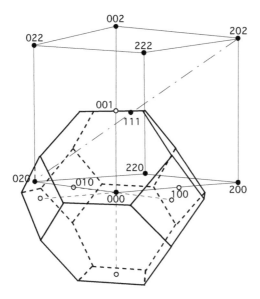

Figure 15.2: Perspective drawing of body-centered cube (reciprocal of fcc), filled circles at its lattice points; corresponding Brillouin zone (or Voronoi polyhedron or Dirichlet region), with 6 of its superstructure points (open circles).

15.3 Landau Approach

The different handling of the bulk free energy $g(c)$ is what distinguishes the Cahn and Hilliard (CH) treatment from that of Landau and Lifshitz, a fairly mild distinction. In CH, the regular solution approximation [see Sec. 7.10, Eq. (7.87)] was used to investigate the form of the diffuse interface, namely the function $c(x)$. In the Landau–Lifshitz approach, the free energy is approximated by a polynomial of the fourth degree (quartic) or of the sixth degree (hextic).[4] In some sense, the regular solution model is superior (without in any sense being exact) because with it, the behavior of the order parameter, or concentration difference, is satisfactory at large values (thanks to the presence of the logarithmic ideal entropy terms). By contrast, order parameters and free energies are unbounded in the Landau–Lifshitz formulation, therefore unrealistic at large values; however, computations are much easier to perform with the polynomial approach of Landau.

In Eq. (14.5) the expression for $g(c)$ is left unspecified, and in Eq. (14.6)

---

[4]The word "hextic" is not in my Oxford two-volume dictionary, but let's use it here anyway; it's a fun made-up word with witchcraft connotations.

the same is true for Δg. In this section, we shall follow the Landau–Lifshitz prescription and use polynomials. In both cases, Sections 14.2 and 15.3, the notation Δg designates a *difference* of free energies, as mentioned briefly in conjunction with Eq. (14.6), a difference between a tangent and a g curve. But what tangent? In the previous section, the tangent was the familiar common tangent of phase equilibrium, as shown in Fig. 14.1. In the examples to follow, the tangent points are *at* the state of disorder of the system, such as the average concentration $\bar{c}$, or the order parameter $\xi = 0$. Hence that value of ξ measures the degree of order of the ordered state. As mentioned previously, there are two main classes of phase transformations: *first order* and *second order*. The Landau approach is particularly well suited for illustrating these two classes and their semi-quantitative properties, as will now be examined.

Initially, Landau in the late 1930's was mainly interested in second-order transitions with applications to superfluidity (for which he was awarded a Nobel prize in 1962, shortly before his death in 1968, from the long sequel of physical problems resulting from an automobile accident in 1960). Landau's formalism was later used by V. L. Ginzburg (1950, Nobel Laureate in 2003) in a theory of superconductivity and superfluidity in which the polynomial expansion was made to include the square-gradient term, as in Sec. 14.2, above. Hence the formalism in question is often referred to as the "Ginzburg–Landau theory", a formalism soon taken up by Devonshire (1949–51) in the latter's theory of ferroelectricity. Since in this section we are concerned with uniform phases only, while discarding interfaces, the gradient term will not be be necessary. So here is the Landau expansion in powers of the (continuous) order parameter ξ

$$\Delta g = A\xi^2 + B\xi^3 + C\xi^4 + \dots . \tag{15.4}$$

This free energy expansion is denoted Δg in keeping with the notation used earlier, and as a reminder that we are investigating the properties of a *difference* in energies. It also means that the expansion, without loss of generality, can start at the quadratic term since the free energy is referred to its tangent at the value zero of the order parameter. All the coefficients of the powers of ξ are themselves function of temperature and pressure (or of other fields). For simplicity, Landau used the following expression for the A coefficient

$$A(P,T) = a(P)\,(T - T_0), \tag{15.5}$$

where, in the original treatment, T_0 was regarded as the second-order transition temperature, and as we shall see more generally, the temperature at which the A coefficient changes sign, thereby initiating an instability. The expansion can go on indefinitely, but Landau initially stopped at the quartic term, as described for example in the 1958 English translation of *Statistical Physics*. In the French translation of a later edition of the Russian text, a footnote informs the reader that there may be cases for which the sixth-order term would be required, for example in studies of ferroelectricity or ferromagnetism, but also in physical metallurgy (in alloy theory). Here then is the expansion needed in this section:

$$\Delta g = a \, (T - T_0)\xi^2 + B\xi^3 + C\xi^4 + D\xi^6 \tag{15.6}$$

with $a > 0$, possibly pressure-dependent. In any case, the expansion must end in an even-order term with positive coefficient, since with large values of the order parameter $|\xi|$ the free energy must always reach large positive values. Different cases suggest themselves:

Second-order transition

Figure 15.3 illustrates the case of expansion parameters $A = a(T - T_0)$, $B = D = 0$ and $C > 0$. The polynomial expansion and its first and second

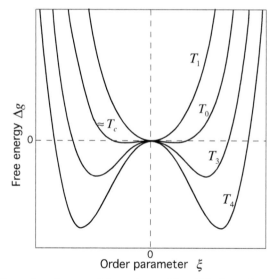

Figure 15.3: Second-order transition, $A = a \, (T - T_0)$, $B = D = 0$, $C > 0$. Curve labeled T_0, is close to critical T_c and instability temperature. Temperature order is $T_1 > T_0 > T_3 > T_4$.

derivatives (dashes) with respect to ξ reduce to

$$\Delta g = A\xi^2 + C\xi^4 \tag{15.7a}$$

$$\Delta g' = 2A\xi + 4C\xi^3 \tag{15.7b}$$

$$\Delta g'' = 2A + 12C\xi^2 \tag{15.7c}$$

with A given by Eq. (15.5). At each temperature, the equilibrium state of the system is given by the minimum of Δg, that is by the vanishing of the first derivative, $\Delta g' = 0$. For $T > T_0$ there is only one real solution: $\xi = 0$, denoting a state of zero order parameter, *i.e.* disorder. For $T < T_0$, $\xi = 0$ is still a solution but now there are also lower minima at $A + 2C\xi^2 = 0$, thus at

$$\xi = \pm\sqrt{\frac{a(T_0 - T)}{2C}}. \tag{15.8}$$

It is thus seen that below T_0, the equilibrium state is that for which the order parameter gradually decreases to the value zero at the transition temperature T_0, which is all at once the critical temperature T_c and instability temperature T_0. At this temperature, all expressions (15.7) take the value zero.

Note that the lower-temperature curves of Fig. 15.3 also feature the "double-well potential" introduced earlier in Fig. 2.2 and elsewhere in this book.

Asymmetric first-order transition

Figure 15.4 illustrates the case of expansion parameters $A = a\,(T - T_0), B \neq 0, C > 0$ and $D = 0$. We have $T_0 < T_1 < T_t < T_3$. At high temperatures, the lowest free energy minimum is located at $\xi = 0$ where the equilibrium state is the disordered one, and at low temperatures the lowest minimum is at $\xi \neq 0$ where the ordered state is stable; between those two temperature regimes there exists a limiting temperature T_t at which the lowest minima have the same free energy difference Δg. It follows that T_t is the transition temperature at which the order parameter jumps discontinuously from $\xi = 0$ to a non-zero value, the order parameter jump at T_t thus indicating that the transition is first-order. Temperature T_0 is that at which the associated free energy curve has vanishing second derivative at $\xi = 0$, *i.e.* for which the disordered state is inherently unstable. The polynomial expansion and its first and second derivatives with respect to ξ reduce to

$$\Delta g = A\xi^2 + B\xi^3 + C\xi^4 \tag{15.9a}$$

$$\Delta g' = 2A\xi + 3B\xi^2 + 4C\xi^3 \tag{15.9b}$$

$$\Delta g'' = 2A + 6B\xi + 12C\xi^2 . \tag{15.9c}$$

The instability temperature T_0 is solution of the equation $\Delta g''(0) = 0$, which implies $A = 0$. What is important in first-order transitions is that T_0 is *not* the equilibrium transition temperature T_t, since actually $T_t > T_0$. The transition temperature itself along with the value of the order parameter is solution of $\Delta G = \Delta G' = 0$, that is to say obtained by solving the simultaneous equations (15.9a) and (15.9b). The solutions of equation

$$\xi \left(2A + 3B\xi + 4C\xi^2\right) = 0 \tag{15.10}$$

are

$$\xi_0 = 0 \quad \text{and} \quad \xi_+ = \frac{-3B + \sqrt{9B^2 - 32AC}}{8C}.$$

The second value (valid for B positive) inserted into equation $\Delta g' = 0$ gives the first-order transition temperature T_t. Representations of first-order transitions are given in Fig. 15.6 and 6.2.

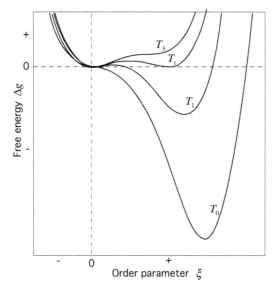

Figure 15.4: Asymmetric First-order transition, $B \neq 0, C > 0, D = 0$. Transition temperature is T_t and T_0 is the instability temperature, much lower.

Symmetric first-order transition

Symmetric first-order transitions, the kind often encountered in ferroelectricity [see Devonshire (1949), Fridkin and Ducharme (2014) and page 229, for example] require the extension to sixth order of the Landau expansion, as shown here with its first two derivatives with respect to ξ:

$$\Delta g = A\xi^2 + C\xi^4 + D\xi^6 \tag{15.11a}$$

$$\Delta g' = 2A\xi + 4C\xi^3 + 6D\xi^5 \tag{15.11b}$$

$$\Delta g'' = 2A + 12C\xi^2 + 30D\xi^4 . \tag{15.11c}$$

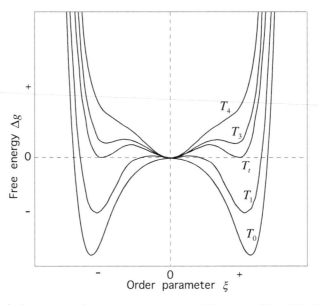

Figure 15.5: Symmetric first-order transition with $A = a(T - T_0), B = 0, C < 0, D > 0$. T_t and T_0 are transition and instability temperatures, respectively.

Figure 15.5 illustrates this case, with order of temperatures being $T_0 < T_1 < T_t < T_3 < T_4$. As above, the instability temperature T_0 is given by $\Delta g''(0) = 0$, which implies $A = 0$. At the transition temperature T_t, the order parameter makes a discrete jump from its equilibrium value 0 above T_t to ξ_t, so that the transition is clearly first-order. As in the previous section, the transition temperature is obtained by solving the simultaneous equations (15.11a) and (15.11b). The solutions of equation

$$2\xi \left(A + 2C\xi^2 + 3D\xi^4 \right) = 0 \tag{15.12}$$

are

$$\xi_0 = 0 \quad \text{and} \quad \xi_+^2 = \frac{-C + \sqrt{C^2 - 3AD}}{D},$$

from which we get the equilibrium order parameter values below T_t

$$\xi_t = \pm\sqrt{\frac{-C + \sqrt{C^2 - 3AD}}{D}} . \tag{15.13}$$

Order Parameter Plots

An instructive way to summarize the various types of transitions is to examine the variation with temperature of the order parameter for the three cases just described. Figure 15.6 shows these transitions in typical cases as calculated by the Cluster Variation Method [R. Kikuchi, unpublished, see de Fontaine (1979)]. In metallic alloys, by far the most common phase transitions are first-order type, including the omnipresent melting transition. There are a few cases for which there exist "coherent" miscibility gaps (MG), *i.e.* where the top of the MG is experimentally accessible, for example in the Au-Ni, Al-Zn and Au-Pd metallic systems.[5] For these alloys and similar ones, there exists a second-order transition at the very top of the MG, first-order everywhere else. There is also a second-order ordering transition in the alloy Cu-Zn around the 50/50 composition, but in very few others in familiar alloys. The simplest analytical model of a miscibility gap is that given by the regular solution model, described in Sec. 9.1.

15.4 Generalized Bragg–Williams Method

In the foregoing, a generalized Bragg–Williams formulation has been introduced whereby each lattice point of a single crystal has been considered as representing a complete sublattice in the sense that the point in question, designated as $\vec{p}$, is occupied by an *average atom* (or molecule). In what follows, the super-arrow will be deleted, with the understanding that the symbol p, and other similar ones recognized by context, should be regarded as triplets of lattice coordinates. That process becomes more useful when certain sublattices are grouped together to form *superlattices* which are proper to the ordering process pertinent to the ordering mechanism in question. Hence the notation $c(p)$ will be used frequently, denoting the *continuous* and *averaged* concentration on sublattice p, of atomic species or electrical or magnetic dipoles, according to what types of transitions are considered. Actually, the procedure described is just the coarse-graining one mentioned earlier, but with the proviso that the lattice site averaging must be performed on those sites belonging to a given sublattice only.

---

[5]By "coherent" we mean that incipient α and β phase regions are separated by interfaces across which lattice planes are always continuous, though perhaps bent, and must not contain dislocations.

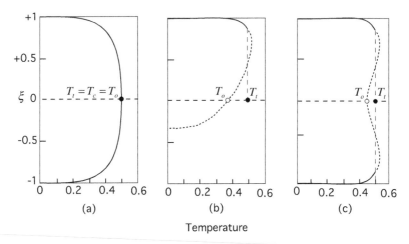

Figure 15.6: Typical order parameters ξ evolution as a function of absolute temperature (calculated by CVM, R. Kikuchi, unpublished): (a) second order; (b) asymmetric first order; (c) symmetric first order. Full lines: equilibrium parameter; dotted lines: non-equilibrium parameter. T_t is transition (filled circles), T_0 is instability (open circles), T_c is critical temperature. In the symmetric cases, both positive and negative order parameters are shown.

Equivalently, one may prefer to use the variable $\xi(p) = c(p) - \bar{c}$, the order parameter associated with the concentration c, as previously defined. For simplicity, only binary systems, A-B, will be considered, with, as usual, average concentrations $c_B = \bar{c}$ and $c_A = 1 - \bar{c}$.

Generalized regular solution

As a first approximation[6] we shall use the the the trusty — but not trustworthy — regular solution approximation (Sec. 7.10), but with the *local, average* (coarse-grained) concentrations $c(p)$ written in place of the sample average c for the free energy of mixing

$$\Delta f = N v_0 \, q_0 + \frac{1}{2} \sum_{p \neq p'} v(p' - p) \, \xi(p) \, \xi(p')$$

$$+ k_B T \sum_p \{ c(p) \ln c(p) + [1 - c(p)] \ln[1 - c(p)] \}. \quad (15.14)$$

---

[6]Much of the material here and in the following sections is based on de Fontaine (1975).

In this equation, $v(r)$ is the effective pair interaction parameter

$$v(r) = \frac{1}{2}[v_{AA}(r) + v_{BB}(r) - 2v_{AB}(r)], \qquad (15.15)$$

just as in Eq. (7.86), but for the case of binaries, where $v_{ij}(r)$ is the interaction parameter for an (i, j) pair with members of the pair separated by the distance $\mathbf{x}(r)$ linking points p and p'; $k_B T$ has its usual meaning. The following notation has also been introduced for brevity

$$q_0 = c_A c_B = \bar{c}(1 - \bar{c}) \qquad \text{and} \qquad v_0 = -\sum_r v(r). \qquad (15.16)$$

Taylor's expansion

In this simple mean field approximation, the free energy has no singularity so can be expanded in a Taylor's series about the disordered state $\xi = 0$. An expansion to fourth order retains the essential features of the model:

$$\Delta f = N f_0 + \frac{1}{2} \sum_{p,p'} f_0^{(2)}(p' - p)\,\xi(p)\,\xi(p') + \frac{1}{3!} f_0^{(3)} \sum_p \xi^3(p) + \frac{1}{4!} f_0^{(4)} \sum_p \xi^4(p),$$

$$(15.17)$$

with coefficients given by

$$f_0 = v_0\, c_A\, c_B + \tau(c_A \ln c_A + c_B \ln c_B) \qquad (15.18a)$$

$$f_0^{(2)}(r \neq 0) = 2v(r) \qquad (15.18b)$$

$$f_0^{(2)}(0) = \tau/q_0 \qquad (15.18c)$$

$$f_0^{(3)} = -\tau\,(c_A^2 - c_B^2)/q_0^2 \qquad (15.18d)$$

$$f_0^{(4)} = 2\tau\,(c_A^3 + c_B^3)/q_0^3, \qquad (15.18e)$$

and with $\tau = k_B T$ for brevity.

Note that the zeroth term, Eq. (15.17) is just the usual regular solution model, the 3rd-order term can have either sign but must vanish for $c_A = c_B$, and the 4th-order term is always positive. Such a result conflicts with the possibility of having a symmetric first-order transition, as seen in the previous section and also experimentally for example in the equiatomic Cu-Au alloy. The failure of the present model to allow such transitions is due to the regular solution approximation which makes the 4th-order term always positive. The CVM allows such transitions because its energy terms contains multi-atom interactions and has a more realistic configurational entropy formulation. The truncated Taylor's expansion introduces approximations as well by allowing the f_0 coefficients to be constants independent of temperature and pressure, which is not realistic. The present formulation is however quite simple and allows one to illustrate at least some of the qualitative features of phase transitions.

The determination of the equilibrium configuration at any temperature and average composition theoretically requires the solution of N coupled transcendental equations resulting from the free energy (15.17) with respect to the N configuration variables $c(p)$. This cumbersome task is simplified by the judicial practice of grouping lattice sites into a small number of sublattices, thus leading to *ad hoc* Bragg–Williams models. Such groupings will not be undertaken here, but can be found in the work of Khachaturyan (1983) and de Fontaine (1979) for example.

Fourier expansion

In the section on the Landau approach (Sec. 15.3), it was not specified about which point the phenomenological expansion was performed; it was merely understood "about the state for which ξ was equal to zero", *i.e.* the disordered state. That means that the phase transformation could be of either clustering or ordering type, depending upon whether the instability in question was due to long wavelength (clustering) or to short wavelength (ordering). The present section will examine these types of instabilities, but the nature of the unstable wave will have to be specified, which means that the wave nature of the process will have to be built-in from the start, which in turn requires expansion to be couched in Fourier waves. This can be accomplished by simply replacing in the expression (15.17) the order parameters $\xi(p)$ by their Fourier expansion $X(h)$ according to Eq. (I.1). The resulting Fourier-space expression for the free energy difference is then found to be

$$\Delta F = \frac{N}{2} \sum_h F_0^{(2)} |X(h)|^2$$
$$+ \frac{N}{3!} F_0^{(3)} \sum_{h,h',h''} X(h)\, X(h')\, X(h'')\, \delta(h + h' + h''|H)$$
$$+ \frac{N}{4!} F_0^{(4)} \sum_{h,h',h'',h'''} X(h)\, X(h')\, X(h'')\, X(h''')\, \delta(h + h' + h'' + h'''|H) + \dots.$$

$$(15.19)$$

The expansion is cut off after the 4th-order term for simplicity, but could be extended as far as required. The coefficient of the 2nd-order term is, by Eq. (15.17)

$$F_0^{(2)} = 2V(h) + \tau/q_0 \qquad (15.20)$$

where $V(h)$ is N times the Fourier transform of the effective pair interaction function $v(r)$. In Eq. (15.19), it is seen that the second-order term is simplified, *i.e* diagonalized with respect to its direct-space counterpart, whereas the higher-order terms are complicated by these discrete delta functions described in Appendix I. Those terms indicate that concentration waves (CW) of indices $h, h', h'', \ldots$ interact in such a way that the delta function vanishes unless the h-indices sum to the indices H of a reciprocal lattice vector. Thus, in the reciprocal space representation the harmonic (2nd-order) contribution term to the free energy is simplified, whereas the anharmonic (3rd-order and higher) contributions complicate matters, but play an essential role, which has not always been appreciated.

Symmetry of V(h)

Of primary importance are the points h° at which the function $V(h)$ takes its minimum values since, by Eq. (15.20), the corresponding wave vectors $\mathbf{k}(h^\circ)$ minimize the harmonic part of the free energy, and that, in that approximation, the second derivative of the configurational entropy does not depend on h. In many cases, these wave vectors generate commonly observed ordered structures. The necessary condition for point h° to be at a minimum of $V(h)$ is given by the set of conditions

$$\frac{\partial V}{\partial h_i^\circ} = 0 \qquad (i = 1, 2, 3) \tag{15.21}$$

where i denotes Cartesian indices. However, the wave vectors satisfying Eqs. (15.21) also yield the locations of maxima, minima and saddle points of $V(h)$. The nature of the extrema at those h° points can be ascertained by expanding $V(h)$ to second order about h° thus:

$$V(h) = V_0 + \frac{1}{2} \sum_{i=1}^{3} \lambda_i(h^\circ) H_i^2 . \tag{15.22}$$

In this equation, the linear term vanishes because the function $V(h)$ is an extremum there, V_0 is the value of the function at its extremum, λ_i are the eigenvalues of the real symmetric matrix of second derivatives of $V(h)$ at h°, and the coordinates H_i are measured along the three eigenvectors of the matrix. The λ_i are then proportional to the curvatures of the surface $V(h)$ along the H_i directions so that if they are all positive, $V(h)$ is concave at that extremum, which is then a minimum, if they are all negative $V(h)$ is convex there, and thus a maximum, and if the eigenvalues are of mixed signs, positive and negative, the extremum is a saddle point. A minimum of $V(h)$ gives rise to stability, a maximum gives rise to instability.

Special points

The function $V(h)$ possesses not only the translational but also the point group symmetry of the reciprocal lattice. Thus, if a solution of Eq. (15.21) is found for some wave vector $\mathbf{k}(h^\circ)$, other extrema of the same type must occur at all other points belonging to the star of the vector $\mathbf{k}$, the star of a wave vector consisting of all those vectors which transform into one another by the operations of the space group of the reciprocal lattice. Also, because of translational symmetry, a topological theorem due to Morse (1935) states that a number of extrema of different types — maxima, minima, saddle points — must always be present and must verify certain inequalities; in particular, the minimum number of extrema is one maximum, one minimum and six saddle points.

If a symmetry element (rotation, rotation-inversion, mirror plane) of the space group in $\mathbf{k}$-space is located at point h, the vector representing the gradient $\nabla_h V(h)$ of an arbitrary function $V(h)$ at that point must lie along or within the symmetry element at that point. If two or more symmetry elements intersect at that point, we must necessarily have

$$|\nabla_h V(h)| = 0 \qquad (15.23)$$

since a non-zero vector cannot lie simultaneously in two intersecting straight lines (or in a line and a plane) having only one point in common. At these so-called *special points*, or high symmetry points, Eqs. (15.21) are thus satisfied by symmetry requirements alone, so that $V(h)$ must present an extremum regardless of the choice of effective pairwise interaction parameters. This universal character of the special points was pointed out by Lifshitz (1942) and by Landau and Lifshitz in their classic text *Statistical Physics* (1958). The special points for all 230 crystallographic space groups can be found in the International Tables for Crystallography under the name of "Wyckoff positions".

In the author's first *Solid State Physics* review paper (1979, page 111), perspective plots of the $V(h)$ potential along (001) and ($\bar{1}10$) surfaces are shown for the fcc lattice for four sets of $v(p)$ parameters out to 3rd nearest neighbors. The four cases illustrated display absolute minima at special points $\langle 000 \rangle, \langle 100 \rangle, \langle \frac{1}{2}\frac{1}{2}\frac{1}{2} \rangle$ and $\langle 1\frac{1}{2}0 \rangle$, respectively. These surfaces also exhibit special point maxima and saddle points, as expected. In that review paper, tables II and III also list the four special points for fcc and bcc Brillouin zones (BZ) and are given here in Table 15.1; there are no other special points for these lattices. For fcc only, the location of such points can be found on the BZ of Fig. 15.2: the actual lattice points of the lattice reciprocal to fcc (*i.e.* bcc) are shown as filled circles, the open circles denoting points of high symmetry. The location of these points are found at the centers of the square faces of the tetrakaidecahedron ($\langle 100 \rangle$, as shown) and for the others at the centers of the hexagonal faces and at the BZ vertices (not shown). For each one of these two lattices, the $v(p)$ interaction parameters were chosen so as to give absolute minima at the four points of symmetry-dictated minima according to the criteria of Eq. (15.22). In terms of diffraction theory, the black points in Fig. 15.2 are the Bragg reflections (at $\langle 000 \rangle$), and the other special points are the superstructure reflections corresponding to ordered structures.

Table 15.1: Special points Miller indices, space groups (International Tables) and symmetry classes for fcc and bcc Brillouin zones. In the review article (DdF, 1979), the indices of the fourth bcc special point were given as $\langle 1\frac{1}{2}0 \rangle$, which is incorrect. They appear correctly here, and in the original publication (DdF, 1975).

fcc			bcc		
Miller	*Internat.*	*Symmetry*	*Miller*	*Internat.*	*Symmetry*
$\langle 000 \rangle$	m3m	cubic	$\langle 000 \rangle$	m3m	cubic
$\langle 100 \rangle$	4/mmm	tetragonal	$\langle 100 \rangle$	m3m	cubic
$\langle \frac{1}{2}\frac{1}{2}\frac{1}{2} \rangle$	$\bar{3}m$	trigonal	$\langle \frac{1}{2}\frac{1}{2}\frac{1}{2} \rangle$	$\bar{4}3m$	cubic
$\langle 1\frac{1}{2}0 \rangle$	$\bar{4}2m$	tetragonal	$\langle \frac{1}{2}\frac{1}{2}0 \rangle$	mmm	ortho.

The importance of these "special points" cannot be overemphasized: they are the basis of symmetry-dictated extrema, which means that absolute minima of any function which has the symmetry of the reciprocal lattice are located at points in reciprocal (or **k**-space) which do not depend on the particular details of the physical parameters (including temperature and pressure) which give rise to these minima. In particular, the special points appear in the Landau rules (see Sec. 15.5) for the existence of phase transitions of second-order. These points, also called symmetry-dictated extrema, also appear as most likely **k**-points associated with phase instabilities and, we reiterate, depend only on symmetry, not on the actual values of interaction parameters. Of course, minima of say $V(h)$ can occur elsewhere in **k**-space as accidental minima, but structures which result from the related CW will tend to be unstable. These points will be discussed further in Sec. 15.5) where it will be argued that, in many respects, ordering and phase separation can be treated within the same framework: in the latter case the special point $\langle 000 \rangle$ (center of BZ) gives rise to unstable concentration waves of infinite wavelength; in the case of ordering, the unstable waves have wave vectors at superlattice positions (for example, open circles on the surface of the BZ, see Fig. 15.2). In either case, the basic arguments are similar.

15.5   Landau–Lifshitz Rules

These rules, first derived by Landau (1937) for the necessary condition of second-order transitions, were completed by Lifshitz (1942) and were expressed together by these two authors in their classic text *Statistical Physics* (1958), and reproduced frequently in more recent publications, all heavy on crystallographic group theory. In what follows, it is shown that it is not essential to dwell to such an extent on group theory, as simpler treatments are available, though perhaps less elegant or complete. To repeat a word of warning: in the past, the question was often asked of spinodal decomposition (see Chapter 18) whether that transition was of first or second order. One possible answer is to state that classical thermodynamics can only take care of *equilibrium* changes *i.e.* occurring *at* (or very near) equilibrium states. Such phase changes we shall call "transitions". Changes occurring far from equilibrium, such as spinodal decomposition, or the glass "transition",[7] we shall simply call "transformations". Hence spinodal decomposition (or spinodal ordering, see below) are merely *transformations* for which the question of first or second order simply does not arise.

First off, in the completely disordered state, and according the "coarse graining" idea, all the "atoms" are "gray".[8] Below a transition, the degeneracy is lifted, meaning that some atoms become "white" (or at least light gray, if there remains some solubility for black atoms), as in Fig. 15.1 for exampled, and some atoms become black (or dark gray). In panel (a) of Fig. 15.1, the atoms on the right are distinguished from those on the left across the phase boundary, which means in crystallographic group theory terms that translational symmetry has been partially lost. In Fig. 15.1 (b) above the transition, if coarse graining averaging is carried out, translational symmetry is lowered because the nearest neighbor distance (along directions at 45° of the main axes) is an allowed translation which is lost in panel (c), at left or right sides, since only translations between the same color are allowed translations. Other symmetry elements are also lost in the ordering case illustrated since mirror planes (actually mirror lines) half-way between rows at 45° exist in the "all-grey" image, whereas mirror lines only exist *through* the diagonal rows of atoms and also in the horizontal and ver-

---

[7]Should be "glass transformation" according to that distinction of mine.

[8]According to a French saying: *La nuit, toutes les souris sont grises* [at night, all the mice are gray, (*i.e.* indistinguishable)].

tical directions in the ordered state, as seen in panel (c). Or, the primitive unit cell of the disordered structure before transition was a square oriented at 45°, after transition it is a larger square, oriented along the vertical or horizontal axes. Accordingly, and indeed quite generally, transitions from disorder to order are associated with loss of symmetry elements, meaning that the symmetry group of the ordered structures must be a subgroup of that of the disordered state. Such is the first Landau rule for second-order transitions. This rule also implies, for the *transformation* to be a true *transition*, that transformed and untransformed region be *coherent* in the sense of the footnote on page 233.

The second rule concerns the "special points" discussed above. It is also couched in terms of symmetry and it states that *stable* ordered structures can only exist for ordered structures whose wave vectors belong to the special points. That rule appears to be due mainly to I. M. Lifshitz (1942). The third rule refers to the phenomenological expansion itself in its Fourier expression; it states that, for a transition to be second-order, the third-order terms in the Fourier expansion must vanish identically, which in turn means that, it must *not* be possible to find a vector sum of three wave vectors which add up to a reciprocal lattice vector. In terms of $\mathbf{k}$-space indices, it means that $(h + h' + h")$ must *not* equal H, in which case the discrete delta function of Eq. (15.19) would not vanish. But we have seen, in the subsection *Symmetric first-order transition* that in that case the third-order does indeed vanish, yet the transition can be definitely first-order because the fourth-order term can be negative ($C < 0$). Hence the Landau–Lifshitz rules must be considered as *necessary* but not *sufficient* conditions. To summarize, the rules allowing a second-order transition may be stated as follows:

Rule I — The space group of symmetry elements of the ordered structure characterized by the local order parameter $\xi(p)$ must be a subgroup of the disordered solution, for which $\xi = 0$.

Rule II — The ordering wave vectors of the ordered structure must be located at the special (high-symmetry) points of the Brillouin zone of the disordered phase, where symmetry elements intersect at a point.

Rule III — It must *not* be possible to find combinations of three members of the ordering wave vectors satisfying rule II which will sum (vectorially) to a reciprocal lattice vector.

It is worth emphasizing the advantages of the Fourier approach to phase changes. For one thing, it is not necessary to know in advance what groupings of sublattices are necessary to treat the phase transition problem, as in the usual Bragg–Williams method; it suffices in principle to let the system take care of itself by searching for a minimum in the $V(h)$ potential. For another, more importantly, the techniques brought to bear on a given transition problem are the same regardless of the softness mode considered. By "soft mode" is meant a point of high symmetry in $\mathbf{k}$-space corresponding to an actual minimum, not simply an extremum (which it always is), a fact which can be ascertained by verifying that the eigenvalues λ_i in Eq. (15.22) are all three positive. The advantages of the Fourier approach are thus those of a unified treatment for different types of ordering processes, including phase separation for which the softest mode is always $\langle 000 \rangle$. However, there are three caveats: (1) minima of $V(h)$ may occur "accidentally" at non-special sites in the BZ, in which case one may encounter ordered domains as per the schematic representation in Fig. 15.7 (to be described later). Another possibility, (2) very frequent this one, is that the phase transformation in question is not of second-order type, or even is not a *continuous* one, not characterized, even in part, by a "softening" of a concentration wave, but triggered by terms in a Landau expansion higher than second order. Finally (3) it is true that the Landau approach gives a very clear, unified phenomenological treatment of phase transformations — which can be used for displacive, ferro– and antiferro– magnetism or electricity — but it is also true that the mean field, coarse-graining approximation cannot do justice to the quantitative aspects of the transitions.[9] Such difficulties will be taken up in the next section.

15.6   Discussion of the Landau Theory

But is the Landau theory "wrong"? It depends on who you ask: the physicist or the materials scientist. The materials scientist is typically interested in "microstructure", in phenomena occurring at the mesoscale, such as small precipitates, ordered domains, phase- and antiphase-boundaries, ... and the transformations which gave rise to them. Most of these processes can be explained by applications of classical thermodynamics, expanded, as we have seen, so as to cover non-uniformity through coarse-graining and so on. It seemed that no surprise could intervene. It was known of course that some phase transformations gave rise to singularities or divergences in thermodynamic potentials. Then around the mid-nineteen hundreds it was found that, for continuous, or second-order transitions, the behavior of some thermodynamic systems near the singularities gave rise to unexpected

---

[9]One materials science instructor at a prestigious university was quoted as saying "since I learned that the Landau theory was *wrong*, then I don't have to learn it. Goody, goody!" Of course, the Landau theory is not "wrong" in that sense.

effects. The singularities, *i.e.* the critical points themselves, could not be investigated, but the approach to criticality could be, and that caused quite a stir in the theorist community. In this book we cannot do justice to the enormous literature which has resulted from current investigation of this puzzling domain, but a few aspects of it may be mentioned. First we recapitulate certain basic points.

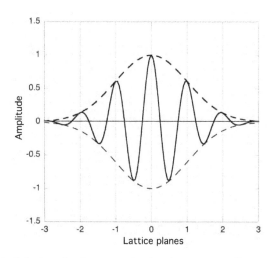

Figure 15.7: Modulation (broken curve) of ordering wave (full curve), leading to ordered domains.

Energy and constitutive expressions

The master equation of classical thermodynamics was given as Eq. (3.18) of Chapter 3 (or in the most condensed form $dU = Y_i dX_i$, using the Einstein convention of implied summation over repeating index). For recall, that fundamental differential form is

$$dU = TdS + \sum_{j=1}^{J} Y_j \, dX_j \,, \qquad (15.24)$$

i.e. a sum of the intensive variables Y_j, times the differential dX_j of its conjugate extensive variable. Despite its "linear" appearance, Eq. (15.24) is "exact" meaning that it is mathematically the total differential of the energy function

$$U = U(S, X_1, \ldots X_J)$$

of its natural extensive variables X_i. But Eq. (15.24) cannot be used in numerical computations because differentials are undefined in magnitude.

Therefore it is essential to "integrate" this differential form, as discussed for instance in Sec. 3.4, a task which can be anywhere from fairly simple to impossibly difficult.

In addition, in order to perform the integration, it is necessary to know the so-called "constitutive equations" such as

$$Y_i = Y_i(S, X_1, \ldots X_J) \qquad \forall\, i$$

which are generally not available. An approximate way to proceed is to obtain such equations from a multivariable Taylor's expansion, as given by Eq. (3.44); repeating it:

$$Y_j = Y_j^o + \sum_{k=1}^{\infty} \frac{1}{k!} \left(\sum_{i=0}^{I} \delta X_i \frac{\partial}{\partial X_i} \right)^k Y_j^o, \qquad (15.25)$$

where the superscript $(^o)$ indicates that the quantity must be calculated in the reference state. Note: this is not a differential form, as was slyly indicated by replacing the symbol "d" of Eq. (15.24) by the (small) finite difference operator δ inside the brackets. In the present application to the neighboring of a critical point, the increments δ are indeed small, and the expansion can be carried out to the first few terms only. That is not true for the investigation of free energies as a function of order parameters, as done in the preceding sections where it is imperative to extend the Landau expansions at least to fourth order otherwise the essential structure of the curves in Figs. 15.3, 15.4 and 15.5 is lost. The integration of the energy differential Eq. (15.24) is here greatly facilitated because the coefficients in the expansion are now constants evaluated at the chosen expansion point, indicated by the superscript $(^\circ)$. What was said above about the internal energy function is of course also valid for various Legendre transforms leading to the familiar potentials of Helmholtz (F) and Gibbs (G) and their respective expansions.

In particular, let us return to the Landau expansion of the Gibbs free energy $g = g(P, T, \xi)$ in the symmetric case introduced above, but let us write it a bit differently. If the function g is a function only of P, T and ξ, we write:

$$g = g_0 + a(P)(T - T_0)\,\xi^2 + C\,\xi^4 , \qquad (15.26)$$

where the expansion is terminated at the 4th-order term since we are investigating the free energy in the immediate vicinity of the critical point. Also, the zeroth-order term g_0 is now placed on the right hand side and, as in Eq. (15.7a), all odd terms are absent since we are investigating continuous transitions which, by symmetry, require their vanishing. In principle the coefficients of the powers of ξ are functions of temperature and pressure (and perhaps other parameters), but the simple form given here suffices for present purposes.

15.7 Critical Exponents

Let us now investigate the immediate environment of the critical point T_{cr}, or T_0. The two temperatures, in present notation, are not identical: T_{cr} is indeed the second-order critical temperature, but T_0 in this context is more generally an *instability* temperature (there will be more on this subject in the next section). The reason for the distinction is that an instability can in principle occur below a first-order temperature, but for that to happen, the latter transformation must be prevented from happening by imposing (formally) certain *constraints*, which in materials science are often kinetic constraints, for example resulting from diffusion too slow to allow the first-order transformation to occur. Thus the effect is to make the instability a *non-equilibrium transformation* (not *transition!*), which means, in turn, that the transformation initiated by the instability takes place away from equilibrium, and that the state of the material just above T_0 is not well characterized: it is history-dependent, meaning that the state of the system depends on how rapidly it is cooled from a high temperature, for example. These may be subtleties, so for present formal purposes the symbol T_0 may be taken to refer either to an instability or to an equilibrium critical temperature.

The distinction between first- and second-order can be further appreciated by comparing the schematic figures 6.1, 6.2 and 6.3. In the first two, the free energies and their immediate metastable extensions (see Fig. 6.1, dashed lines) are composed of two distinct curves representing the two phases α and β. Their intersections do give rise to singularities in the form of delta functions in the second derivative, but that is an artifact of their relative positions in thermodynamic space; each function, by itself is perfectly regular, and reaches the delta function at T_0 quite abruptly, with no warning as to what is about to happen, that is: a singularity in the total equilibrium free energy curve. The second-order case is quite different: to the eye, the free energies show no sign of a forthcoming divergence, it is in some sense hidden, but appears in the first and second derivatives, and the plots of these derivatives, on either sides of the singularity at T_0, show strong deviations from their regular behavior farther away from T_0. That tells us that free energy functions must be non-analytic.[10] Clearly then, the Landau theory must break down for second-order transitions, the very

---

[10] A function is non-analytic if it cannot be expanded in a Taylor's series because it is not infinitely differentiable. See also definition on page 143.

situation that the theory was designed to represent.

The divergences can be studied by examining the temperature dependence of the energy derivatives in the immediate vicinity of the singularity.[11] For that, let us construct power series of various thermodynamic quantities, such as that given by Eq. (15.26), the classical Landau expansion of the function $g(P, T, \xi)$, suspected of presenting singularities. Note however that in this expression, the variables P, T and ξ play slightly different roles: pressure and temperature may be assigned at will, but the order parameter's value must be arrived at by seeking the minimum of the free energy, i.e. by setting equal to zero the derivative of g with respect to ξ:

$$\frac{\partial g}{\partial \xi} \equiv 0 + 2a(P) \, \Delta T \, \xi + 4g_4 \, \xi^3 + \cdots = 0 \qquad (15.27)$$

where

$$\Delta T = (T - T_0) . \qquad (15.28)$$

So let us look at the temperature dependence of thermodynamic functions near the critical point, in the form of the exponent of ΔT in the expansion of the thermodynamic function of interest, in their classical (Landau) formulation. For the function g itself, in the classical description, there is no singularity, so the exponent is simply 0.

Coexistence curve

The coexistence curve, i.e. the line of equal free energies of the two phases involved in the phase transition, is given, in the second-order case, by the vanishing of the derivative of the free energy, Eq. (15.27), for $\xi \to 0$. In the limit, all powers of ξ go to zero faster than the lowest term in the expansion, so that we have, from Eq. (15.27),

$$a \, \Delta T \, \xi + 2g_4 \, \xi^3 = 0 \qquad (15.29)$$

and that for $T > T_0$ there is only one solution: $\xi = 0$, since g_4 must be positive for overall stability at large order parameter, and a must be positive by *fiat*.

For $T < T_0$, the extremum at $\xi = 0$ is actually a maximum, but there are two additional solutions which are symmetrically placed minima given by

$$\xi^2 = \frac{a \, (T_0 - T)}{2g_4} \qquad i.e. \qquad \xi = \pm \sqrt{\frac{a}{2g_4}} \, |\Delta T|^{\beta} , \qquad (15.30)$$

that is to say that the coexistence curve should behave as $(T_0 - T)^{\beta}$, our first example of a "critical exponent", with $\beta = \frac{1}{2}$ in the vicinity of the critical

---

[11] The following text closely follows Callen (1985, pp. 263-275).

point. The same exponent $\beta = \frac{1}{2}$ is found at the top of the miscibility gap obtained by the regular solution model, for example. All simple classical treatments coexistence curves lead to $\beta = \frac{1}{2}$ whereas experimental values give the range of values $0.3 < \beta < 0.4$, significantly at variance from the classical value. Why a "range"? Because exact values of exponents (and other thermodynamic quantities) related to the second-order phase transitions cannot be obtained for even the simplest three-dimensional model systems (the three-dimensional Ising model, for instance), as was mentioned earlier, so that experimental results must be used. How could that discrepancy have come about?

Cahn and Hilliard on their classical paper on non-uniform systems (1958), surveyed here in Sec. 14.1, mention the problem, explaining that the coexistence curve should be parabolic (exponent $\frac{1}{2}$), but is often encountered in practice better fitted to exponent $\frac{1}{3}$. The authors add (in 1958) "So far there appears to be no satisfactory explanation for this anomaly". But now there is an explanation; this is how Callen (1985) states the problem:

> The reader may well ask how so simple, direct and general an argument as that of the preceding section can possibly lead to incorrect results. Does the error lie within the argument itself, or does it lie deeper, at the very foundations of thermodynamics? That puzzlement was shared by thermodynamicists for several decades. Although we cannot enter here into the renormalisation theory that solved the problem, it may be helpful at least to identify the source of the difficulty. To do so we return to most central postulate of thermodynamics — the entropy maximum postulate. In fact that "postulate" is a somewhat over-simplified transcription of the theorems of statistical mechanics. This over-simplification has significant consequences only when fluctuations become dominant — that is, in the critical region.

The problem seems to lie in the fact that the classical derivations implicitly rely on a statistical approximation, made in Sec. 8.2, which replaces the full distribution of states by the value of the most probable distribution. Such an approximation is current in the field and is normally acceptable because the optimal distribution in most instances is very sharp, a narrow Gaussian, practically a delta function. Very near the critical point of a second-order transition that is not the case, however: the thermodynamic potential is now very shallow and, as seen for instance in the potential curve T_0 of Fig. 15.3, no longer symmetric about its minimum point, so that average energies, say, calculated in accordance with this approximation, are no longer correct. In conclusion, it is interesting that an apparently mild approximation turns out to be the cause of serious discrepancies in the description of second-order transitions which had baffled thermodynamicists for decades. Renormalisation group methods of calculating free energies,[12] still approximate, confirm the cause of the discrepancies. Consequences of this work led to notions of "scaling" and "universality" with the conclusion that the values of critical exponents depended solely on the dimensionality of the system (normally 3) and on the multiplicity of the order parameter, not on the physical nature of the system in question [see Callen (1985) for example].

Susceptibility

Figure 6.2 shows that even for first-order transitions singularities are expected for the second derivatives of free energies with respect to its parameters at panels (c) and (f) for, respectively, heat capacity c_P and compressibility κ_T. The same is seen to be true for second-order transitions as seen in Fig. 6.3. In general, the vanishing of second derivatives of thermodynamic potentials give rise to generalized instabilities χ. In a one-dimensional model, we have

$$\chi^{-1} \equiv \frac{\partial^2 g}{\partial \xi^2} = 2a\,\Delta T + 12g_4\,\xi^2 + \text{terms of order higher in powers of } \xi$$
(15.31)

from which it follows, with $\xi \to 0$ and $T > T_0$, in the limit, that

$$\chi^{-1} = 2a\,(T - T_0)^\gamma$$
(15.32)

with $\gamma = 1$, the second example of a critical point. For $T < T_0$, the order parameter ξ becomes non-zero, then inserting the value for ξ from Eq. (15.30) into Eq. (15.31) we find

$$\chi^{-1} = \frac{\partial^2 g}{\partial \xi^2} = 2a\,\Delta T - 12g_4\left(\frac{a}{2g_4}\right)\Delta T + \cdots = 4a\,(T_0 - T) + \ldots \quad (15.33)$$

so that here also the classical critical exponent is $\gamma' = 1$, the accent on the symbol γ meaning that the approach to singularity (phase α, say) might differ from the exit from singularity (phase β). On the schematic diagram of Fig. 6.3 (b), that would refer to the curved lines on either side of the singularity. In classical theories the two curves are the same; but experimentally we find $1.2 < \gamma < 1.4$ and $1 < \gamma' < 1.2$, significantly different from the classical value of 1. As for the singularity itself, it is called here (generalized) susceptibility, but in Fig. 6.3 it is the heat capacity at constant

---

[12]Kenneth Wilson was awarded a solo Nobel prize for that work in 1982.

pressure c_P, or it could be the isothermal compressibility κ_T as in panel (f) of the figure just quoted, and so on. At singularities, those χ parameters are the inverses of the generalized susceptibilities, so that when these tend to zero, their opposites tend to infinity, and vice versa.

References

Born, M. and Huang, K. (1962), *Dynamical Theory of Crystal Lattices*, Oxford, Clarendon Press.

Brillouin, L. (1946). *Les Tenseurs en Mécanique et en Elasticité*, Dover.

Cahn, J. W. and Hilliard, J. E. (1958), *J. Chem. Phys.*, **28**, 258.

Callen, H. (1985), *Thermodynamics and an Introduction to Thermostatistics; Second Edition*, John Wiley & Sons, Inc.

de Fontaine, D. (1975), *Acta Metall.*, **23**, 553.

de Fontaine, D. (1979), *Configurational Thermodynamics of Solid Solutions*, in *Solid State Physics*, Ehrenreich, H., Seitz, F. and Turnbull, D., **34**, p. 73-274, Academic Press.

de Fontaine, D. (1994), *Cluster Approach to Order-Disorder Transformations in Alloys*, in *Solid State Physics*, Ehrenreich, H., Seitz, F. and Turnbull, D., **47**, p. 33-176, Academic Press.

Devonshire, A, F. (1949), *Philos. Mag.* **40**, 1040; (1951) *Philos. Mag.* **42**, 1065.

Ducastelle, F. (1991), *Order and Phase Stability in Alloys*, North Holland.

Fridkin, V. and Ducharme, S. (2014), *Ferroelectricity at the Nanoscale*, in Nanoscience and Technology, Springer-Verlag, Berlin Heidelberg.

Ginzburg, V. L. (1945), *Zh. Eksp. Teor. Fiz.* **15**, 739; (1946) *J. Phys. USSR* **10** 107 (translation).

Ginzburg, V. L. and Landau, L. D. (1950), *Zh. Exsp. Teor. Fiz.* **20**, 1064 (in Russian).

International Tables for Crystallography (see 8 Volumes on the Web, URL http://it.iucr.org/).

Khachaturyan, A. G. (1973), *Phys. Status Solidi* **B 60**, 9.

Khachaturyan, A. G. (1983), *Theory of Structural Transformations in Solids*, J. Wiley & Sons, New York.

Kikuchi, R. (1951), *Phys. Rev. B*, **17**, 2926.

Krivoglaz, M. A. and Tikhonova, E. A. (1958), *Ukr. Fiz. Zh.* **3**, 297

Landau, L. D. (1937), *Phys, Z. Sowjet.*, **11**, 26; 545.

Landau, L. D. and Lifshitz, E. M. (1958), *Statistical Physics*. Addison-Wesley, Reading, MA. Also: Landau, L. D., Lifshitz, I. M. and Pitaevskii, L. P.(1980). *Statistical Physics*, Third Edition, Pergamon Press (English versions).

Landau, L. D. (1965), *Collected papers*. Oxford, Pergamon Press, p. 546.

Lifshitz, E. M. (1942), *J. Phys. USSR* **7**, 61 and 251.

Morse, M. (1935), *Functional Topology and Abstract Variational Algebra* in *Mem. Sci. Math.*, Fascicule 92, Gauthier Villars, Paris.

Nye, J. F. (1957), *Physical Properties of Crystals*, Oxford, Clarendon Press.

Onsager, L. (1944), *Phys. Rev.*, **65**, 117.

Sanchez, J. M., Ducastelle, F. and Gratias, D. (1984), *Physica A*, **128**, 334.

Wilson, K. G. (1975), *Rev. Mod. Phys.* **47**, 4, 773.

Chapter 16

Thermodynamic Stability of Crystals

Strictly speaking, classical thermodynamics is limited to the study of systems in equilibrium; but the *loss of stability* of systems can be investigated, by graphical and/or algebraic means, as was done for example in the previous chapter in the context of the Landau theory or equivalent formulations.[1] In a complete study of loss of stability, certain important aspects must be looked into: (A) is the transition from the homogeneous state continuous or discontinuous? (B) is the order parameter of dimension 1 (as it was in Ch. 15 mostly) or is it of higher dimension, *i.e.* does ξ have several structural or compositional components, (C) is the loss of equilibrium due to an infinitesimal departure ($\delta\xi$) from the initial homogeneous state ($\xi = 0$), or from a large departure taking place locally. Point (A) has already been discussed but will be brought up again here, point (B) will taken up in the present chapter, and point (C) will be discussed in Ch. 17, pertaining to nucleation theory, then also in Ch. 18 in an attempt to unify infinitesimal and discontinuous transitions.

16.1 Transformation Mode

In this book, transformation temperatures have been designated by a variety of symbols. Usually, the context will define the meaning, but often a subscripted upper-case T will designate transformations, as shown in Table 16.1: Almost 150 years ago, in his famous treatise Gibbs described loss

---

[1]Consider indeed the title of Gibb's treatise, *"On the Equilibrium of Heterogeneous Substances"*.

of equilibrium of a given system by two apparently very different processes: through perturbations small in intensity but large in extent, and perturbations large in intensity but small in extent. In the materials science lingo, the first mode today is called spinodal decomposition, the second one nucleation and growth. The two appear to be incompatible: it's one of the other. Actually, thanks in part to fluctuations, the two go smoothly into one another, as will be described in Ch. 18, by a transformation not known to Gibbs at the time. Considering the two cases separately, spinodal decomposition would be a continuous process, so that its transformation temperature would be denoted by the symbol T_o, whereas nucleation and growth would be a discontinuous process with transformation temperature T_t characteristic of a first-order transition. Practically, a first-order transformation — occurring normally at temperature T_t — can often be initiated by a continuous transformation at a lower temperature $T_o < T_t$ if the equilibrium transition is suppressed due to sluggish atomic diffusion, a common occurrence in material science. In such cases the instability temperature T_o will be history dependent, *i.e.* dependent, to some extent at least, on how fast was the cooling rate, how deep, and from what starting temperature.

An example will be given in Ch. 18, Fig. 18.3. It is shown there, and in original references quoted therein, that over the whole NiMo phase diagram a $\langle 1\frac{1}{2}0 \rangle$ ordering wave becomes unstable at concentration-dependent T_o and subsequently triggers the transformation to the equilibrium ordered phases, whose ordering waves are crystallographically related, but not necessarily identical to the unstable wave. By analogy with the well-known mechanism of spinodal decomposition, we shall denote by the term "spinodal ordering" this transformation initiated through an ordering instability.

Table 16.1: Symbols for transformation temperatures.

Symbol	Meaning
T_t	transformation or transition, first-order or triple point
T_o	transformation, general or instability or spinodal
T_c	critical, second-order transition
T_g	glass transformation

The symbol T_o should not be confused here with its use in what physical metallurgy called the "T_o-line", *i.e.* the locus of intersection of two free energy curves as in Figs. 9.3, 10.4 and 10.6, denoting metastable (diffusionless) equilibrium between two different phases.

Extension of instability theory to multicomponent order parameters will be taken up in following sections.

16.2   Multivariable Instabilities

Up to now, we have considered only one-dimensional order parameters, *i.e.* parameters having only a single degree of freedom, such as the independent concentration in a binary system, or a vectorial order parameter constrained to lie in one direction. But an order parameter may be a vector with three independent components (in three space dimensions), or a tensor of second rank (two indices), third and fourth rank (three or four indices) and so on. The general ideas concerning multicomponent instabilities are the same as in the one-component case, but the formalism now includes some linear algebra and tensor calculus, here limited to Cartesian tensors, *i.e.* referred to a Cartesian coordinate system. The formalism will be given mainly for the case of infinitesimal stability, *i.e.* stability with respect to very small departures from equilibrium. Stability with respect to large departures instead requires, in principle, knowledge of the entire energy (or potential) curve and will be addressed in Chapter 18. As will be seen, multicomponent order parameters allow consideration of many different types of ordering processes with different physical properties, compositional, magnetic, electronic, displacive, . . . in crystalline materials for example.

Consider again Fig. 2.2 which depicts graphically the equilibrium conditions for a single-variable system. As was stated, at the point ξ indicated by a filled circle, the system is stable, it is metastable at the open circle, and unstable at the starred circle. For now, we shall be concerned only with "stability in the small", *i.e.* for infinitely small departures from the initial position on the curve, so that the system is shown to be stable in the small at the open and closed circle positions — since there the energy along the curve is raised — and unstable at the starred position — since there the energy is lowered for any small departure from the point in question. More generally, for any initial point situated on the curve between the inflection points, the system will be unstable. The limiting points of the negative curvature region of the curve, as we have seen, are the spinodal points of an energy surface, or "potential"; they are indicated as open circles on Figs. 9.1 in Ch. 9 and 14.1, the vanishing-curvature points on the "double-well" potential curve. The general conclusion concerning stability "in the small", for the single (independent) component case, is thus that for points of the curve inside the open circles (spinodals) there is infinitesimal instability. Outside the negative-curvature region, there is stability in the small. Between open and closed circles — the latter being the common tangency points — there is still stability in the small, but instability with respect to perturbations "in the large", large in intensity but small in extent (requiring nucleation events).

This chapter, however, is concerned with multidimensional order param-

eters; practically that means that the curve in Fig. 2.2 becomes a surface
and tangents become (hyper)-planes. Stability in the small therefore consists in constructing planes tangent to the potential surface at the point
representing the system. By analogy with the single component case, infinitesimal stability will result if the surface lies entirely above the tangent
plane in the immediate vicinity of the contact point, will be partially unstable otherwise. By "partially", it is meant that some portions of surface may
lie "under" the tangent plane along certain directions in parameter space;
others, not. Indeed there are not merely two cases to consider, positive
curvature and negative curvature; at a given point, there may be some directions in parameter space presenting positive, others negative curvature.
Fortunately, there are straightforward mathematical processes for handling
stability for n-dimensional order parameters, for arbitrary n.

Quadratic Forms

Mathematically, the infinitesimal stability analysis consists in expressing
the energy potential by a quadratic form, *i.e.* expanding the free energy
function in a Taylor's expansion to second order. That will suffice since we
are interested here in infinitely small displacements from a "point" state.
The form is then transformed into a sum of squares to see if it contains one
or more negative terms. The system will be unstable to small perturbations
unless all square terms are positive. The matrix of the second derivatives
of the energy potential surface is the Hessian introduced previously. Diagonalization of the Hessian matrix produces the eigenvalues and eigenvectors
which are respectively the coefficient of the sum of squares of the diagonalized quadratic form and the directions in space of the instabilities, but can
also be performed in different ways (see below).

   Let us write the quadratic form associated with a given potential energy
function $\delta u(x)$, *i.e.* as the second-order term of a Taylor's expansion, using
"content-free" mathematical notation:

$$\delta u(x) = \frac{1}{2} \sum_{i,j=1}^{n} \left(\frac{\partial^2 u}{\partial x_i \partial x_j} \right)^{\circ} \delta x_i \, \delta x_j \qquad (16.1)$$

where the lower-case x and y do not necessarily have the physical meaning
of extensive and intensive variables they had previously, u is the energy
potential or density and the superscript $\circ$ indicates that the derivatives
must be calculated at the tangency point. The symbols δ denote small
but finite differences, not differentials, with the difference δu indicating the
distance from a point on the tangent at point $\mathbf{x}$ to the curve at the same
abscissa. For intrinsic stability, the quadratic form δu must then be positive

for *any* infinitesimal change in the variables x_i. It is therefore required to examine the matrices

$$
Q = \begin{pmatrix} u_{,11} & u_{,12} & \cdots & u_{,1n} \\ u_{,21} & u_{,22} & \cdots & u_{,2n} \\ \cdots\cdots\cdots\cdots\cdots \\ u_{,n1} & u_{,n2} & \cdots & u_{,nn} \end{pmatrix} = \begin{pmatrix} \frac{\partial y_1}{\partial x_1} & \frac{\partial y_1}{\partial x_2} & \cdots & \frac{\partial y_1}{\partial x_n} \\ \frac{\partial y_2}{\partial x_1} & \frac{\partial y_2}{\partial x_2} & \cdots & \frac{\partial y_2}{\partial x_n} \\ \cdots\cdots\cdots\cdots\cdots \\ \frac{\partial y_n}{\partial x_1} & \frac{\partial y_n}{\partial x_2} & \cdots & \frac{\partial y_n}{\partial x_n} \end{pmatrix} , \qquad (16.2)
$$

where we have defined

$$
y_i \overset{\text{def}}{=} \frac{\partial u}{\partial x_i} = u_{,i} \qquad (16.3)
$$

and

$$
\frac{\partial^2 u}{\partial x_i \partial x_j} = \frac{\partial y_i}{\partial x_j} = u_{,ij} . \qquad (16.4)
$$

It is seen that the matrix Q is symmetric by construction (it is a Hessian matrix, see page 104), so that $\frac{\partial y_i}{\partial x_j} = \frac{\partial y_j}{\partial x_i}$, which are just the *Maxwell equations* in the appropriate variables. Also, the number of independent first derivatives is $n(n+1)/2$, provided that the x_i variables themselves are independent. Since Q is symmetric, it has real eigenvalues, but the computation thereof can be challenging for $n \geq 3$. Alternately, one can use various mathematical techniques of "completing the square" to transform the quadratic form in a sum of squares, which leads directly to the "rule of minors", again introducing n conditions.[2] Whatever the technique, the conclusion is always the same: as soon as one of the terms of the sum becomes negative (for instance as one of the principal minors vanishes as temperature or concentration or pressure changes, say), an instability sets in. The most elegant method for testing stability is to use Jacobian determinants (see Appendix K and Appendix- E for a brief summary of properties of Jacobians), as follows: first, one notices that the determinant of the second matrix Q in Eq. (16.2) is the *functional determinant*, or Jacobian of the transformation of the of the of the x_i into the y_i variables:

$$
\text{Det}(Q) \equiv J = \frac{\partial(y_1, y_2, y_3, \dots y_n)}{\partial(x_1, x_2, x_3, \dots x_n)} . \qquad (16.5)
$$

The requirement that a quadratic form be positive for any variation of its independent variables [see Eq. (16.1)] may now be written in matrix

---

[2]The rule of minors states that that a real symmetric square matrix is positive definite if all of its leading principal minors are positive. A leading principal minor of an $n \times n$ determinant is an $m \times m$ determinant, with $m \leq n$, extracted from the $n - m$ rows and columns of the original determinant. Some such minors are indicated schematically in Fig. 16.1.

form as follows:

$$\delta u = \frac{1}{2}\,\mathbf{x}'\mathbf{Q}\,\mathbf{x} > 0 \tag{16.6}$$

with the symbol $\mathbf{x}'$ designating the transpose of the column vector $\mathbf{x}$. Equation (16.6) is written out in full at Fig. 16.1 (for clarity, dropping the commas indicating derivatives). The Jacobians J_k may be used to transform the quadratic form to a sum of squares according to the Jacobi formula:

$$u = \sum_{k=1}^{n} \frac{J_k}{J_{k-1}}\, z_k^2 \qquad (\text{with } J_0 = 1) \tag{16.7}$$

where z_k $(k = 1, 2, \ldots n)$ are the transforms of the x_i variables. The proof is given in Appendix K and makes use of the properties of determinants given in Appendix E.

$$\begin{array}{cc} & \begin{array}{cccccc} J_1 & J_2 & J_3 & \cdots & J_n & \end{array} \\ \delta u = \frac{1}{2}\left(x_1\, x_2\, x_3 \ldots x_n\right) & \begin{pmatrix} u_{11} & u_{12} & u_{13} & \cdots & u_{1n} \\ u_{21} & u_{22} & u_{23} & \cdots & u_{2n} \\ u_{31} & u_{32} & u_{33} & \cdots & u_{3n} \\ \vdots & \vdots & \vdots & \ddots & \vdots \\ u_{n1} & u_{n2} & u_{n3} & \cdots & u_{nn} \end{pmatrix} \begin{pmatrix} x_1 \\ x_2 \\ x_3 \\ \vdots \\ x_n \end{pmatrix} > 0 \end{array}$$

Figure 16.1: Expansion of Eq. (16.6) showing row and column vectors and matrix Q with 3 leading principal minors J_1, J_2, J_3 and J_n, the latter being the full minor, equal to the determinant J of the matrix Q itself.

Linear Stability Conditions

The ratios of Jacobians in Eq. (16.7) are *not* eigenvalues, but are simply the coefficients of a sum of squared variables (z_k) in this particular method of diagonalizing a quadratic from. However, a theorem due to Sylvester states that the number of positive, negative and null terms in the sum of squares is always the same, regardless of the algorithm used for diagonalization. It follows that instability sets in if any of the determinants J_k vanish, including the determinant $J_n \equiv J$ of the matrix Q itself.

The rule of minors may be expressed as follows: where use has been made of Jacobian properties 2, 5, and 7 of Appendix E. In the latter expression, and in preceding ones, the partial derivatives just to the right of the arrow signs are written with straight "d" (Roman letter) for a purpose, which is to emphasize the importance of the subscripts $x_i, y_j, \ldots$ indicating which variables are being held constant. Indeed, the very presence of these subscripts outside the parentheses tell the reader that a partial derivative operation is being carried out, so that the expressions $\left(\frac{dy_2}{dx_2}\right)_{y_1,x_3,\ldots x_n}$ and $\frac{\partial y_2}{\partial x_2}$, for example, have exactly the same meaning, except that the former one is more explicit. In particular, in the inequality (K.18) the set of subscripts indicating the variables being held constant is a "hybrid" one, containing both x and y variables. That in turns means that the Jacobian method also performs virtual Legendre transformations.

Equation (K.18) and preceding ones may appear quite formidable, but an example for a simple binary alloy (mixture) A-B will hopefully clarify the situation. The independent extensive variables (actually molar densities) in this case are s, v and $c \equiv c_B$, the latter being the concentration of the B species; the dependent variable $c_A = 1 - c$ does not appear. The intensive variable conjugate to c is what we have called the "chemical field" $\mu = \mu_B - \mu_A$, as was explained in the section on independent variables (Sec. 7.6, page 90). The total differential of the energy density is thus

$$\mathrm{d}u = T\,\mathrm{d}s - P\,\mathrm{d}v + \mu\,\mathrm{d}c \qquad (16.8)$$

with quadratic form matrix

$$Q = \begin{pmatrix} \frac{\partial T}{\partial s} & \frac{\partial T}{\partial v} & \frac{\partial T}{\partial c} \\ \frac{-\partial P}{\partial s} & \frac{-\partial P}{\partial v} & \frac{-\partial P}{\partial c} \\ \frac{\partial \mu}{\partial s} & \frac{\partial \mu}{\partial v} & \frac{\partial \mu}{\partial c} \end{pmatrix} \qquad (16.9)$$

from which we deduce the Maxwell relations

$$\left(\frac{\partial T}{\partial v}\right)_{s,c} = -\left(\frac{\partial P}{\partial s}\right)_{v,c}, \quad \left(\frac{\partial T}{\partial c}\right)_{s,v} = -\left(\frac{\partial \mu}{\partial s}\right)_{v,c}, \quad -\left(\frac{\partial P}{\partial c}\right)_{s,v} = \left(\frac{\partial \mu}{\partial v}\right)_{s,c}$$

between independent partial derivatives of which there are $\frac{3\cdot4}{2} = 6$. The stability conditions are thus:

(1) $\left(\frac{\partial T}{\partial s}\right)_{v,c} > 0 \qquad \Rightarrow c_v > 0$ \hfill (16.10a)

(2) $\dfrac{\partial(T,-P)}{\partial(s,v)} = \dfrac{\partial(T,-P)}{\partial(T,v)}\dfrac{\partial(T,v)}{\partial(s,v)} = \dfrac{T}{v\,c_v\kappa_T} > 0 \qquad \Rightarrow \kappa_T > 0$ \hfill (16.10b)

(3) $\dfrac{\partial(T,-P,\mu)}{\partial(s,v,c)} = \dfrac{\partial(T,-P,\mu)}{\partial(T,-P,c)}\dfrac{\partial(T,-P,c)}{\partial(T,v,c)}\dfrac{\partial(T,v,c)}{\partial(s,v,c)} \qquad \Rightarrow \left(\frac{\partial \mu}{\partial c}\right)_{T,P} > 0.$ \hfill (16.10c)

These conditions, which are explicit expressions of the formulas associated with of Eq. (K.18), are quite sensible: number (1) tells us that the heat capacity must be positive, otherwise a substance would cool on receiving heat, number (2) tells us that the compressibility must be positive as well, otherwise a substance would expand upon being pressurized,[3] number (3) tells us that the average concentration c must lie outside the *spinodal*.

Recall that in a binary system the intensive variable conjugate to the independent concentration variable c is the "chemical field" μ (see page 90), so that we have, after a few Legendre operations

$$\mathrm{d}g = -s\,\mathrm{d}T + v\,\mathrm{d}P + \mu\,\mathrm{d}c. \tag{16.11}$$

Therefore, at constant T and P we have

$$\mu \overset{\text{def}}{=} \mu_B - \mu_A = \left(\frac{\partial g}{\partial c}\right)_{T,P} \tag{16.12}$$

and

$$\left(\frac{\partial \mu}{\partial c}\right)_{T,P} = \left(\frac{\partial^2 g}{\partial c^2}\right)_{T,P} \equiv g'', \tag{16.13}$$

as expected. It follows that condition (16.10c) is equivalent to specifying, for infinitesimal stability, that the system of average concentration $\bar{c}$ must lie outside the spinodal indicated by a dashed line in Fig. 9.1, and defined by $g'' = 0$. The stability condition at T and P constant for a binary (two-component) system is thus $g'' > 0$, meaning that the solution is unstable with respect to concentrations located inside that region of the free energy curve $g(c)$, between the spinodal concentrations.

Multicomponent Compositional Instabilities

What if the alloy system contains more than two components (atomic or molecular), say $m \geq 3$? Then the composition variable c becomes an $(m-1)-$vector and the quadratic form matrix Q can be subdivided into two principal minors, one involving the diagonal elements 1,1 and 2,2, the same as those of the previous matrix of Eq. (16.9), and another principal minor, of order $(m-1) \times (m-1)$, in place of the 3,3 element $\partial \mu / \partial c$ of the binary case. Having ascertained that the determinant of the 2×2 matrix is positive, there remains the requirement of the positive character of the remaining $(m-1) \times (m-1)$ minor, call it $\mathbf{M}$, which introduces $m-1$ conditions, to be examined as was done in the general treatment of the Jacobi formula, Eq. (16.7). Or one can search for the eigenvalues and eigenvectors of the concentration minor $\mathbf{M}$. The vanishing of those $m-1$ eigenvalues introduces $m-1$ surfaces in phase diagram space, these *instability surfaces* generalizing the dashed spinodal curve of Fig. 9.1, mentioned above. To each one of these instability surfaces corresponds an eigenvector which defines a most unstable direction in composition space. Note that some of these surfaces may be located at

---

[3]as does cork partially, I am told.

$T < 0$, and are thus unphysical; note also that some may mutually intersect. As is the case for most thermodynamic considerations in dimensions greater than two, the geometrical representation of the problem is challenging, but an attempt was made to summarize the results for ternary systems in a review paper already mentioned (de Fontaine, 1979), which cites the relevant original literature.

To summarize, in the general multicomponent case of linear instability, one should expand the free energy to second order in the $(m-1)$ *independent* concentration variables to construct the resulting Hessian matrix (see page 104). It is preferable to express the matrix **M** in Fourier space so as to treat both clustering and ordering within the same framework, as described in Sec. 15.2 and in Appendix I. Thus, one picks the correct underlying lattice for the solid solution in question (fcc or bcc, for example), and for each one constructs the **M** matrix for the high-symmetry crystallographic points [or *special points* ordering wave vectors $\mathbf{k}(h)$, for instance $\langle 000 \rangle$ for clustering and $\langle 100 \rangle$ (or others ...) for ordering]. There remains the task of determining values for the effective pair interactions Ω_{ij}, defined in Eqs. (7.86) and (7.87). There are several ways of doing that: fitting to experimental data *à la* CALPHAD or fitting to the calculated energy of compounds (real or not), as is available in computer codes such as ATAT of Axel van de Walle (http://www.brown.edu/Depatment Engineering Labs/avdw/atat). So, even if a model free energy is given, at least to second order, the stability determination can be rather extensive, although such complete tests are rarely required. An example of such calculations is given in Sec. 10.4 in connection with high-entropy alloys.

16.3 Tensor Formalism

Now that the general formalism is in place, we can apply it also to many different situations, in particular to various sets of physical variables in crystalline materials. To simplify matters we shall consider only uniform systems. So what this section proposes is an extension of the basic thermodynamics covered so far — non-uniformity excluded — applied to multidimensional order parameters, *i.e.* physical variables which are not simply scalars but also to vectors and tensors in crystalline media. It will be seen that the form taken by these vectors and tensors will depend on the symmetry of the crystals under study.

Definitions

We start with some definitions (the treatment given here follows that given by J. F. Nye (1957, reprinted many times).[4] As mentioned previously, we shall consider only so-called *Cartesian tensors, i.e.* those whose components are referred to Cartesian coordinate systems $[x_1, x_2, x_3]$, the axes of which are mutually orthogonal. The definition is then as follows: *a tensor is a physical quantity which, with respect to a given set of axes, has components which transform — upon a change of coordinate system — according to well-defined rules.* These rules are given in Table 16.2.[5] In this table,

Table 16.2: Tensor definitions for 0-rank to 4-rank tensors.

Objects	Rank	Transformation laws	Examples
Scalar	0	$t' = t$	Temperature
Vector	1	$t'_i = a_{ij} t_j$	Electric field
Tensor	2	$t'_{ij} = a_{ik} a_{jl} t_{kl}$	Elastic strain
Tensor	3	$t'_{ijk} = a_{il} a_{jm} a_{kn} t_{lmn}$	Piezoelectric coefficient
Tensor	4	$t'_{ijkl} = a_{im} a_{jn} a_{kp} a_{iq} t_{mnpq}$	Elastic constants

all the symbols t indicate the original tensors and the accented t' are the transformed ones. The indices $i, j, k, l, \ldots$ take on the values 1, 2, and 3 since we assume that we are working in crystalline 3-space. The rank of a tensor is simply the number of its indices (subscripts) so that, in this nomenclature, a tensor of rank 0 is a scalar and a tensor of rank 1 is a vector. The parameters a_{ij} are the directional cosines of axes $[x'_1, x'_2, x'_3]$ projected onto the original ones $[x_1, x_2, x_3]$. These parameters form the 3×3 transformation matrix

$$\mathbf{A} \overset{\text{def}}{=} [a_{ij}] = \begin{pmatrix} a_{11} & a_{12} & a_{13} \\ a_{21} & a_{22} & a_{23} \\ a_{31} & a_{32} & a_{33} \end{pmatrix}. \tag{16.14}$$

---

[4] Just as is the textbook by Goldstein (1950), mentioned earlier, the delightful text by Nye (1957) is also "user-friendly" and highly recommended for a more complete treatment of the topics presented in this section. See also the more recent textbook by Newnham (2005).

[5] For oblique or curvilinear coordinate systems, a more complete definition is required, where coordinate indices are indicated as subscripts and superscripts (co-variant and contra-variant), but these will not be needed here. Were we to describe vectors in oblique coordinate systems for crystals of low symmetry say, the direct space vectors would be contra-variant vectors of rank one, whereas vectors in reciprocal space would be rank-1 co-variant tensors.

It is important to note that these 2-index parameters form a matrix, *not a tensor*. Indeed, this array of numbers belongs in a sense to two systems of axes, the old and the new, whereas the physical objects (tensors) t_{ij} belong always to a single coordinate system. The inverse transformation from the $[x'_1, x'_2, x'_3]$ to the $[x_1, x_2, x_3]$ is obtained by permuting the indices i, j in the matrix of Eq. (16.14).

Note also that, since the transformation parameters are directional cosines, the following relations can be shown to hold

$$a_{ik}a_{kj} \overset{\text{def}}{=} \sum_{k=1}^{3} a_{ik}a_{kj} = \delta_{ij} = \begin{cases} 1 & (i = j) \\ 0 & (i \neq j) \end{cases}. \tag{16.15}$$

Hence the matrix of the Kronecker deltas is the unit matrix. Equation (16.15) makes use of the so-called "Einstein convention" whereby the summation sign is suppressed from sums over a repeating dummy index, k in the present example.

Cubic Symmetry

The case of cubic symmetry is the simplest to treat and also one of the most important due to the frequent observation of this class of symmetry in nature. Along with other symmetry elements, it is characterized by three mutually orthogonal axes of 4-fold rotation, whereas in the tetragonal class, there is but one 4-fold rotation axis. The presence in the cubic class of three 4-fold axes results in a further reduction of independent tensor components from 5 in the tetragonal class [see Eq. (J.2) in Appendix J] to only 3 in the cubic: C_{11}, C_{12}, and C_{44}, which is the case for elastic constants, for example (also called stifnesses). The notation just used is the so-called *Voigt notation* which replaces pairs of indices of tensors of rank greater than 3 by single indices but of range $1, 2, 3, 4, 5, 6$. The Voigt notation is explained in detail in Appendix J; it is used in what follows, where we look at the case of crystals of cubic symmetry. That of tetragonal symmetry is mentioned in Appendix J. The matrix representing the fourth rank tensor for cubic symmetry can be derived in similar ways:

$$\text{Cubic} \equiv \begin{pmatrix} C_{11} & C_{12} & C_{12} & 0 & 0 & 0 \\ C_{21} & C_{11} & C_{12} & 0 & 0 & 0 \\ C_{21} & C_{32} & C_{33} & 0 & 0 & 0 \\ 0 & 0 & 0 & C_{44} & 0 & 0 \\ 0 & 0 & 0 & 0 & C_{44} & 0 \\ 0 & 0 & 0 & 0 & 0 & C_{44} \end{pmatrix}, \text{ with } C_{IJ} = C_{JI}. \tag{16.16}$$

In this case, one thus obtains a reduction of tensor components from 81 (in tensor notation for the general case of minimal symmetry) to just 3 (in

Voigt matrix notation), as just stated which is significant to say the least. Thanks to this simplification, the condition of mechanical stability in cubic crystals can be simply expressed. As was stated previously, it is that the matrix of Eq. (16.16) be positive definite, that is to say that its principal minors be positive. It therefore follows that C_{11} and C_{44} must be positive along with the principal minors of the 3×3 upper minor of the full matrix, that is (matrix symmetry has ben taken into account):

$$\begin{vmatrix} C_{11} & C_{12} \\ C_{12} & C_{11} \end{vmatrix} = C_{11}^2 - C_{12}^2 > 0 \quad \Rightarrow \quad C_{11} > |C_{12}| \tag{16.17}$$

and

$$\begin{vmatrix} C_{11} & C_{12} & C_{12} \\ C_{12} & C_{11} & C_{12} \\ C_{12} & C_{12} & C_{11} \end{vmatrix} = (C_{11} - C_{12})^2 \, (C_{11} + 2|C_{12}|) > 0$$

$$\Rightarrow \quad \text{same condition plus}: \ C_{11} > -2\,C_{12} \quad \text{if} \quad C_{12} < 0. \tag{16.18}$$

In summary, if $C_{12} > 0$ (the usual situation), the stability conditions can be summarized by the familiar criteria: $C_{11} > 0$, $C_{44} > 0$ and $C' \equiv (C_{11} - C_{12}) > 0$. If only one of these inequalities fails, the crystalline material will soften and ultimately transform to another phase. If this failure occurs at equilibrium, the transition is of second-order type at T_c. If the failure occurs below the temperature T_t of a suppressed first-order transition, the phase change results from an instability T_o and is simply called a *transformation*, such as a Martensitic transformation, much studied experimentally in metallurgy because of its importance in the the fabrication of steel.

16.4   Physical properties

In this subsection, new physical variables will be introduced, of the kind that one may encounter in crystal physics [see for example the textbooks by Nye (1957) and Newnham (2005)]. In general these new variables will be inherently multidimensional, that is, will be tensors of rank 1, 2, 3, or 4. In particular, let us take as a typical example the extensive variables appearing in the energy

$$U = U(S, \mathcal{P}_k, \epsilon_{ij}), \quad \text{with} \quad i, j, k = 1, 2, 3, \tag{16.19}$$

namely: entropy S, electric polarization $\mathcal{P}_k$,[6] and elastic strain ϵ_{ij}. These variables interact with one another in ways which may appear rather com-

---

[6]For simplicity, we shall make no distinction between polarization $\mathcal{P}$ and electric displacement $\mathcal{D}$.

plex, but which can be represented graphically as shown in Fig. 16.2 where the symbols for the independent extensive variables are indicated inside shaded circles at the vertices of an equilateral triangle in the figure. The

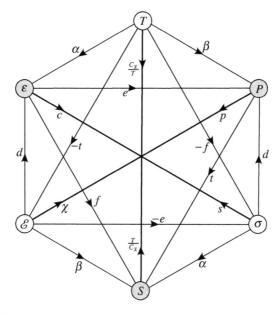

Figure 16.2: Graphical representation of interaction between intensive variables T, E and σ (open circles) and conjugate extensive variables $S, \mathcal{P}$ and ϵ, respectively (shaded circles). Other interactions are listed in Table 16.3.

corresponding intensive variables: temperature (T), electric field $(\mathcal{E}_k)$ and elastic stress (σ_{ij}) are indicated inside open circles also at the vertices of an equilateral triangle. Note the use of scripts $\mathcal{E}$ and $\mathcal{P}$ to distinguish these symbols from those of energy and pressure. The symbols in the columns of Table 16.3 are tensors, but the relevant subscripts are not indicated, for clarity. The physical meanings of these symbols are given in books on crystal physics (see References), with sometimes different notations. Of course other physical parameters could be included in the energy expression of Eq. (16.19), for example variables pertaining to magnetic fields, but the thermodynamic treatment would be formally exactly the same as that for the electric field. One could also consider mole numbers N_i or concentrations (mole fractions) c_i as additional variables. Symbols are defined in Tables 16.3 and 16.4.

It is often preferable to perform a Legendre transformation in order to express the energy in terms of independent intensive instead of extensive

Table 16.3: Physical meaning of symbols used in Fig. 16.2.

Symbol	Meaning	Symbol	Meaning
T	Temperature	S	Entropy
$\mathcal{E}$	Electric field	$\mathcal{P}$	Polarisation
σ	Elastic stress	ϵ	Elastic strain
$\frac{C_X}{T}$	Heat capacity	$\frac{T}{C_X}$	Inv. heat capacity
χ	Susceptibility	p	Inv. permittivity
s	Elastic compliance	c	Elastic stiffness
α	Thermal expansion	t	Electrothermal eff.
β	Pyroelectric eff.	f	Heat of deformation
d	Piezoelectric eff.	e	Electromechanical eff.

variables (which are experimentally awkward to use), thus

$$\Psi = \Psi(T, \mathcal{E}_k, \sigma_{ij}), \tag{16.20}$$

where the symbol Ψ designates free energy obtained in a manner similar to that leading to the Gibbs free energy G. The corresponding differential form is

$$d\Psi = -S\,dT - \mathcal{P}_k\,d\mathcal{E}_k - \epsilon_{ij}\,d\sigma_{ij}, \tag{16.21}$$

an exact expression, with implied summation over the indices. As usual, the coefficients are given by partial derivatives such as those included in Table 16.4 where the variables held constant are indicated as superscripts on the symbols for partial derivatives. Concentration is held constant implicitly. In Table 16.4, the rule for attributing subscripts to the tensors of the right hand side of the identity signs is a simple one: the indices which appear are all those which are found on the left hand side of the identity, *i.e.* as subscripts in the partial derivatives. The order is immaterial, but must be observed consistently.

The differential expressions — or differential "work" terms of the type $y\,dX$ in the fundamental form (16.21) — correspond symbolically to the diagonals of the hexagon of Fig. 16.2 with one (tensor) extensive variable at one end and the corresponding *conjugate* intensive variable at the other end. Other links between extensive-intensive variables correspond to the external edges of the hexagon, such as the one labeled β linking the (non-conjugate) variables: temperature T and polarization $\mathcal{P}$, for example. Together these (3+6) nine links symbolize the terms which express the differentials of 3 extensive tensors situated at the nodes of Fig. 16.2 in terms of the 3 intensive ones.[7]

---

[7]The (main) diagonal (conjugate) elements of table 16.4 also correspond to a diagonal

Table 16.4: Partial derivatives of extensive variables as a function of intensive ones commonly used in crystal physics. Variables held constant are indicated as superscripts only on physical parameters themselves.

$\dfrac{\partial S}{\partial T} \equiv C^{\mathcal{E},\sigma}$	$\dfrac{\partial S}{\partial \mathcal{E}_k} \equiv p_k^{T,\sigma}$	$\dfrac{\partial S}{\partial \sigma_{ij}} \equiv \alpha_{ij}^{T,\mathcal{E}}$
Specific heat	Electrocaloric	Piezocaloric
$\dfrac{\partial \mathcal{P}_l}{\partial T} \equiv p_l^{\mathcal{E},\sigma}$	$\dfrac{\partial \mathcal{P}_l}{\partial \mathcal{E}_k} \equiv \chi_{lk}^{T,\sigma}$	$\dfrac{\partial \mathcal{P}_l}{\partial \sigma_{ij}} \equiv d_{lij}^{T,\mathcal{E}}$
Pyroelectric	Susceptibility	Piezoelectric
$\dfrac{\partial \epsilon_{ij}}{\partial T} \equiv \alpha_{ij}^{\mathcal{E},\sigma}$	$\dfrac{\partial \epsilon_{ij}}{\partial \mathcal{E}_k} \equiv d_{ijk}^{T,\sigma}$	$\dfrac{\partial \epsilon_{ij}}{\partial \sigma_{kl}} \equiv s_{ijkl}^{T,\mathcal{E}}$
Thermal expansivity	Piezoelectric	Elastic compliance.

The terms appearing in these two tables were chosen historically to indicate the variables or processes involved; for example the word "pyro-electric" relates to the effect of temperature on electric polarization, from the Greek word *pyro*, meaning "fire" or, by extension "heat". Similarly, the word "piezoelectric" refers to the effect of pressure (from the Greek *piezo*, to "press") on electric polarization. By "electric polarization" is meant the existence of electric dipole in the material such that the center of positive charge in the crystal does not coincide with the center of negative charge, and can be measured by a vector, or first-rank tensor, $\mathcal{P}_j$. The polarization vector itself is related to the dielectric displacement vector, but we shall not be concerned with that complication, particularly since these two vectors are collinear in all but two symmetry systems, the two of lowest symmetry. In other systems, the polarization vector, or axis, is generally taken to be a rotation axis of high order. Further descriptions of the physical meaning of the parameters indicated in the tables are given in the textbook by Nye (1957).

---

of Fig. 16.2. The distinction between "X-tensive" and "Y-tensive" conjugate variables, which we have stressed all along, shows up here again.

Application to Ferroelectricity

Ferroelectricity[8] is to ferromagnetism what electric dipoles are to magnetic dipoles, which means that from a strictly thermodynamic sense, they can both be treated within the same formal framework. Here we select the electrical case since it was featured in Fig. 16.2. Ferroelectricity (or ferro-magnetism) is also to compositional clustering what dipoles are to types of atoms (or ions), A and B. Likewise, anti-ferroelectricity (or magnetism) is analogous to ordering in the binary compositional case, meaning a ten-dency for individual elements to surround themselves with unlike neighbors in various ordered arrangements, so that again the same thermodynamic formalism, as treated in this chapter, may be applied.

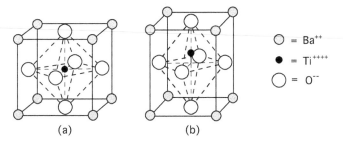

(a)   (b)

Figure 16.3: Unit cell of a piezoelectric crystal such as $BaTiO_3$: (a) unstressed, (b) stressed. In configuration (a), the Ti ions *on average* occupy the central position of the cell, in configuration (b), the Ti ions are in positions displaced along one of the $\langle 100 \rangle$ cube directions.

Many elements and compounds are ferromagnets or ferroelectrics such as the familiar oxide barium titanate ($BaTiO_3$) whose unit cells are de-picted in Fig. 16.3. At low temperature — below the so-called Curie point at which the substance transforms from cubic to tetragonal symmetry — the Ti ions occupy positions along a 4-fold axis in the z-direction (say), causing a tetragonal distortion of the unit cell and of the oxygen ion oc-tahedra which surround it [Fig. 16.3 (b)]. These ionic displacements cause electrical dipoles to form along the 4-fold axis. Above T_c, all interstitial positions along the three 4-fold axes are equivalent, so that the Ti ions occupy, *on average* the central positions, as shown in Fig. 16.3 (a). We shall be concerned here with the ferroelectric *transition*, which implies that we are investigating not only a problem in crystallography, but a thermo-dynamic one, in particular one in which a cooperative phenomenon occurs

---

[8]The prefix "ferro" comes from the latin *ferrum* since iron (whose chemical symbol is of course Fe) was the metallic element in which magnetism was first observed.

such that dipoles spontaneously align themselves parallel to their neighbors or, in the case of anti-ferroelectricity, in antiparallel (ordered) fashion.

Landau, Lifshitz, Ginzburg Formulation

The treatment of the ferroelectric transition that follows is that based on the Landau theory (1937), supplemented as we have seen by Lifshitz (1958), then applied more specifically to ferroelectrics by Ginzburg (1945, 1949), sometimes referred to as the LLG theory. Applications more specifically geared to ferroelectricity were treated by Devonshire (1949), shortly after the publication of Ginzburg's work (1945–6). Quantitatively the LLG treatment is not quite satisfactory theoretically, but its simplicity is very appealing in a qualitative manner, and we have already covered its basic aspects, so that we can immediately apply it to the ferroelectric case, as well as to the general case of ordering transitions in the binary alloys, or to ferromagnetism, or to the "antiferro" equivalents, or indeed to the somewhat different phenomena of atomic displacements which we may denote "ferro-(or antiferro)-displacive transitions" (also known as soft-mode transitions). Recalling Chapter 15, we may once again insist on the unity of the concentration wave approach coupled to the Landau formalism: one need not treat *clustering* (ferro-) and *ordering* (antiferro-) differently since they are variations on the same theme. Simply put, various ordering instabilities correspond to different special-point wave vectors, with clustering instabilities corresponding to the $\langle 000 \rangle$ wave vector, *i.e.* an infinite wave length. Ordering is more diverse because there are more ordering waves available than the sole clustering wave $\langle 000 \rangle$.

In all of these applications, the basic formalism is always the same, and is also the one sketched previously in Chapter 3 of this text. In it we saw that the starting point is the fundamental differential form for the energy in terms of extensive variables X

$$dU = T\,dS + Y\,dX\,, \tag{16.22}$$

the last term of which indicates symbolically a sum of differentials of extensive variables (all the ones deemed to be necessary for the problem at hand) times the corresponding intensive variable y. Since we know, from the basic laws of thermodynamics (particularly the first law, which actually creates the energy function) that the differential form on the right side of Eq. (16.22) is an exact differential, dU, integration of the energy differential is always possible, without having to specify the relationship between entropy S and the other extensive variables.

The integration can then proceed by use of Eq. (3.44), or some version thereof, so it is quite obvious that expressions of the variables T and the $Y's$ as a function of the extensive variables: $Y_i = Y_i(S, X_1, X_2, \ldots X_n)$, $i = 0, 1, 2, \ldots$ are required. As was stated in Sec. 3.6, one way of proceeding is to write those relations as Taylor's expansions in the X variables, often stopping at the first-order term if it is ascertained that the linear approximation suffices. In what follows, we shall see that the linear approximation often is not sufficient; in fact the expansion must generally go to fourth- or sixth-order terms in what will be recognized as Landau expansions in one or more extensive variables, which will also be recognized as order parameters ξ.

Much of what has been said here can also apply to thermodynamic potentials obtained through Legendre transformations. So, it is quite permissible to exchange the roles of intensive and extensive variables to create new thermodynamic potentials which have the desired independent variables, the way the Gibbs free energy is formed from the internal energy for example. The procedure described is also the one illustrated in Fig. 6.2 with successive derivatives of the Gibbs energy as a function of T and of P used, by differentiation, to produce entropy and volume and then, again by differentiation, to derive the measurable quantities: heat capacity and compressibility. In an experimental procedure, one follows the opposite method, *i.e* one of *integration* of measured quantities, such as the ones found in the boxes of Table 16.4. There is another method, a more recent one, which is the reverse of the one just mentioned, that of *derivation*, consisting of starting from a computationally constructed thermodynamic potential and taking mathematical derivatives. Here the difficulty lies in deriving thermodynamic potentials practically from first principles.

So much for similarities; now let us go to our example of ferroelectricity as an illustration. It should be made clear for the case of solids, crystals in particular, is far more complicated than that of fluids, even if the basic thermodynamics is similar. The main difference is that the so-called constitutive relations, $Y(X)$ are much more complex: scalars become tensors and thermodynamic potentials depend much more critically on the shape of materials investigated. That means that thermodynamic variables are strongly position-dependent (non-uniform), so that methods such as coarse-graining must be applied, and that position-dependence must be obtained by solving the partial differential equations of boundary valued problems. But since the present book is not one in applied mathematics, we shall retain only the barest of uniform thermodynamics for illustration.

The Ferroelectric Transition

To simplify further, let us leave out the variables stresses and strains, leaving as thermodynamic "potential" the function $\Phi(T, \mathcal{P})$. After having performed the required Legendre transformations we find, from Eq. (16.22), the fundamental total differential for the problem to be:

$$d\Phi = -S \, dT + \mathcal{E}_k \, d\mathcal{P}_k. \quad \text{(summation implied)} \tag{16.23}$$

One-Dimensional, Second-Order Case

In this first application, we consider only a one-dimensional transition, meaning that the high-temperature phase, the more symmetric, disordered one, is biased in such a way that only one crystallographic direction is allowed for aligning the electric dipole (pyroelectric) vectors. In principle, this can be achieved by, for instance, squeezing laterally the structure shown in Fig. 16.3(a) to the (b) configuration, or by considering a crystal structure which is, say, tetragonal rather than cubic, with pyroelectric vector along the tetragonal axis z.

We assume that above a transition temperature T_c, known as the Curie temperature, the vector in question lies along the positive or negative direction in equal proportions, so that the net dielectric (average) dipole moment has value zero. Below the transition (Curie) temperature — which is also, in a second-order transition, the instability T_o — the +/- symmetry is broken and one finds a net dielectric dipole pointing either "up" or "down". This up/down degeneracy causes some regions of crystal space to have a majority of up dipoles, and other regions to have a majority of down dipoles. These regions are called ferroelectric *domains*, separated from one another by domain walls, a situation which parallels that of antiphase domains in the order-disorder case illustrated schematically in Fig. 15.1. In this section we do not take domains into account so that the math refers really to an infinitely large domain, an approximation which, for thermodynamic purposes, is generally adequate unless the domains are really small.

Having made all these simplifying assumptions, we see that the problem maps neatly into the Ising model which was treated earlier with the polarization $\mathcal{P}_z$ in the z direction playing the role of a Landau order parameter ξ. All of the math introduced in that section can thus be used without modification, with the free energy curves of Fig. 15.3 representing those expected for the one-dimensional second-order ferroelectric case, such as that seen in Fig. 16.4(a), traced for a particular temperature τ, and having constant- T Landau expansion in the z direction

$$\Delta\Phi = A \, \mathcal{P}_z^2 + C \, \mathcal{P}_z^4 \tag{16.24}$$

with A and C being in general (weakly) pressure and temperature dependent, and with $\mathcal{P}_z$ as the order parameter, being zero above T_o and non-zero

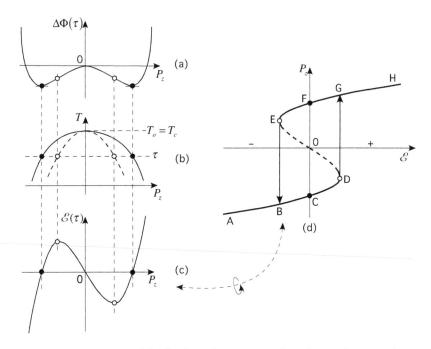

Figure 16.4: Free energy (a) of a ferroelectric crystal in the unidirectional, symmetric, second-order case; (b) phase diagram T versus inverse susceptibility P_x; (c) derivative of field $\mathcal{E}$ with respect to P_x (analogous to chemical potential); (d) same as (c) but with axes inverted and rotated.

below. Recall also that the coefficients A and C are calculated at $\mathcal{P}_z = 0$, so that Eq. (16.24) is actually a truncated Taylor's expansion, whereas in the differential form (16.23), which *appears* to be "linear", the coefficients in question are still a function of the pyrolytic dipole moments. As was shown before, the instability temperature, which is also the critical temperature, is reached when the quadratic coefficient $A = a(T - T_o)$ becomes negative, *i.e.* when $T < T_o$. A is then a (one-dimensional) inverse dielectric susceptibility

$$A = \chi^{-1}. \tag{16.25}$$

Let us however reiterate another word of caution: The original Ising model considered elementary dipoles, or *spins* $\sigma = +1$ or -1, and the solution in one or two dimensions did not appeal to the mean field approximation, in other words the spins kept their discrete nature, hence the extreme difficulty of finding a mathematical solution to the problem. Here however, we implicitly make the coarse graining approximation, so that the dipoles in question, dielectric or magnetic are averaged quantities, the average being taken over a region including a few unit cells in the clustering, of ferro-case, or a sublattice average in the ordering, or antiferro-case. Hence, in the following, the dipole variable $\mathcal{P}$ will be a continuous variable, implying a mean field treatment of the thermodynamic functions.

Figure 16.4 also shows, in panel (b), the locus of the equilibrium points T as a function of $\mathcal{P}_z$ (filled circles) and the associated spinodal curve, shown as a dashed curve, locus of the inflection points of the free energy curve (open circles) in the (a) panel.[9] The third panel (c) is a plot, at temperature τ of the electric field

$$\mathcal{E}_z = \left(\frac{\partial \Delta \Phi}{\partial \mathcal{P}_z}\right)_T, \tag{16.26}$$

also as a function of the polarization. The resulting curve, turned over on its side, is plotted in panel (d) and is seen to resemble the typical one of polarization exhibiting hysteresis. This conclusion is merely suggestive, it is not meant to produce an actual hysteresis curve since that relies on non-equilibrium thermodynamics which cannot be handled in classical thermodynamics. But in actual practice the first-order reaction consisting of flipping spins (or electric dipoles) from "up" to "down" can be rather sluggish, forcing the system into non-equilibrium states. Indeed, the curve A to H in panel (d) contains a non-equilibrium portion D to E, shown dotted, whose shape is history-dependent (time-dependent), and the width of the resulting hysteresis loop depends on how the domain walls are pinned by imperfections, hindering their motion. Still, qualitatively the progression of electric polarization (or of magnetization) will be guided by the mechanism described here: at high negative electric field A the polarization is also high and negative, then decreases in magnitude B, reaches equilibrium at C, then, instead of jumping to point F as it should do in an equilibrium first-order transition, continues to decrease smoothly until reaching the "spinodal" (unstable) point D, at which it jumps to point G, then finally continues on an equilibrium curve to H and beyond. If the field is reversed the trajectory will be H-F-E-B-A. So, equilibrium (classical) thermodynamics can indeed give us at least a framework for understanding *real* non-equilibrium material behavior.

One-Dimensional, First-Order case

In this case, we must use the (even) polynomial expansion (15.11a), out to sixth power in the order parameter $\mathcal{P}_z$

$$\Delta \Phi = A\,\mathcal{P}_z^2 + C\,\mathcal{P}_z^4 + D\,\mathcal{P}_z^6 \tag{16.27}$$

---

[9]Note similarity with Fig. 9.1.

again with $A = a\,\Delta T$ and $\Delta T = (T - T_o)$. Of course T_o is no longer the transition temperature: with $C < 0$ (and $D > 0$) there is a first-order transition at $T_t > T_o$. Whereas, in the second-order case, the order parameter varies with temperature according to the curve of Fig. 15.6(a), in the symmetric first-order case the order parameter varies according to the curve of Fig. 15.6(c). In both cases the curves are symmetric about the temperature axis since, given the crystallography of the problem, in the absence of an applied field, positive and negative dipoles are equally probable.

Cubic Crystals, Second-Order Case

With three-dimensional order parameters, the coefficients A and C of Eq. (16.24) become respectively second- and fourth-rank tensors, which complicates matters quite a bit. The Landau expansion now looks like this:[10]

$$\Delta\Phi = A_{ij}\,\mathcal{P}_i\mathcal{P}_j + C_{ijkl}\,\mathcal{P}_i\mathcal{P}_j\mathcal{P}_k\mathcal{P}_l\,, \qquad (16.28)$$

with summation convention implied. In the case of second-order transitions, it is not necessary to extend the expansion beyond the fourth power, as we have seen. Fortunately, crystal symmetry comes to the aid: in cubic symmetry a second-rank tensor reduces to just one diagonal component and the fourth-rank tensor C_{ijkl} reduces to just 3 independent components in the Voigt matrix notation, i.e it behaves like the elastic stiffness tensor with elements C_{11}, C_{12}, C_{44}. Actually, in the present case, an additional symmetry reduces the number of components to just two, a "diagonal" one, C_1, proportional to C_{11} (in present notation) and an off-diagonal one C_2, proportional to $(C_{12} + 2C_{44})$ so that the required expansion can also be written

$$\Delta\Phi = a\,\Delta T\,(\mathcal{P}_x^2{+}\mathcal{P}_y^2{+}\mathcal{P}_z^2){+}C_1(\mathcal{P}_x^4{+}\mathcal{P}_y^4{+}\mathcal{P}_z^4){+}2C_2(\mathcal{P}_x^2\mathcal{P}_y^2{+}\mathcal{P}_y^2\mathcal{P}_z^2{+}\mathcal{P}_z^2\mathcal{P}_x^2)\,. \qquad (16.29)$$

According to arguments given in Chapter 14, stability requires that the quadratic term of Eq. (16.29) be positive definite, which reduces here to $T > T_o$, as expected, and that the quartic term be positive definite. To make that condition explicit, it is useful to rewrite that term as a product of three matrices (here C_1 and C_2 are *not* elastic stiffnesses):

$$C_{ijkl}\,\mathcal{P}_i\mathcal{P}_j\mathcal{P}_k\mathcal{P}_l = \begin{pmatrix} \mathcal{P}_x^2 & \mathcal{P}_y^2 & \mathcal{P}_z^2 \end{pmatrix} \begin{pmatrix} C_1 & C_2 & C_2 \\ C_2 & C_1 & C_2 \\ C_2 & C_2 & C_1 \end{pmatrix} \begin{pmatrix} \mathcal{P}_x^2 \\ \mathcal{P}_y^2 \\ \mathcal{P}_z^2 \end{pmatrix}. \qquad (16.30)$$

---

[10]The following example is from Landau and Lifshitz (1995).

Through the rule of minors applied to the C-matrix, starting at the upper left hand corner, we find the conditions

1. $C_1 > 0$

2. $(C_1 + C_2)(C_1 - C_2) > 0$

3. $(C1 - C_2)^2(C_1 + C_2) > 0$

which reduce to $C_1 > 0$ and $C_1 > |C_2|$ for $C_2 > 0$, the usual case.

It is also possible to determine what are the most unstable directions of polarization in crystal space by examining the quartic term (the quadratic term is direction-independent). Let the directional cosines of the pyrolytic vector **n** be n_1, n_2, n_3. Then dividing Eq. (16.29) through by the magnitude squared of the vector Φ we get

$$\text{normalized quartic term} = C_1(n_x^4 + n_y^4 + n_z^4) + 2C_2\, N^2(\mathbf{n})$$
$$\text{where} \quad N^2(\mathbf{n}) = (n_x^2 n_y^2 + n_y^2 n_z^2 + n_z^2 n_x^2) \quad \text{or}$$
$$\text{normalized quartic term} = C_1 + 2(C_2 - C_1)\, N^2(\mathbf{n}). \tag{16.31}$$

The objective is to find the optimal orientation of the unit vector **n** for a given choice of the difference $(C_2 - C_1)$. The optimal directions are thus

- For $C_2 > C_1$, minimize expression N, hence choose $\mathbf{n} = \langle 100 \rangle$, yielding a tetragonal distortion,

- For $C_2 < C_1$, maximize expression N, hence choose $\mathbf{n} = \frac{1}{\sqrt{3}}\langle 111 \rangle$, yielding a trigonal distortion.

This instructive example, given by Landau and Lifshitz (1995), demonstrates that classical thermodynamics, in coarse-graining mode, combined with elements of crystallography can tell us a great deal about the workings of electrodynamics of crystals, for instance, leading to useful applications. Other applications concern piezoelectricity, for example, describing the influence of stress on polarization. Applications abound: *all* modern watches and clocks use piezoelectric crystals for regulation of time intervals, and a novel application to energy production is to be tested: cars running on highways generate pressure; pressure on piezoelectric crystals embedded in the road surface can, in principle, generate electric current. It remains to be seen if the method will turn out to be industrially useful.

References

Brillouin, L. (1946). *Les Tenseurs en Mécanique et en Elasticité*, Dover.

de Fontaine, D. (1972), *J. Phys. Chem. Solids* **33**, 297.

de Fontaine, D. (1979), *Configurational Thermodynamics of Solid Solutions*, in *Solid State Physics*, Ehrenreich, H., Seitz, F. and Turnbull, D., **34**, p. 73-274, Academic Press.

de Fontaine, D.(1994), *Cluster Approach to Order-Disorder Transformations in Alloys*, in *Solid State Physics*, Ehrenreich, H., Seitz, F. and Turnbull, D., **47**, pp. 33-176, Academic Press.

Devonshire, A. F. (1949), *Philos. Mag.* **40**, 1040; (1951) *Philos. Mag.* **42**, 1065; (1954) *Philos. Mag. Suppl.* **3**, 85-139.

Ginzburg, V. L. (1945) *Zh. Eksp. Teor. Fiz.* **15**, 739 [(1946) *J. Phys. USSR* **10**, 107 (translation)]; (1949) *Zh. Eksp. Teor. Fiz.* **19**, 36.

Goldstein, H. (1950). *Classical Mechanics* Addison-Wesley Publishing Co., Reading, MA.

Landau, L. D. (1937). *Zh. Exsp. Fiz.* **7**, 627; (1949) *J. Phys. USSR* **10**, 107.

Landau, L. D. and Lifshitz, E. M. (1958), *Statistical Physics*, Addison-Wesley, Reading, MA. Also: Landau, L. D., Lifshitz, I. M. and Pitaevskii, L. P.(1980). *Statistical Physics*, Third Edition, Pergamon Press (English versions).

Landau, L. D. and Lifshitz, I. M. (1995), *Electrodynamics of Continuous Media*, Oxford, Butterworth–Heinemann, Boston, MA.

Newnham, R. E. (2005), *Properties of Materials*, Oxford University Press, UK.

Nye, J. F. (1957), *Physical Properties of Crystals*, Oxford, Clarendon Press.

Chapter 17

Nucleation and Growth

Nucleation is one of the most important topics in the field of phase transformations, but has often been misused.[1] A theory of nucleation is required to describe the appearance, in the parent phase (α), of an embryo of the "new" phase (β) below a (first-order) $\alpha \rightarrow \beta$ phase transition. It will be shown that the $\alpha \rightarrow \beta$ transformation can proceed at a finite rate only if the transforming system is maintained below the equilibrium transition temperature, *i.e.* if the system is sufficiently *undercooled*.

Nucleation theory is therefore a *non-equilibrium* one which, however, we shall attempt to handle by techniques borrowed from classical equilibrium thermodynamics. In this way we shall be able to derive expressions for the *work of nucleation*. The way of doing this was given by Gibbs, over a hundred years ago, but his rigorous treatment has not always been handled satisfactorily by later continuators. Here, we shall first appeal to statistical mechanics in order to derive an expression for the rate of nucleation.

17.1 Statistics of Embryo Formation

For simplicity, consider fluid systems (gas, liquid) whose properties are isotropic and uniform. Suppose its high-temperature phase, α, to be unstable below T_t to the formation of a new phase, call it β. Suppose further

---

[1]The following treatment is based partially on a graduate course given by the late John Hilliard at Northwestern University in the 1960's. It follows joint work with John Cahn (1958), following, of course, the classic publications by Gibbs (1875) and Landau and Lifshitz (1958). Only the classical theory is covered in the present work. Actually, a more general theory, covering a wider range of phase changes should be used, as will be shown briefly in the next chapter. For an earlier attempt at generalization, see de Fontaine (1081).

that a small quantity of the new phase, roughly spherical in shape, will constitute an *embryo* of the new phase β [the Gibbs theory does not require the embryo to be strictly spherical]. When the embryo is in (unstable) equilibrium with α it will be called a *nucleus*, and the nucleus associated with the minimum work of formation will be called *critical nucleus*.

For the sake of the derivation, we shall set up a somewhat idealized system, which, however, entails no loss in generality. The system proper and its surroundings will be enclosed by rigid, impermeable adiabatic walls. The resulting isolated *supersystem* was already depicted schematically in Fig. 4.1. The system proper is open and contains a *fluctuation* (region of different density, composition, structure...) here in the form of a β embryo, indicated by a shaded circle. The complement of the system proper is assumed to be very large in extent and uniform in its properties; that portion of the supersystem can thus be regarded as a *reservoir*, the intensive variables of which ($\overline{T}, \overline{P}, \overline{\mu}_1, \overline{\mu}_2$, in a "binary" example) remain sensibly constant for any change occurring in the system proper.

According to macroscopic (classical) thermodynamics, once the supersystem has reached equilibrium, no further change can be brought about since, by construction, it is *isolated*. On a microscopic scale, however, *fluctuations* are perpetually taking place. Classical thermodynamics does not neglect these fluctuations, merely, their properties cannot be measured at equilibrium since these fluctuations are not sufficiently long lived. We may still handle fluctuations theoretically by the following artifice however: once a fluctuation of the desired type has formed, quite by chance, let us maintain it in existence by imposing on the system a set of appropriate microscopic constraints. Such internal constraints cannot be applied in practice; nevertheless, we can, in principle, enumerate the number W^* of individual microstates available to the supersystem in the particular fluctuating state envisaged (*i.e.* containing a localized embryo), compatible with given fixed values of internal energy U_s, volume V_s, masses N_s, ... Likewise, let W_s be the total number of microstates available to the system proper, under the same macroscopic conditions, but for *any* fluctuation state. We have $W_s \gg W^*$, since there are relatively so few embryo-containing microstates compared to the number of all possible states.

The probability of finding the supersystem in an embryo-containing state is given by the ratio of the number of embryonic microstates to the total number of states

$$\mathcal{P} = \frac{W^*}{W_s}. \tag{17.1}$$

From the Boltzmann (8.13), valid for the present case of an isolated system, we obtain the entropy change associated with the formation of an embryo:

$$\Delta S_s = S^* - S_s = k_B \ln W^* - k_B \ln W_s = k_B \ln \frac{W^*}{W_s} = k_B \ln \mathcal{P}.$$

Hence, we have

$$\mathcal{P} = e^{\frac{\Delta S_s}{k_B}}. \tag{17.2}$$

Thus, we have obtained an expression for the probability of finding a system containing a specified embryo fluctuation as a function of a total entropy change, of which its entropy change is in fact negative: $\Delta S_s < 0$. It is now necessary to express ΔS_s in terms of measurable thermodynamic quantities.

17.2 Creating an Embryo

Let the extensive variables of the system proper be denoted by U, S, V, N (N written for the set of N_i) and those of the "reservoir" by the same symbols barred ($\overline{U}$, etc.). Do likewise for the intensive variables T, P, μ (μ written for the set μ_i). Since the supersystem is isolated, we have the conservation relations, valid for any change Δ in the system,

$$\Delta U_s = \Delta U + \Delta \overline{U} = 0, \tag{17.3a}$$
$$\Delta V_s = \Delta V + \Delta \overline{V} = 0, \tag{17.3b}$$
$$\Delta N_s = \Delta N + \Delta \overline{N} = 0, \tag{17.3c}$$
$$\Delta S_s = \Delta S + \Delta \overline{S} \geq 0. \tag{17.3d}$$

For the (infinite) reservoir, the internal energy differential

$$d\overline{U} = \overline{T}d\overline{S} - \overline{P}d\overline{V} + \overline{\mu}d\overline{N} \tag{17.4}$$

(μdN written for $\sum_i \mu_i dN_i$) may be integrated directly since the intensive variables are regarded as constant (equivalently, Euler integration may be used on Eq. (17.4)) yielding

$$\Delta \overline{U} = \overline{T}\Delta\overline{S} - \overline{P}\Delta\overline{V} + \overline{\mu}\Delta\overline{N} = 0. \tag{17.5}$$

By Eqs. (17.3) and (17.5) we have

$$\Delta \overline{U}_s = \Delta U + \overline{T}(\Delta S_s - \Delta S) + \overline{P}\Delta V - \overline{\mu}\Delta N = 0$$

or

$$-\overline{T}\Delta S_s = \Delta U - \overline{T}\Delta S + \overline{P}\Delta V - \overline{\mu}\Delta N. \tag{17.6}$$

This hybrid expression, containing the extensive variables of the system and the intensive variables of the reservoir, is inconvenient to use. Let us then specify particular boundary conditions on the system-reservoir complex by choosing for their common boundary a permeable, diathermic but rigid wall. Hence, we have

$$\Delta V = 0, \quad T = \bar{T} = \text{constant}, \quad \mu = \bar{\mu} = \text{constant}.$$

Then Eq. (17.6) becomes

$$-T\Delta S_s = \Delta U - T\Delta S - \mu\Delta N = \Delta\Omega \tag{17.7}$$

where we have defined

$$\Omega = U - TS - \mu N, \tag{17.8}$$

which is the omega (grand) potential already defined by Eq. (4.25). Finally, the probability (17.2) is given by the Boltzmann-type expression

$$\mathcal{P} = e^{\frac{-\Delta\Omega}{k_B T}} \tag{17.9}$$

where Ω is given, by Eq. (17.8), in terms of measurable quantities, as required.

17.3 Work of Formation of a Nucleus

For the specific case of nucleus formation, $\Delta\Omega$ can be given a simple form. For that purpose, consider the "embryo" just the way we did for the Gibbs treatment of interfaces (see Sec. 13.1): enclose the embryo β by an infinitely thin, sharp interface and let the properties of phase α (the phase outside the embryo) be uniform right up to the constructed interface of total area A. Now use the Gibbs interface stratagem idea, expressed by Eq. (13.6) to define effective *interfacial* quantities u^I, etc. Also define *densities* by $u^\alpha = \frac{U^\alpha}{V^\alpha}$, $\rho^\alpha = \frac{N^\alpha}{V^\alpha}$, etc. Then the difference between the Ω function after and before nucleation is

$$\Delta\Omega = V^\alpha(u^\alpha - Ts^\alpha - \mu\rho^\alpha) + V^\beta(u^\beta - Ts^\beta - \mu\rho^\beta)$$
$$+ A(u^I - Ts^I - \mu\gamma) - (V^\alpha + V^\beta)(u^\alpha - Ts^\alpha - \mu\rho^\alpha), \tag{17.10}$$

expressing the following contributions: the omega potential for the material outside the embryo, plus that inside the embryo, plus the omega potential of the α/β interface minus the potential for the material α, before the appearance of the embryo β, occupying the total volume $V_\alpha + V_\beta$. For the bulk phases we have for the omega (or PV) potential, from Eq. (4.25), after dividing through by the volume V:

$$u - Ts - \mu\rho = -P \tag{17.11}$$

and for the interface, by Eq. (13.18), with appropriately modified notation,

$$u^I - Ts^I - \mu\gamma = \sigma. \tag{17.12}$$

Hence Eq. (17.10) becomes

$$\Delta\Omega = A\sigma - V^\beta \Delta P \tag{17.13}$$

where

$$\Delta P = P^\beta - P^\alpha. \tag{17.14}$$

For a spherical embryo region (β) of radius R we must have:

$$V^\beta = \frac{4}{3}\pi R^3, \qquad A = 4\pi R^2. \tag{17.15}$$

Inserting these (geometrical) values into Eq. (17.13) yields an expression for the omega potential in terms of the nucleus radius R:

$$\Delta\Omega = 4\pi R^2 \sigma - \frac{4}{3}\pi R^3 \Delta P. \tag{17.16}$$

Since the *critical nucleus* must be in (unstable) equilibrium with the parent phase α we may set the derivative of expression (17.16) with respect to R to zero (as was done for Eq. (13.12)) to obtain the compact result for the *critical* radius R^*

$$R^* = \frac{2\sigma}{\Delta P}, \tag{17.17}$$

known as the Gibbs–Thomson equation (spherical special case). Inserting that expression into Eq. (17.13) then yields the important formula for a spherical critical nucleus

$$\Delta\Omega^* = \frac{16}{3}\pi \frac{\sigma^3}{\Delta P^2}. \tag{17.18}$$

This is the Gibbs critical nucleus result. It is an *exact* result, given the starting assumptions.

Let us now see how to replace the pressure difference ΔP by free energy differences, in an *approximate* manner. In general, consider a bulk phase (β) isothermally compressed from pressure P^α to P^β. From the differential of the molar Gibbs energy

$$dg = -sdT + vdP, \tag{17.19}$$

as previously seen at Fig. 6.1 we have, by integration and at constant T,

$$g(P^\beta) - g(P^\alpha) = \int_{p^\alpha}^{p^\beta} v \, dp. \tag{17.20}$$

If the substance (β) is practically incompressible, then Eq. (17.20) becomes

$$g(P^\beta) - g(P^\alpha) \cong v\Delta P. \tag{17.21}$$

The molar Gibbs energy difference must now be evaluated for various cases of interest.

17.4 Single-component System

Denote by Δg the Gibbs energy difference

$$\Delta g = g^\beta(P^\alpha) - g^\alpha(P^\alpha) \tag{17.22}$$

or

$$\Delta g = \mu^\beta(P^\alpha) - \mu^\alpha(P^\alpha) \tag{17.23}$$

between chemical potentials of α and β phases at the ambient pressure P^α. The critical nucleus must be in equilibrium with the parent α so that

$$\mu^\alpha(P^\alpha) = \mu^\beta(P^\beta). \tag{17.24}$$

Hence

$$\Delta g = \mu^\beta(P^\alpha) - \mu^\beta(P^\beta) \tag{17.25}$$

Expression (17.25) is just the negative of the difference in Eq. (17.21) so that

$$\Delta g \cong -v\Delta G_v \tag{17.26}$$

where, by definition, ΔG_v is the difference between Gibbs energies *per unit volume* between the bulk β and bulk α phases, both *at ambient pressure* P^α. This quantity ΔG_v, in turn, can be obtained from a (molar) Gibbs free energy plot at constant P, as shown in Fig. 17.1, where $\Delta g(P^\alpha)$ is the difference between the two g-curves at pressure P^α, at some temperature below the equilibrium transition temperature T_t (compare to Fig. 6.1). In the vicinity of the latter point, the two curves can be regarded almost as straight lines, signifying that the enthalpy and entropy changes, Δh and Δs, at the first-order transition T_t, vary little with temperature. Hence, with this new approximation, we have at the transition

$$\Delta g = \Delta h - T_t \Delta s = 0 \tag{17.27}$$

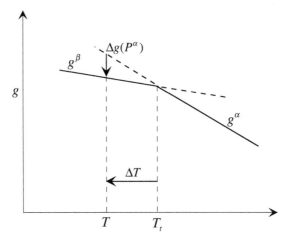

Figure 17.1: Intersection of two free energy curves at the first-order transition temperature T_t. At undercooling ΔT, the reduction of free energy is $\Delta g(P^\alpha)$.

hence

$$\Delta s = \frac{\Delta h}{T_t}, \tag{17.28}$$

so that, close to T_t,

$$\Delta g = \Delta h(1 - \frac{T}{T_e}) = \frac{\Delta h}{T_e}\Delta T \tag{17.29}$$

where ΔT is the *undercooling* $T_t - T$. Nucleation cannot occur unless the enthalpy change is negative.

The critical radius, R^*, can be expressed as

$$R^* = -\frac{2\sigma}{\Delta G_v} \tag{17.30}$$

showing that $R^* \to \infty$ as the transition temperature T_t is approached (from below).

17.5 Multicomponent System

The introduction of concentration variables complicates things a bit: the critical nucleus (associated here with the symbol (*)), despite being under pressure P^* higher than that of the surrounding (denoted here as $\overline{P}$) is also at different concentration c^*, and may be of a different type of phase, α, let us say. For a critical nucleus in (unstable) equilibrium with the surrounding multicomponent solution, the condition replacing Eq. (17.24) is

$$\bar{\mu}_i(\overline{P}, \bar{c}) = \mu_i^*(P^*, c^*) \quad (\text{all } i = 1, 2, \ldots n),$$
(17.31)

the chemical potentials being respectively evaluated at the composition $\bar{c}$ of the parent phase α and at the composition c^* of the nucleating phase β, all at ambient pressure $\overline{P}$. Note: here the symbol c without the subscript i stands for all concentrations c_i, $i = 1, 2, \ldots n$.

The counterpart of Eq. (17.22) for the Gibbs energy difference δg due to an increase in pressure of the nucleating phase from $\overline{P}$ to P^* is then

$$\Delta g = g^*(P^*, c^*) - g^*(\overline{P}, c^*),$$
(17.32)

where the single symbol "c" has been written for all concentrations $c_A, c_B \ldots$. With these simplifications, and with the use of the standard equation $g = \sum \mu_i c_i$, we have:

$$\Delta g \equiv v\Delta P = \sum_i \mu_i^*(P^*, c_i^*) c_i^* - \sum_i \mu_i^*(\overline{P}, c_i^*) c_i^*,$$

or, by substituting (17.31) into the latter equation,

$$\Delta g = \sum_i \bar{\mu}_i(\overline{P}, \bar{c}_i) c_i^* - \sum_i \mu_i^*(\overline{P}, c_i^*) c_i^*.$$
(17.33)

The first term after the equal sign in Eq. (17.33) is a hybrid expression featuring chemical potentials at one concentration multiplying concentrations at another. The related sum is thus *not* a partial molar quantity, but with the chemical potential held constant at $\bar{\mu}$, it is a linear function of the concentrations c, and hence, according to the intercept rule (and its somewhat elaborate proof in Sec. 7.5), it represents a tangent in binary systems, a tangent plane in ternaries, a tangent hyperplane in multicomponents in general, the tangency point having coordinates here indicated as $\bar{c}$. Thus the difference Δg given by Eq. (17.33) is measured by the distance on the g axis from the tangent plane on the free energy surface g^α at point $\bar{c}$ to the Gibbs energy surface α at point c^*. This tangent construction is illustrated for a binary solution at Fig. 17.2 for the case of maximum rate of nucleation, that is corresponding to minimum work of nucleation, hence to the maximum Δg. That requirement in turn corresponds to taking the critical nucleus composition at a point on the β curve located on the tangent parallel to the $(\bar{\mu}_A, \bar{\mu}_B)$ tangent. Such is the famous *parallel tangent rule*.[2]

---

[2]It must be emphasized that critical nuclei are elusive creatures: sitting on top of an energy hill, as shown symbolically in Fig. 2.2, they decay exponentially fast: if they have been "observed", they were probably mere *preprecipitates*.

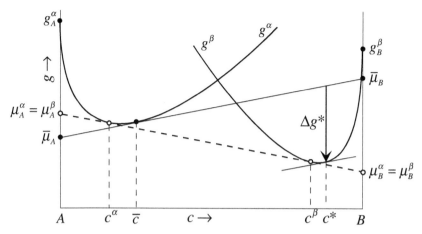

Figure 17.2: Schematic illustration of the "tangent rule" for the binary case. Starred (*) symbols refer to the critical composition, overbar symbols to the ambient phase.

17.6 Growth of a New Phase

After the new phase (β, say) has nucleated, it must grow to its equilibrium amount as specified by the lever rule. In doing so, at least for the case of more than one component, it must adjust its composition from that given by the "tangent rule" (as given in the previous section) to that specified by the common tangent rule (indicated by a dashed line in Fig. 17.2). The problem of "growth" is thus a fairly complex one, but here we shall be concerned only with the simplest case: that of spherical particles of β growing in the parent α phase with no change in concentration or molar volume, and such that the β phase completely replaces the α.

Let the total volume V of "old" (α) and "new" (β) phases remain constant

$$V = V^\alpha + V^\beta = \text{const.}$$

The *volume fraction transformed* is defined by

$$X_v = \frac{V^\beta}{V^\alpha + V^\beta}.$$

The problem is to find an approximate expression for the evolution $X_v = X_v(t)$ as a function of time. To obtain such a function we shall use the exceedingly clever trick developed independently by Kolmogorov, Avrami and Johnson and Mehl (KJAM, see for example Fultz, 2014). The idea is the following: define a fictitious volume fraction, the *extended volume fraction* which is *the volume which the particles of the new phase would have occupied if they had continued to nucleate and grow unimpeded everywhere in the system, including in the previously transformed regions.*

We thus have

$$0 \leq X_v \leq 1$$

but

$$0 \leq X_e < \infty .$$

At any moment, the fraction untransformed is $(1 - X_v)$, and is actually available for transformation. At time t, during time interval δt, an actual volume fraction δX_v is transformed. The corresponding incremental *extended* volume fraction δX_e is related to the real volume by

$$\delta X_v = (1 - X_v)\, \delta X_e .$$

In the limit, we have

$$\frac{\mathrm{d}X_v}{1 - X_v} = \mathrm{d}X_e .$$

This differential equation is readily solved for initial conditions $X_v = X_e = 0$ at time $t = 0$. Thus, with $x = (1 - X_v)$,

$$\int_1^{1-X_v} \frac{\mathrm{d}x}{x} = - \int_0^{X_e} \mathrm{d}y$$

yielding the desired result:

$$X_v = 1 - e^{-X_e} . \tag{17.34}$$

It is seen that $X_v \to 1$ as $X_e \to \infty$.

Phase β growing under steady state conditions locally increases its volume by capturing atoms (or molecules) on the element ΔA of the α/β interface. In time δt, the volume change $\delta \Delta V$ is given by the net number of atoms arriving *and sticking* to the interface in that time interval times the atomic volume Ω_o:

$$\delta \Delta V = \Delta A\, \Gamma\, \Omega_o \delta t$$

where Γ is the rate of arrival for unit area for unit time. For a spherical particle of β, of radius R, growing unimpeded in α we then have

$$\delta V = 4\pi R^2 \delta R = 4\pi R^2 \Gamma \Omega_o \delta t$$

or

$$\delta R = C \delta t \tag{17.35}$$

where we have defined $C = \Gamma\Omega_o$ as the *linear growth rate*, also given by

$$C = \frac{\mathrm{d}R}{\mathrm{d}t} \tag{17.36}$$

which will be assumed constant for simplicity (not a bad assumption for particles growing unimpeded, *i.e.* for X_e).

The linear growth law (17.35) can then be integrated for a particle growing from $R = 0$ at time $t = 0$ to a given R at time t. Of course, we do not expect linear growth (with constant C) to be valid down to vanishing radius, particularly since, at the start of the growth process, the initial radius R must have reached finite supercritical size R^*. However, assuming that observed radii are $R \gg R^*$, we can extrapolate expression (17.35) to $R = 0$ so that $R(t) = Ct$ holds approximately for a particle nucleated at time $t = 0$.

For a particle nucleated at time $t = t_o > 0$ we have

$$R(t) = C(t - t_o)$$

so that the corresponding volume at time t is

$$\frac{4}{3}\pi R^3 = \frac{4}{3}\pi C^3 (t - t_o)^3. \tag{17.37}$$

The total volume transformed for *all* particles nucleated between time t_o and $t_o\delta_o$, is thus $I_e(t_o)$ larger, where $I_e(t_o)$ is the nucleation rate in unit volume of "extended" β at time t_o.

The corresponding extended volume fraction at time t due to particles nucleated during the time interval $(t_o, t_o + \delta t_o)$ is thus

$$\delta X_e(t, t_o) = I_e(t_o)\frac{4}{3}\pi C^3 (t - t_o)^3 \delta t_o. \tag{17.38}$$

To obtain the total extended volume fraction at t, due to nucleation events at all times prior to t, we must sum all contributions of the type (17.38) for $t_o = t_1, t_2, \ldots t_k$ ($t_k \le t$). In the limit, the sum is replaced by an integral:

$$\int_0^{X_e} \mathrm{d}X_e = \frac{4}{3}\int_0^t I_e(t_o)C^3 (t - t_o)^3 \, \mathrm{d}t_o.$$

With (an assumed) constant growth rate we have:

$$X_e(t) = \frac{4}{3}\pi C^3 \int_0^t I_e(t_o)(t - t_o)^3 \, \mathrm{d}t_o. \tag{17.39}$$

The actual volume fraction X_v is obtained by substituting Eq. (17.39) into Eq. (17.34).

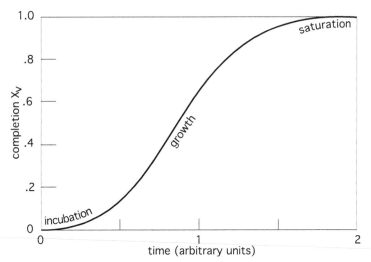

Figure 17.3: Plot of Eq. (17.42) (here with $K = 1$ and $n = 3$), known as the "Avrami equation", is the main result of the KAJM analysis. The shape of the curve is practically model-independent.

Some limiting cases are of interest:

I = Const.

Thus all ("extended") particles are nucleated at a constant rate and the volume fraction becomes, upon integrating (17.39)

$$X_v(t) = 1 - \exp(-\frac{\pi}{3}IC^3t^4). \tag{17.40}$$

I = $I_\circ\delta(t)$

Thus all particles are nucleated at $t = 0$ with instantaneous rate $I_\circ$, $\delta(t)$ being an infinitely sharply peaked function centered at $t = 0$ (delta function). Integration yields

$$X_v(t) = 1 - \exp(-\frac{4}{3}\pi I_\circ C^3t^3). \tag{17.41}$$

Other cases may be examined, the general conclusion being that the volume fraction may be expressed as an exponential function of a power of time:

$$X_v(t) = 1 - e^{-Kt^n}, \tag{17.42}$$

with K being a constant containing nucleation and growth parameters and n being an appropriate exponent, not necessarily equal to an integer.

Equation (17.42) is the main result of the Kolmogorov–Avrami–Johnson–Mehl analysis. X_v as a function of t follows a "sigmoid" curve, as shown in Fig. 17.3. Typically, three regimes can be observed: an "incubation" stage during which the volume fraction initially grows very slowly, then an approximately "linear growth" stage, followed by a "saturation" stage during which X_v slowly tends to unity, denoting completion of the transformation.

References

Cahn, J. W. and Hilliard, J. E. (1958), *Journal of Chemical Physics*, **28**, 258.

de Fontaine, D. (1981), *Thermodynamics and Kinetics of Phase Separation* in USA/China Bilateral Metallurgy Conference: Treatises in Metallurgy, Beijing, PRC, edited by J. Elliot and J.K. Tien (TMS/ASM).

Fultz, B. (2014), *Phase Transitions in Materials*, Cambridge University Press.

Gibbs, J. W. (1875), *On the Equilibrium of Heterogeneous Substances. Trans. Connecticut Acad.*, pp. 108-248; (1877), *ibidem*, pp. 343-524. See for example (1993), *The Scientific Papers of J. Willard Gibbs*, Ox Bow Press.

Landau, L. D. and Lifshitz, I. M, (1958), *Statistical Physics*, Addison–Wesley.

Chapter 18

Kinetic Aspects

The present book is about thermodynamic phase equilibrium, *not* about kinetics or dynamics, yet topics such as nucleation and *spinodal decomposition* are so often part of the relevant tropics treated in classical thermodynamics that brief descriptions are worth mentioning, hence the inclusion of Chapters 17 and 18. Moreover, materials science is a recent discipline which originally sprung out of so-called physical metallurgy. In the latter discipline, it turned out that the steel industry (and others) was highly concerned with phase transformations, particularly with *phase separation* whereby a disordered high-temperature phase gives rise on cooling to two or more new phases. The basic mechanism invoked was then "nucleation and growth", which was covered in the previous chapter. It then followed that Gibbsian thermodynamics, with its concept of "phases" (defined in the Introduction on page 3) was of primary importance. It was then a more or less tacit assumption that all phase changes were governed by nucleation, and it took two papers by John W. Cahn (1960, 1961) on spinodal decomposition to partially overthrow the nucleation dogma.[1] That change of paradigm came none too soon: although nucleation theory applied to fluids is quite successful, it often fails for solids, partly because it is often used in regions of phase diagrams where it had no business being applied, partly because the theory features the exponential of the cube of interfacial energies, the latter being usually very poorly known, especially in solids. Also, nucleation theory was often unreasonably applied to displacive transformations, such as martensite.

There were other problems with the traditional treatments: it was often believed that the only type of phase transformations worth considering

---

[1] John Cahn passed away on March 14, 2014 in Seattle, WA, as this book was being written.

was that of phase separation, but what about order-disorder? Should it not be treated on a roughly equal footing? Typically, phase separation was handled in metallurgy departments, and order-disorder in physics departments, but in the form of ordering of magnetic spins, as in the Ising model. And yet, the two mechanisms – if we disregard the nature of the materials involved – clustering and ordering are not that dissimilar. In a sense, both involve "phase separations" of sorts: in the former case, objects (atoms or molecules) tend to surround themselves by *like* objects; in the latter case they tend to surround themselves by *unlike* objects. Thus in the former case, like objects tend to cluster together in separate regions of the lattice, in the latter, like objects tend to occupy one lattice, or sets of lattices. Whereas clustering can arise in pretty much one way only, objects can surround themselves by unlike objects in many crystallographically different ordered ways. Or, to put things differently, in the ordering case, a phase separation does occur, but the occupied regions (lattices) are interpenetrating; in clustering, they are located in separate regions of space. The two cases of "phase separation" were illustrated previously in a 2-dimensional model in Fig. 15.1; at panel (a), white objects and black ones cluster in separate regions, separated by a phase boundary, at panel (c), white and black objects occupy distinct sublattices, separated by antiphase boundaries.

To emphasize the analogy between ordering and clustering, consider the concentration wave description of the processes and the Landau approach: in clustering, the related special point instability is at the origin $\langle 000 \rangle$ of reciprocal space (or of the Brillouin zone), in ordering, the instabilities are located at special points other than the origin of reciprocal space. However the mathematics of instabilities is the same in both cases: that of spinodal transformations clustering and ordering. The first one is relevant to Cahn's original breakthrough, spinodal decomposition (or spinodal clustering), the second one came later, spinodal ordering.

It may be objected that ordering often involves the nucleation of new phases which in no way resemble the parent phase (in crystallographic parlance, the space groups of the new phase would not be a subgroup of that of the parent phase, as required by the first Landau rule) but it is frequently observed in practice that a first-order phase change is suppressed by sluggish kinetics, so that the first hint of an incipient phase transformation is an instability, *i.e.*: a spinodal reaction occurring for instance at lower temperatures. Therefore in this chapter we shall dwell upon the spinodal (instability) concept before surveying the more complex one of (first-order) phase separation.

To summarize, the topic of "phase transformations" is a varied one, which is not really part of equilibrium thermodynamics, but closely related to it. Applied to solids (in particular to crystals) it can be broken up in various sub-topics, which recall the dichotomies listed at the start of Chapter 14:

(a) *coherent vs. incoherent.* By *coherent* it is meant that, in crystals, both original and product phases have the same underlying lattice structures, or, in the terminology of the Landau theory, that the space group of the product phase should be a subgroup of the original one. Or again, it means that the interface between original and product phases should contain no lattice discontinuities, such as dislocations, at least in the initial stages of the transformation. In an *incoherent* transformation no such restrictions are present. Actually, in practice incoherent transformations are often initiated at grain boundaries where lattice discontinuities are already present.

(b) *replacive vs. displacive.* The former transformation describes replacing atom "A", say, by atom "B", or "spin up" by "spin down". The latter refers to coordinated movements of atoms (or dipoles), as in martensitic transformations.

(c) *phase separation vs. order-disorder.* This classification has just been covered and need not be repeated.

(e) One more distinction should be mentioned, one that is not standard: that between *transition* and *transformation*, the former being a transformation occurring (practically) *at* equilibrium, the latter referring simply to *any* transformation.

For example, the spinodal process is a (non-equilibrium) phase separation (spinodal decomposition) *transformation* or ordering (spinodal ordering) *transformation*.

18.1 The Spinodal Concept

As mentioned above, spinodal decomposition was the brainchild of John W. Cahn, originating mathematically from the linearization of a non-linear partial differential equation, itself resulting from the free energy functional derived by Cahn and Hilliard (1958), and mentioned in Chapter 14 [Eq. (14.1)] and also derived, actually earlier, by Landau and co-workers (1958). Soon thereafter, experimental work was undertaken at Northwestern University (which created the first Materials Science Department) in the graduate student research group of John Hilliard.[2] Thus was born a new mechanism of phase separation from a disordered solid solution, a transformation which previously had been abandoned almost exclusively to nucleation and growth, often resulting in failure of nucleation theory. Historically though, the spinodal theory was anticipated in 1956 by Mats Hillert in his doctoral

---

[2]that I was a part of, hence my knowledge of the early developments of the field.

dissertation at MIT, which was followed only later by a publication by that author (Hillert, 1961).

At first, the initiators of the field of spinodal decomposition were not quite sure of how to interpret their new mechanism of phase separation; was it a competing mechanism to nucleation, or was it applicable to only certain regions of the solid solution miscibility gap? At present we can answer somewhat facetiously "none of the above", as will be discussed in the final section of this chapter where it is shown that a full non-linear theory must be considered, combining both nucleation and spinodal approaches, and including fluctuations. As we shall see, the linear simplification is not really a reliable kinetic theory (in particular, it diverges at late times); it is actually merely a study of instabilities of the free energy of disordered systems, as already mentioned previously. The difference is that in the kinetic formulation the time variable is introduced.

As noted by Cahn (1968),"the origin of the word *spinodal* has been somewhat of a mystery". Its linguistic root is the Latin word *spina*, meaning *thorn*. Mathematically, the spinodal should then be the locus of sharp points or cusps, the latter word defined as *a double point at which two tangents to a curve are coincident; Syn: spinode* [James & James (1960)]. The French mathematical term is more descriptive: *point de rebroussement*, or *the point at which one retraces one's steps*. One therefore expects the spinodal to be a locus of cusps, but to quote Cahn again, "nowhere in the modern theory [...] is there a clue to the cusp that gave origin to the name". In the present context, it is shown that the double-well thermo-dynamic potential that characterizes the free energy of a phase separating binary system, usually plotted as a function of concentration, when plotted versus an intensive variable, such as a chemical potential, gives rise to a locus of cusps, or spinodes. Figure 18.1, taken from Kikuchi and de Fontaine (1976), is a plot of the grand potential $g = f - \mu\xi$, or omega potential [see Sec. 4.3(d)], as a function of μ, the chemical field [see Eq. (7.66) and general definition on page 90] and temperature T, and where f is the normalized Helmholtz free energy and ξ is an order parameter (or concentration difference). Ridge lines are clearly seen as loci of cusps characteristic of spinodal features. The figure also represents the *swallowtail catastrophe* as defined by René Thom (1973).

18.2   The Kinetic Equation

The kinetics of spinodal decomposition follow a standard diffusion equation which, in words, states that the time rate of change of concentration c (we are treating a binary solid solution only) is equal to minus the divergence

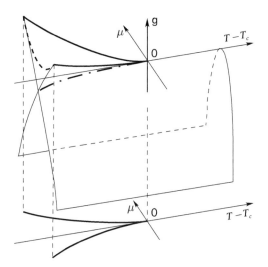

Figure 18.1: The *Swallowtail Catastrophe* according to R. Thom (1973) applied to the spinodal mechanism.

of the flux J:

$$\frac{\partial c}{\partial t} = -\operatorname{div} J \qquad (18.1)$$

where here and in subsequent equations the numerical factor N_V, the number of atoms per unit volume, has been left out for simplicity. The flux J itself is proportional to the gradient of a suitable potential μ:

$$J = -M \operatorname{grad} \mu, \qquad (18.2)$$

where M is a positive mobility term. The physics of the problem is specified by the potential μ, more correctly by the chemical field, to be defined by appeal to the Cahn–Hilliard functional, used as well in Sec. 14.2 for the equilibrium of a nonuniform system. In that section, the Euler expression provides the form of the potential which is required here; the potential was given by Eq. (14.8), labeled there as λ, instead of μ, thus we have:

$$\mu = \frac{\partial g}{\partial c} - 2\kappa \nabla^2 c, \qquad (18.3)$$

in which g is, as usual, a free energy as a function of concentration, and κ is, as in Chapter 14, the gradient energy coefficient of the square-gradient approximation. The latter three equations, augmented by a suitable double-well bulk free energy, combine to produce a diffusion equation, valid for solving the conserved concentration variable c

$$\frac{\partial c}{\partial t} = \mathrm{div}\left\{ M \mathrm{grad}\left[\frac{\partial g}{\partial c} - \frac{\partial \kappa}{\partial c}(\nabla c)^2 - 2\kappa \nabla^2 c \right] \right\}, \tag{18.4}$$

in which the elasticity term has been left out. The double well potential g may be obtained from the regular solution model (see Fig. 9.1) or from a fourth degree polynomial (quartic), as was done initially by Landau in Sec. 15.3, less accurate, but more convenient for performing numerical calculations, required for the non-linear equation, as shown in Sec. 18.4.

Such is the famous equation by which John Cahn (1960) introduced *spinodal decomposition* to the field of materials science. Surprisingly, Cahn appeared at the time not to have grasped the importance of his equation, which after all also ushered in the *phase field technique*, as he left this equation unnumbered, which forced continuators to refer to it as "the last equation before Eq. (18) in John Cahn's 1960 paper on spinodal decomposition". Present tradition now refers to this equation as the "Cahn–Hilliard equation", which is not quite correct since it was first presented in a paper authored by Cahn alone. It would appear that Cahn, recognizing that his equation was a highly non-linear partial differential equation which could not be solved by direct analytical techniques, immediately turned his attention to its linearized version, to be investigated next. Attempts at numerical solutions of the equation will also be mentioned in later sections.

18.3   Linear Solution

As mentioned above, Cahn in his introductory paper to spinodal decomposition (1960) turned his attention directly to the linearized version of Eq. (18.4), which is (with constant concentration gradient coefficient κ and without the elastic term)

$$\frac{\partial c}{\partial t} = M(g_0'' \nabla^2 c - 2\kappa \nabla^4 c), \tag{18.5}$$

where g_0'' is the second derivative of the free energy evaluated at the average concentration c_0. The linear partial differential equation (18.5) can be Fourier transformed into an ordinary differential equation for the concentration wave amplitude $A(\mathbf{k}, t)$, the solution of which is found to be

$$A(\mathbf{k}, t) = A(\mathbf{k}, 0)^{\alpha(\mathbf{k})t}, \tag{18.6}$$

where $\mathbf{k}$ is the wave vector for the concentration wave of initial amplitude $A(\mathbf{k}, 0)$ and wave amplification factor[3]

$$\alpha(\mathbf{k}) = -Mk^2(g_0'' + 2\kappa k^2). \tag{18.7}$$

Note that strictly speaking the linear solution is incorrect since it is unbounded: the amplitude A grows without limit as time t increases.[4] It follows, as Cahn pointed out, that the linear solution should be used only at the very first instants of the spinodal transformation, but how short is that stage of validity? It later turned out that in practice that this stage is often too short to provide a meaningful representation of the morphology of spinodal microstructures, as will be discussed later.

The linear solution, given here by Eqs. (18.6) and (18.7) is still useful, as a perturbative one however, indicating which concentration waves are at first subjected to maximum amplification, namely those which maximize the amplification factor given by Eq. (18.7). In some cases, the optimal amplitudes dominate the amplitude spectrum, therefore leading to quasi-periodic morphology. More generally, even at very early stages, the non-linear terms intervene and much more complex non-periodic morphology ensues, resembling ones that might have arisen from nucleation and growth. It follows that spinodal decomposition does not necessarily lead to periodicity, nor does periodicity necessarily imply a spinodal mechanism. That conclusion helps explain the confusion which often followed early experimental work on spinodal decomposition.

Despite such reservations, the linear theory is also useful for the study of compositional instabilities in solid solutions since it can be generalized to order-disorder transformations. Formally, all one needs to do is to recognize that the formalism is still valid if, instead of examining instabilities in $\mathbf{k}$-space around high-symmetry point $\langle 000 \rangle$, one were to investigate other high-symmetry points in the Brillouin zone of the crystal (see Sec. 15.2 for examples of such points). One more important modification to the linear equation needs to be performed, however, before it can be used in the order-disorder case: the nature of the crystal lattice must be taken into account. Formally, that means that the nabla (∇) vector of Eq. (18.5) must be replaced by finite differences noted as (Δ). Now these operators are structure dependent, that is, a second difference Δ^2 on an fcc lattice is not the same as one on a bcc lattice, for example. The resulting equation

---

[3]The notation given here differs slightly from that of Cahn; it is that given in de Fontaine (1975).

[4]Once I left out inadvertently the non-linear terms in a numerical calculation of the solution of Eq. (18.4) and the computer immediately returned the value 10^{300} for the amplitude A, which should have been confined to the interval -1 to +1. Turning the non-linearities back on restored the correct limits to the amplitudes.

now reads

$$\frac{\partial c}{\partial t} = M(g_0'' \Delta^2 c - 2\kappa\Delta^4 c)\,, \tag{18.8}$$

in which the concentration variable c must now be obtained by a mesoscopic coarse-graining procedure outlined in Chapter 15. Just as in the continuum case of Eq. (18.5) a Fourier transform is used to change Eq.(18.8) into an ordinary differential equation in the concentration wave amplitude A. In this case, however, the Fourier transform introduces lattice periodicity, which feature was absent in the continuum case. In particular, this periodicity transforms the squared wave vector k^2 into the trigonometric expression $1 - \cos ka$, responsible for introducing lattice periodicity into the solution.

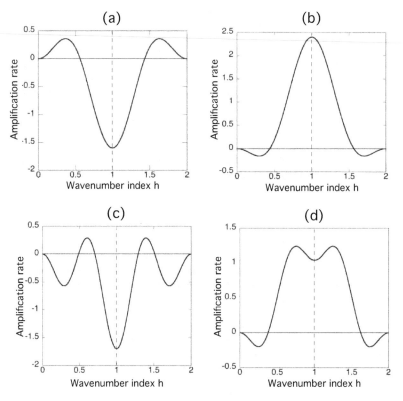

Figure 18.2: Four cases of kinetic amplification factors in 1-dimensional reciprocal space (vertical scale in arbitrary units): (a) spinodal decomposition, (b) spinodal ordering at [100] position, (c) continuous ordering for wave vector *not* belonging to special-point ordering, (d) spinodal ordering for long-period modulation of [100] ordering wave, leading to ordered domains, as in Fig. 15.7.

For illustration consider Fig. 18.2. Amplification factors are plotted in one dimension (for instance along the $\langle 100 \rangle$ direction in cubic crystals) for various cases of $\alpha(h)$ around phase separation or ordering high symmetry reciprocal space points (see caption for explanation). Panel (a) shows an amplification factor profile for the case of an instability in the vicinity of special point $\langle 000 \rangle$, *i.e.* for the clustering case (but also for the crystallographically equivalent points $\langle 200 \rangle$ and so on, in diffraction terms). The curve at the left portion of the figure resembles the corresponding amplification curve for the case of spinodal decomposition, as expected, but the additional feature here is that this curve repeats around all equivalent points in **k**-space. In this case, as in the classical spinodal theory, the long wavelength concentration waves, in the vicinity of reciprocal space index $h \approx 0.5$ (or equivalently 1.5) are being amplified, with the one receiving maximum amplification having $\mathbf{k}(h)$-vector at $h \approx 0.35$, in the case of Fig. 18.2.

For panel (b), the amplification curve resembles that of panel (a) but flipped vertically, due to the fact that, in Eq. (18.8) the kappa gradient coefficient has *negative* sign, a feature not encountered in the original continuum theory. It follows that in this case the short wavelength concentration waves in the vicinity of the ordering point $\langle 100 \rangle (h = 1)$ are preferentially amplified. Panel (c) illustrates the case of amplification of ordering waves of wave vector away from high symmetry points, and likewise for panel (d). In the latter case, the wave of maximum amplification can be regarded as a $\langle 100 \rangle$ wave amplitude-modulated by a wave of wavelength equal to the reciprocal of the **k**-space distance between the maximum (at about $h \approx 0.8$ in the figure) and the center line at $h = 1$. The overall result is shown (not to scale) in Fig. 15.7, which can be interpreted as one member of a row of equally spaced ordered precipitates.

Hence, it is seen that, by combining Cahn's original spinodal ideas with some borrowing from the Landau ordering theory, a simple linear finite-difference equation can provide useful information about the very early stages of phase transformations, *i.e.* of spinodal clustering and spinodal ordering. An example of application will now be given.

Example

Spinodal clustering or ordering can often act as a precursor to first-order phase transitions or transformations. Typically, a transformation starts with the amplification of the most unstable concentration waves — those located in the positive regions of the amplification curves seen in Fig. 18.2, for example — but later, as the non-linear contribution to the diffusion equation takes over, the waves interact with one another by anharmonic

coupling, leaving finally the waves with $\mathbf{k}(h)$ which produce the correct
equilibrium crystal structure, or very close to the equilibrium ground state
structures. Wave coupling is awkward to treat algebraically, so that real
space considerations for studying ground state problems are required. However, reciprocal space methods are helpful for qualitative descriptions, as
will now be shown in the case of ordering in the NiMo alloy, treated in
more detail in the original paper (de Fontaine, 1975).

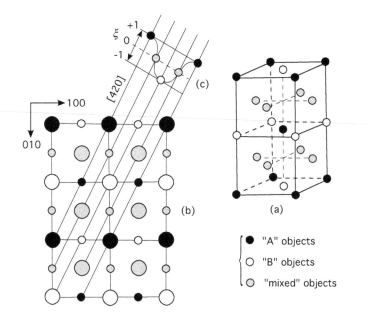

Figure 18.3: (a) Unit cell of mesoscopic AB alloy; (b) projection of averaged
AB cell (large circles are in plane of projection, small ones are one-half lattice
parameter away); (c) [420] concentration wave [after de Fontaine (1979)].

Quenching fcc solid solutions from the disordered state over a wide range
of compositions from about 8 to 33 at. % Mo produces, in X-ray and elec-
tron diffraction patterns, diffuse intensity peaks centered about point $\langle 1\frac{1}{2}0 \rangle$
in reciprocal space, a point which coincides with only one of the superlat-
tice positions of the equilibrium structures of the stable ordered phases in
this alloy, namely Ni_2Mo, Ni_3Mo and Ni_4Mo. Thus the spinodal ordering
wave $\langle 1\frac{1}{2}0 \rangle$ is seen as a precursor to the transformation to the more stable
ordered structures in that alloy, the ordered structures peaks still being
related to the $\langle 1\frac{1}{2}0 \rangle$ peak by lying all in the $\langle 4\,2\,0 \rangle$ direction in reciprocal
space. Figure 18.3 shows an fcc lattice decorated by a $\langle 1\frac{1}{2}0 \rangle$ concentration wave acting alone and producing Ni-rich and Mo-rich "coarse-grained

atoms" and (grey) "average atoms". A way of describing the particular ordered structure associated with the $\langle 1\frac{1}{2}0 \rangle$ ordering wave is to refer to these as belonging to the $\langle 1\frac{1}{2}0 \rangle$ "ordering wave family" (de Fontaine, 1975).

18.4 The Non–Linear Equation

Equation (18.4), known today as "the Cahn–Hilliard equation", despite its general significance — in particular as the starting point for phase field modeling — was almost ignored initially; it was impossible to solve analytically and therefore thought, even by its initiators, to be of little use, except as a necessary step towards a linear solution. So it took a while for the present author, then a graduate student, to risk searching for numerical solutions of this non-linear equation. The trouble with numerical solutions, of course, is that every calculation is that of a special case. Nevertheless, such a solutions was undertaken, but because of the author's inexperience, in a rather clumsy way.[5] Many such calculations have been carried out since.[6]

The numerical approach essentially requires replacing the derivatives of Eq. (18.4) by finite differences, though not those of Eq. (18.8) which take the lattice structure into account. Here, the calculation proceeds by stepping forward in time with very short time steps at the early stages of the transformation and large ones towards the end. The general objective was to mimic, as far as possible, the experiments on aging of AlZn alloys inside the miscibility gap being performed at the same time by Rundman and Hilliard (1967); hence the choice of alloy concentration selected in Figs. 18.4 and 18.5. The AlZn alloy system features a partially accessible miscibility gap, slightly depressed in temperature because of quasi-isotropic elastic coherency strains. The initial condition for all three compositions — 20, 22.5 and 37.5 at. % Zn — was taken to be the same in all three cases: a very small "random" fluctuation, not shown at the relevant figures because its amplitude was so small.

Despite the fact that all starting points were identical, further development of the composition amplitude profile show marked differences, even qualitative differences, and even though all average concentrations are lying inside the same miscibility gap (MG) at the annealing temperature of 100 C. The first one, at 20 at.% [Fig. 18.4 (a)] is very close to the limit of

---

[5]de Fontaine, (1967) and (1975). That way involved only a one-dimensional space (plus two cases in two-dimensional space) calculations and unnecessary Fourier-space transformations.

[6]A very recent publication is entitled *Random scalar fields and hyperuniformity*, by Ma, Z. and Torquato, S. (2017). Despite the sophisticated language, it actually describes, in part, an application of the Cahn–Hilliard equation in two dimensions.

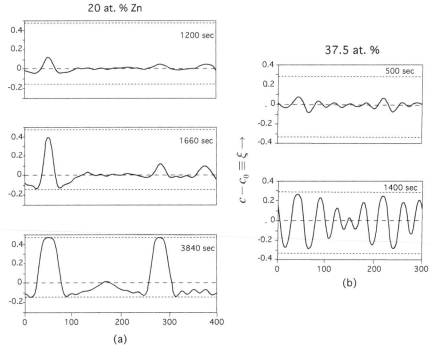

Figure 18.4: One-dimensional composition profiles of annealing at 100 C of AlZn alloys of the indicated compositions in a 400 Å (a) and (300) Å (b) domain with periodic boundary conditions and with random initial condition of small amplitude (de Fontaine 1967 and 1975). Vertical axes denote concentration differences $c - c_0 \equiv \xi$, equivalent to an order parameter, as previously defined. The dashed line denotes the average concentration and the dotted lines denote the concentrations of the coherent miscibility gap.

the coherent AlZn MG, the second one at 22.5 at.% [Fig. 18.5 (a) and (b)] is a little more inside, and the third at at. 37.5 % [Fig. 18.4 (b)] is close to the center of the MG. It is immediately apparent that the central average composition (37.5 %) exhibits the amplitude modulated periodic profile character expected for a typical spinodal morphology, but quite irregular for those short annealing times.

The off-center compositions however, do not resemble the (classical) spinodal morphology predicted by the linear theory. Further growth of the composition variation requires a lengthening of the pseudo-period, which does not take place readily due to the extreme resistance to coarsening of such regular periodic structures, even after long annealing times, as the extrema of the composition profile move slowly towards the MG boundaries (dotted lines). The resulting profiles of annealing at a composition nearer

22.5 at. % Zn

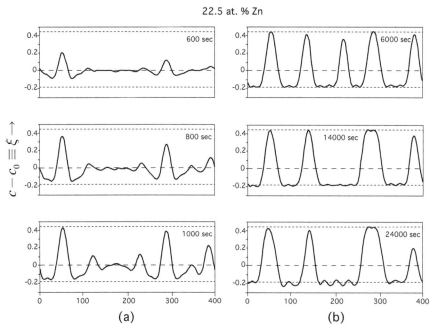

Figure 18.5: One-dimensional composition profiles of annealing at 100 C of AlZn alloys of the indicated composition in a 400 Å domain with periodic boundary conditions and with random initial condition of small amplitude (de Fontaine 1967 and 1975); (a) at early and (b) at later times. Vertical axes denote concentration differences $c - c_0 \equiv \xi$, equivalent to an order parameter, as previously defined. The dashed line denotes the average concentration and the dotted lines denote the concentrations of the coherent miscibility gap.

the Zn-poor side of the MG (22.5 %), but still well inside the spinodal, are depicted in Figs. 18.5. After a 600 sec anneal at 100 C the same small amplitude profile as that taken as starting point for the 37.5 case develops morphologies consisting of Zn-rich peaks (two of them in this case) flanked by depleted regions. Such profiles have been denoted as "Guinier zones". Bonfiglioli and Guinier (1966) correctly observed that those profiles resemble a single period of a periodic function whose wavelength is practically equal to that of the wavelength predicted by solution of Cahn's linear equation [Eq. (18.5)]. That profile grows in amplitude until the "zone" amplitude reaches the MG composition indicated by the dotted lines in the figure. Later (around 6000 sec) a periodic set of peaks develops, but the period is notably larger than that predicted by the linear equation. Later still (around 14000 sec, and later), a coarsening reaction causes the larger peaks to grow at the expense of the smaller ones, which progressively disappear, due to the requirement of minimization of interfacial area. Such is the so-called "coarsening" stage.

Finally, Fig. 18.4 (a), for an average composition very close to, but still inside the spinodal composition, the same low amplitude profile also gives rise to Guinier zones, but later on in the annealing process, a periodic structure of peaks does not develop, leaving only isolated particles, as per a typical nucleation morphology. These calculations clearly show that *spinodal decomposition* and *nucleation — and growth* are not separate mechanisms, but result in fact from the same process given by the same so-called *Cahn–Hilliard equation* (18.4), which is shown to be applicable for the whole MG, regardless of average composition. It is just that the classical spinodal mechanism is an approximation valid for central compositions while classical nucleation is valid very near an MG boundary. The spinodal line itself does not represent a singularity, as was initially believed, but is "ignored" in the non-linear solution, the fluctuations "carrying" the solution over the spinodal. Actually, as was pointed out earlier, the spinodal is not an equilibrium concept, and spinodal decomposition is neither a first-order, nor second-order transition, but a mere *transformation*. John Hilliard had the last word: "it is rather remarkable that a single equation can depict the complete life cycle, from birth to death, of a particle" (1970). It was known, of course, that regardless of the approximation employed, phase separation always shows the transformation to proceeded as indicated schematically in Fig. (**??**), qualitatively described previously as consisting of *incubation, growth* and *coarsening*, and requiring three separate theoretical models. In fact, there is only one mechanism, as shown here, at least for a *coherent* transformation.

Note that such calculations, and phase field calculations in general, may be used not only for phase separation — with instabilities located at $\langle 000 \rangle$ — but for order-disorder transformations, as mentioned in the previous section, with instabilities located at other high-symmetry points in $\mathbf{k}$-space. And of course the objects undergoing phase separation or ordering may be other than atoms of various species, such as spins or dipoles, etc. Recently, attempts have been made to make the calculations more realistic by means of a more quantitative theory which we shall briefly review here.

18.5   General Quantitative Theory

Standard treatments of thermodynamic states have relied, often tacitly, on the *coarse-graining* approximation, convenient for continuous, mesoscopic descriptions of reality, be it gas, liquid or solid. It should be clear, however, that by so doing, much physics is being "swept under the rug", so to speak. The actual discrete nature of atoms, spins, dipoles, etc. is neglected for obvious reasons, since taking correctly into account such short wavelength phenomena in general leads to very serious mathematical and computational difficulties. For example, as we have seen, phase transitions which take the discrete atomistic nature of thermodynamic systems into account (Ising models) lead to no transition in one -dimension, an extremely intricate mathematics in two dimensions, and (currently) no exact solution at all in three dimensions, whereas the regular solution model model, or its order-disorder counterpart (Bragg–Williams model), in any number of dimensions is quite straightforward.

The coarse-graining (mesoscopic or mean field or phase field) models — associated with the names of Bragg–Williams, Landau, Cahn–Hilliard, and many others — are indeed very useful qualitatively, but are they reliable quantitatively? That depends: in some cases, that approximation works quite well, in other cases, the mean field model is not even qualitatively correct. Hence, it might be of interest to construct a theoretical model which combines both microscopic and mesoscopic approaches. Such is precisely what Finel and co-workers have proposed in the framework of the phase field model [Bronchart *et al.* (2008)]. The mathematical treatment is too advanced to be described here, so only a brief summary of it will be given.

The authors envisage a binary alloy evolving towards phase separation — a *clustering* reaction — and, as an example, perform the calculations on a simple cubic lattice, without loss of generality as they hasten to add, with lattice parameter a. Indeed, since ordering is not considered, crystallography is not a concern in this case. The general derivation starts with a (microscopic) master equation for the evolution of an alloy configuration, the atomic transition rates being regarded as a kinetic process based on the saddle point mechanism. In the example chosen, atomic interactions are attractive between first neighbors ($J_1 < 0$) of like atoms, in keeping with the clustering reaction. For coarse-graining purposes, the thermodynamic system under study is subdivided into cubic cells of linear size d, over which the compositional average is performed, the dimensionless parameter a/d thus characterizing the coarse-graining process. Local thermodynamics inside the individual cells are performed by Monte Carlo simulations, which is very rapid since the cells are small compared to the bulk material.

The calculation proceeds with averages taken over cells, leading to a Langevin equation which is equivalent to the Cahn–Hilliard (C-H) equation (18.4) plus a "Gaussian noise" term which provides fluctuations. This is an essential point: the C-H equation is deterministic but Langer (1975) has pointed out the necessity of including random fluctuations, absent in the

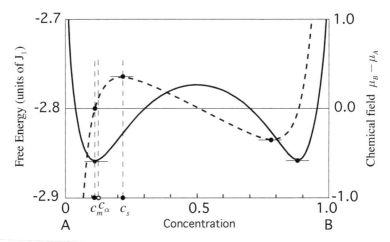

Figure 18.6: Chemical potential difference (dashed line) and corresponding free energy calculated by Bronchart *et al.* (2008). Indicated atomic concentrations denote: common tangent construction c_m, equilibrium concentration c_α, and spinodal concentration c_s. First neighbor interaction is $J_1 = -k_B/4$. Coarse-graining linear dimension is $d = 8a$, where a is the lattice parameter.

original computational treatment of the non-linear equation (de Fontaine 1967).[7] Treating fluctuations is absolutely essential as it helps drive concentration changes over energy barriers and also helps to ignore the spinodal stability limit. In particular, the full treatment outlined here allows nucleation to be handled as well as the spinodal process, with exactly the same mathematics, as was mentioned in Sec. 18.4. As in the C-H formalism, the coarse-graining theory arrives at a diffusion equation through a potential which is the functional derivative of an integral (or summation) over coarse grains, or cells of a uniform free energy and a square-gradient term, but here the terms in question are obtained by taking the microscopic physics into account. Not surprisingly then, the resulting bulk free energy, function of the coarse-grained concentration c, has the aspect of a standard regular solution model energy, as in Figs. 2.2 and 9.1, but now, its non-equilibrium portions are physically (rather, thermodynamically) meaningful, having been calculated from a microscopic basis and as a result turn out to depend on the coarse-graining cell size dimension d. An example of this resulting free energy is shown as a full curve in Fig. 18.6 along with the associated chemical field (dashed line). An interesting feature of the new formulation concerns the well-known *common tangent rule* (see Sec. 7.9): in

---

[7]Actually, in the latter calculation, a short wavelength cutoff was included for practical purposes, but was physically unreal.

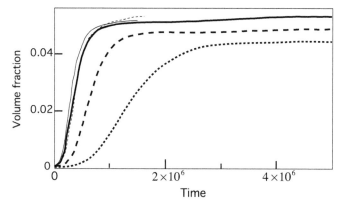

Figure 18.7: Volume fraction of precipitates as a function of time for same interaction parameter as in Fig. 18.6 for three different average concentrations, all of them inside the miscibility gap but outside the spinodal concentrations: $\bar{c} = 0.16$ (heavy dotted line), $\bar{c} = 0.165$ (heavy dashed line), and $\bar{c} = 0.17$ (heavy full line). For the latter concentration, the two light lines, full and dotted, are calculated for $d = 10a$ and $d = 6a$, respectively, and are practically superposed on the full curve computed with $c = 8a$, indicating quasi invariance under change of coarse-graining size d.

Fig. 18.6 it is seen that the calculated equilibrium concentration c_α differs from that given from the common tangent concentration c_m, though not by much. The reason for that difference is the neglect, in the standard formulation, of the concentration fluctuations which are explicitly taken into account in the newer theory.

As a critical test of the general theory, the authors (Bronchart *et al.*, 2008) examine solid-state precipitation predicted from their formulation for three different average concentrations $\bar{c}$, all located inside the calculated miscibility gap, but *outside the spinodal concentration* c_s. This key fact is emphasized because the present formulation is by no means "a new theory of spinodal decomposition". Indeed, its strength is its ability to apply to both nucleation and growth and to spinodal decomposition; in the present case, it was shown that starting from a C-H model, usually regarded as the basis of spinodal decomposition, the present theory can be tested for its applicability for nucleation-and-growth. The results of such a test are indicated in Fig. 18.7 where the volume fraction of precipitates, starting from very small values, are plotted as heavy curves for coarsegraining of cell-size $d = 8a$, and for average concentrations indicated in the figure caption. As expected, the volume fraction curves show the classical *incubation-growth-coarsening* shown in Eq. (17.42) in the previous chapter on classical nucleation.

In addition, Fig. 18.7 indicates in fine lines calculations pertaining to cell sizes $d = 6a$ and $d = 10a$. It is seen that those curves are practically superimposed on the curve for $d = 8a$, all of them for average concentration $\bar{c} = 0.17$, near the solubility limit of the solid solution (and outside the spinodal). This coarse-graining size independence is a very reassuring result, and is somewhat surprising given the clear d-dependence of the various steps of the derivation. Is it then necessary to derive such a complex theory if it simply returns the classical results? The answer is yes because, as we have said, the coarse-graining formulation gives quantitative meaning to the metastable or unstable portions of the free energy curves, which is essential for treating the critical nucleus, or the location of the spinodal, in short for a variety of non-equilibrium phenomena. Of even greater interest, we repeat, the general hybrid theory provides a unique formulation continuously valid all the way from average concentrations at very small supersaturations (near the edge of the MG), where nucleation theory is approximately valid (at early stages), to large supersaturations (near the center of the MG), where linear spinodal theory is approximately valid (at early stages). The combined theory is of course valid for all average concentrations in between those two extremes. We may summarize this important result by claiming:

> The classical nucleation model is approximately valid for starting states just inside the MG, i.e. for very *shallow* quenches, and the linear spinodal model is approximately valid for starting states near the center of the MG, i.e. for very *deep* quenches. For concentrations covering the whole MG, for shallow, deep and intermediate quenches, a full non-linear formulation with fluctuations and coarse graining is required. Such a general theory, covering both nucleation and spinodal decomposition, is now available (Bronchart *et al.* 2008).

Note that the penultimate chapter of this book on *classical* thermodynamics (Ch. 18) treats *non-equilibrium* thermodynamics; but this chapter was included because it may be regarded as an application of the strictly classical theory. The type of procedure referred to in the previous paragraph, a very common one in experimental work in materials science, can be described as one that consists in preparing an equilibrium high-temperature disordered solid solution, and then quenching it (generally very rapidly) into a MG where the solution is unstable (or metastable), and is then allowed to evolve into final phase-separated states or, if an ordering reaction is expected, into separate sublattices. Applications thereof are legion. A general conclusion of the revised view of "spinodal decomposition" can be formulated thus:

The difference between spinodal and nucleation mechanisms is that the former is triggered by an instability small in degree and large in extent, and the latter, by an instability large in degree and small in extent. Such was the (correct!) early view, although in practice — theoretically and experimentally — this neat distinction is actually masked by perpetually present fluctuations. *Critical exponents are not an issue with ordinary T_o transformations (resulting from instabilities). As a result, a spinodal mechanism does not necessarily imply periodicity, and periodicity does not necessarily imply a spinodal mechanism (in displacive transformations, for example.)*

For second-order transitions the situation is not so clear. With mean field theories (using analytical free energies), there is really no difference between an instability and the transitions proper. With better non-analytic free energies, the transition must exhibit singularities, which are mentioned only briefly in Sec. 15.7.

References

Bonfiglioli, A. and Guinier, A. (1966), *Acta Met.* **14**, 1213.

Bronchart, Q., Le Bouar, Y. and Finel, A. (2008), *Phys. Rev. Lett.* **100**, 015702.

Cahn, W. (1960), *Acta Met.* **8**, 554; (1961), *Acta Met.* **9**, 795.

Cahn, J. W, (1968), *Trans, AIME,* **242**, 179.

Cahn, J. W. and Hilliard, J. E. (1958), *J. Chem. Phys.,* **28**, 258.

Chandrasekhar, S. (1957), *Introduction to the Study of Stellar Structure*, Dover Publications; Original edition by the University of Chicago Press (1939).

de Fontaine, D. (1979), *Configurational Thermodynamics of Solid Solutions*, in *Solid State Physics*, Ehrenreich, H., Seitz, F. and Turnbull, D., **34**, pp. 73-274, Academic Press.

de Fontaine, D. (1967), *A Computer simulation of the Evolution of Coherent Composition Variations in Solid Solutions; Ph. D. Dissertation*, Northwestern University, Evanston, IL.

de Fontaine, D. (1975), *Treatise on Solid State Chemistry, Vol. 5: Changes of State*, N. B. Hannay, Ed. (Plenum, N.Y.) p. 129.

de Fontaine, D, (1975), *Acta Metall.*, **23**, 553.

Hillert, M. (1956), *A theory of Nucleation of Solid Metallic Solutions, Sc. D. Dissertation*, Mass. Int. Tech.; and (1961), *Acta. Met.* **9**, 525.

Hilliard, J. E. (1970), *Phase Transformations*, Aaronson, H. I. ed. American Society for Metals, Metals Park, Ohio, pp. 497-569.

James, G. and James, R. C. (1960), *Mathematical Dictionary*, Van Nostrand and Co., Princeton, N. J.

Kikuchi, R. and de Fontaine, D. (1976), *Scr. Metall.*, **10**, 995.

Langer, J. S., Bar-On, M. and Miller, H. D. (1975), *Phys. Rev. A*, **11**, 1417.

Ma, Z. and Torquato, S. (2017), *J.A.P.* **121**, 244904.

Rundman, K. B. and Hilliard, J. E. (1967), *Acta Met.*, **15**, 1025.

Thom, R. (1973), *Stabilité Structurelle et Morphogénèse*, Benjamin, Reading, MA.

Chapter 19

Summary and Conclusion

This book consists of two parts; Part I introduces the basic laws of classical thermodynamics and the resulting mathematical apparatus which follows from the *fundamental energy differential*. Part II describes various applications of interest to materials science, as summarized below.

The fundamental internal energy differential dU can be written as

$$dU = TdS + \sum_{j=1}^{n} Y_j \, dX_j \,,$$

introduced in Sec. 3.4 by Eqs. (3.18) or more compactly by (3.19), or else more explicitly by Eq. (3.66). All of these (equivalent) expressions are linear (Pfaffian) differential forms of "work terms" such as $Y \, dX$, involving an intensive variable times the differential of the conjugate extensive one. The expressions just cited differ from corresponding ones in classical mechanics or electromagnetism by the adjunction of the "thermal" term $T \, dS$ which transforms these classical formulations, as if by magic, into thermodynamic formulations. Of course, important and new concepts are introduced as well, those of *entropy S* and *absolute temperature*. For the subject of solution thermodynamics, there is also a "chemical work" term, $\mu_i \, dN_i$, to include for the N_i ($i = 1, 2, \ldots n$) chemical species and which introduces the *chemical potential μ*. The various contributions to the fundamental energy differential are made more explicit in Eq. (3.66), which is of course of the type (3.18).

It is useful to express energy potentials in terms of *natural variables* — those which are differentiated in the differential form — that are of the intensive rather than of extensive kind, more convenient to hold constant experimentally, for instance. This can be accomplished by Legendre transformations, as explained in Sec. 4.3. Regardless of the number and types of variables, the thermodynamic differential forms are integrable, thanks to the equality of certain partial second derivatives which are expressed in the Maxwell relations, given for example by Eq. (3.40). It is not always emphasized that, because the condition of integrability is verified, there is an explicit and general formula — Eq. (3.41) and Appendix A — for the expression of the Pfaffian differential forms in terms of integrals. These integrals are usually challenging to compute, primarily because the relationship between the variables X and Y are generally unknown. Then much of practical thermodynamics consists of finding suitable approximations for such relationships.

A simple example of a differential form for a closed system (*i.e.* one for which $dN_i = 0, \forall i$) is given in Sec. 3.4, yielding the Gibbs free energy $dg = -s\, dT + v\, dP$ expressed in terms of (intensive) temperature and pressure as natural variables and the *densities* entropy and volume. This important differential form of natural variables T and P is also pictured in Fig. 6.2, and discussed in Sec. 6.2. Figure 6.2 also illustrates schematically two approaches to performing thermodynamic calculations: (A) given the potential g as a function of T and P [in figures (a) and (d) respectively], obtain by differentiation the corresponding entropies and (molar) volumes [in figures (b) and (e) respectively], then differentiate once more to obtain the specific heat (c) and compressibility (f). Or else (B), start with experimentally determined or tabulated data, as represented schematically at panels (c) and (f), proceed by integration to entropies and volumes, then to the Gibbs function itself, as indicated schematically in the upper panels of Fig 6.2. It is of course simpler to differentiate than to integrate, and integration requires also the introduction of constants of integration, so that it would appear that method (A) would be far preferable to method (B), were it not for the fact that Gibbs energies are not readily available in most cases. Today however, it may be possible to calculate such potentials *ab initio*, *i.e.* from "first principles", as briefly shown in Sec. 9.4 in connection with Eq. (9.16). Thus, (A) is seen as a microscopic (better: mesoscopic) "top-down approach", whereas (B) is a macroscopic "bottom-up approach", the choice resting on what is available or feasible. Either way, the final objective is to integrate the differential form of a suitable potential from which just about everything follows.

As was mentioned in the Introduction, the general topic is that of *Classical* Thermodynamics, the qualifier signifying that only *macroscopic* objects are considered, and furthermore that, strictly speaking, only *homogeneous* fluids are examined. However, the treatment is enlarged in Part II of this book since certain sections do treat briefly the crystalline state, non-homogeneous (non-uniform) systems and also microscopic effects (in a brief introduction to statistical mechanics). The influence of stresses and strains in elastic solids is not covered at all. Kinetics and dynamics are not treated, except briefly in the next-to-last chapter (Ch. 18) in order to introduce the topic of *phase transformations*. Finally, and very importantly, classical thermodynamics implies the study of *equilibrium* phenomena, which may however be enlarged to *constrained* equilibria so as to include metastable states.

A full chapter is reserved for the important topic of *solutions*, and care is taken to distinguish between dependent and independent concentration variables. The use of n (unspecified) concentration variables may appear confusing at first, but it actually simplifies the mathematics by making the equations more symmetric and the notation more symbolic and general. Several graphic constructions are introduced which are required for temperature-composition phase diagrams, presented in Chapter 9. This topic, often regarded as the heart and soul of materials thermodynamics, is preceded by a short introduction to statistical mechanics, as a nod towards the microscopic realm.

Non-equilibrium systems, or glasses, present another departure from purely "classical" thermodynamics. The topic is presented in a "Macroscopic" manner, as indeed no general microscopic theory of the glass transition exists at present.[1] Classical chemical reactions come next, with applications to point defects, treated as chemical elements in crystals.

Thermodynamic systems are normally non-uniform, *i.e.* their local composition varies from place to place. To handle that non-uniformity, Gibbs took the important step of defining the notion of *phase*, portions of matter which are themselves uniform, in equilibrium with one another, and separated by phase boundaries. Gibbs was well aware that near interphases, and near other imperfections, concentration would vary, and that such non-uniformity would have to be taken into account. He therefore adopted the view that actual interphases could be replaced by "mathematical" surfaces of separation which would in some sense include the inhomogeneities through their own 2-dimensional equilibrium thermodynamics. That viewpoint is taken up in Chapter 13.

Another, quite different approach is to make the extensive variables of the system concentration-dependent. But real materials are composed of atoms and molecules, and the notion of local *concentration* reduces to atomistics. An intermediate solution is therefore required, one we may characterize as *mesoscopic*, *i.e.* in between the macro- and microscopic domains. In order to define a meaningful concentration *at a point* it is then necessary to define operationally a local averaging procedure, obtained

---

[1]In this book the word "transition" is replaced by the less specific word "transformation" for the simple reason that the so-called *glass transition* is not an *equilibrium* transition.

by taking a local (atomic) average in a small region about the point in question, the region being large enough to include several lattice points (we are talking about crystals here) and small enough to describe the main features of the inhomogeneities. That mesoscopic averaging is called *coarse-graining*, and is taken up in Chapters 14 and 15.

The latter chapter introduces the notion of *order-disorder* effects and that of the Landau theory of second-order transitions, but in a more general manner than that of the original formulations. In particular, the formalism includes both long-wavelength and short-wavelength instabilities, through the concept of *concentration waves*, which also requires coarse-graining, with the averaging performed on the neighboring points of the same lattice of the point selected (for clustering) or that of the same sublattice; equivalently, lattice Fourier transforms can take care of the averaging procedure. Chapter 16 concerns crystals, but mostly from the viewpoint of symmetry.

Chapter 17 deals with the topic of classical nucleation and growth, a non-equilibrium phenomenon made quasi-classical by the notion of the *critical nucleus* in metastable (actually unstable) equilibrium with the untransformed solution. Another aspect of the kinetics of phase transformations is presented in Chapter 18 with the introduction of spinodal decomposition and its ordering counterpart spinodal ordering. At one time, spinodal-decomposition and nucleation-and-growth were regarded as competing transformations, although they are in fact two aspects of the same mechanism according to more recent studies (Finel and co-workers, 2003 and 2008). The problem with older formulations lies with classical nucleation and the linear spinodal theories themselves, both of which are approximations only valid over narrow ranges of concentration of the untransformed solid solutions.

The aim of this book was thus to start from a restrictive, quasi-axiomatic treatment of classical thermodynamics, and to extend the formalism in Part II to illustrate non-uniformity, kinetics, order-disorder reactions, crystals and metastable materials (glasses). These non-classical topics were treated rather superficially, but were included to show the applicability of Gibbsian thermodynamics to areas not imagined by early practitioners of that venerable branch of learning. These extensions allowed bridging of length scales from the very large (light years!) to nanometers; but going further down to the atomic scale requires statistical mechanics plus a transitory step into the mesoscopic scale attained by coarse-graining. In the calculations proposed by Finel and co-workers (2003, 2008), the thermodynamics inside the averaging regions themselves are not neglected, but are performed by Monte-Carlo simulation with atomic interaction parameters obtained from *ab initio* quantum mechanical computer codes. Much more work is still required along those lines, but the general procedure for bridging length scales is well understood. In such advanced calculations, the overall guiding principle is still classical thermodynamics.

Appendices, mostly of mathematical character, are included in an effort to make the book reasonably self-contained.

References

Bronchart, Q., Le Bouar, Y. and Finel, A. (2008) *Phys. Rev. Lett.* **100**, 015702.

NATO Science Series II conference proceedings *Thermodynamics, Microstructures and Plasticity* (2003), Finel, A., Mazière, D. and Veron, M., Eds.

Appendix A

Pfaffian Differential Forms

A Pfaffian differential form is a linear combination of the differentials of independent variables x, y, z, ... with coefficients which are themselves differentiable functions of those variables, thus:

$$dw = X(x, y, z, \ldots)\, \mathrm{d}x + Y(x, y, z, \ldots)\, \mathrm{d}y + Z(x, y, z, \ldots)\, \mathrm{d}z + \ldots\,.^{1} \quad \text{(A.1)}$$

In Sec. 3.5, the fundamental theorem for insuring that a Pfaffian form was an *exact* differential was given for the case of two independent variables. The necessary condition was derived explicitly, but the sufficient condition and method of integration is given here, at first for the case of two independent variables x and y, as taken from the textbook of de La Vallée Poussin (1949).

For the sufficient condition, assume that the criterion of "exactness" (3.40) is fulfilled. Choose a function such that

$$\frac{\partial f}{\partial x} = X(x, y)\,.$$

Now choose an arbitrary but fixed constant a. The definite integral

$$\int_{a}^{x} X(\alpha, y)\, \mathrm{d}\alpha$$

is a particular solution of this differential equation. The general solution is obtained by adding a constant of integration, which however, must still in

---

[1] Note the use here of standard mathematical d (italicized) as distinct from the straight d used to indicate the operator $\mathrm{d} \equiv$ differentiation, introduced below, in anticipation that the function which we seek is integrable, as in $\mathrm{d}f(x, y)$, indicating an *exact* differential. The expression dw is then merely a label for the differential form, integrable or not.

general be a function of the remaining variable y; call it $K(y)$. Hence we have the general solution

$$f(x,y) = \int_a^x X(\alpha, y)\, \mathrm{d}\alpha + K(y)\,. \tag{A.2}$$

Let us now further impose the condition

$$\frac{\partial f}{\partial y} = Y(x,y)\,. \tag{A.3}$$

By taking the derivative of both sides of Eq. (A.3) with respect to y we obtain

$$\frac{\partial f}{\partial y} = \int_a^x \frac{\partial X}{\partial y}\, \mathrm{d}\alpha + \frac{\mathrm{d}K}{\mathrm{d}y} = Y(x,y)\,.$$

We now make use of the criterion (3.40) to obtain

$$\frac{\partial f}{\partial y} = \int_a^x \frac{\partial Y}{\partial \alpha}\, \mathrm{d}\alpha + \frac{\mathrm{d}K}{\mathrm{d}y} = Y(x,y) - Y(a,y) + \frac{\mathrm{d}K}{\mathrm{d}y}\,.$$

Since Eq. (A.3) holds, we must have

$$\frac{\mathrm{d}K}{\mathrm{d}y} = Y(a,y)$$

and the solution of that ordinary differential equation is given by the indefinite integral

$$K(y) = \int Y(a,y)\, \mathrm{d}y + C\,,$$

where C is a constant of integration. Inserting that latter expression into Eq. (A.2) gives the required solution

$$f(x,y) = \int_a^x X(\alpha, y)\, \mathrm{d}\alpha + \int Y(a,y)\, \mathrm{d}y + C \tag{A.4}$$

in terms of an indefinite integral and a constant of integration C. This solution is just the one given by Eq. (3.41) in terms of definite integrals and with the constant of integration taking care of the lower limit b. Likewise, for three independent variables, the most general Pfaffian form

$$\mathrm{d}w = X(x,y,z)\, \mathrm{d}x + Y(x,y,z)\, \mathrm{d}y + Z(x,y,z)\, \mathrm{d}z \tag{A.5}$$

is an exact differential, which therefore can be integrated without specifying a path of integration, when the following conditions hold:

$$\frac{\partial X}{\partial y} = \frac{\partial Y}{\partial x}\,, \quad \frac{\partial Y}{\partial z} = \frac{\partial Z}{\partial y}\,, \quad \frac{\partial X}{\partial z} = \frac{\partial Z}{\partial x}\,. \tag{A.6}$$

It will be recognized that Eq. (A.6) is just the condition that the curl of a vector $\mathbf{F}$, of components X, Y and Z, vanishes; in vector notation:

$$\nabla \times \mathbf{F} = 0 .$$

If Eq. (A.6) holds, then the differential form (A.5) is an exact differential whose integrated form, written with definite integrals, is

$$f(x, y, z) = \int_a^x X(\alpha, y, z) \, \mathrm{d}\alpha$$

$$+ \int_b^y Y(a, \beta, z) \, \mathrm{d}\beta + \int_c^z Z(a, b, \gamma) \, \mathrm{d}\gamma , \quad (A.7)$$

where a and b are arbitrary but fixed parameters, and c takes care of the constant of integration. Note carefully the placement of a and b in the integrands of the second and third integrals. These results can obviously be generalized to an arbitrary number of independent variables.

Example

Let us consider the following example, arbitrarily chosen, but with the proviso that the integrability condition (3.40) holds:

$$X(x, y) = 2x + y, \qquad Y(x, y) = x + 4y \qquad (A.8)$$

for which the integrability condition simply gives $1 = 1$.

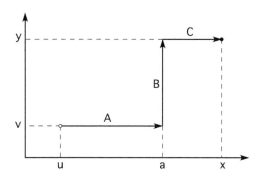

Figure A.1: Example of path of integration.

General method of integration

According to (3.41), we must use here explicitly

$$f(x, y) = \int_u^x (2\alpha + y) \, \mathrm{d}\alpha + \int_v^y (a + 4\beta) \, \mathrm{d}\beta , \qquad (A.9)$$

and integrate between initial point of coordinates (u, v) and final point with running coordinates (x,y) (see Fig. A.1). After performing the two required simple integrations (α and β and are dummy variables of integration), and after choosing an arbitrary but fixed parameter a (arbitrarily placed here on the x axis), we have

$$f(x, y) = \left[\alpha^2 + y\alpha\right]_a^x + \left[a\beta + 2\beta^2\right]_v^y \tag{A.10}$$

which, after canceling and rearranging yields the required result

$$f(x, y) = (x^2 + xy + 2y^2) - (a^2 + av + 2v^2). \tag{A.11}$$

Interestingly, only two bracketed expressions are left, the first one containing only upper limit variables, the second one only lower limit parameters, a property resulting from the canceling of the ay terms, the only ones which contained "mixed" parameters. Thus the difference Δf between final and initial values of the integrated differential form depend only on those two states, as is required for "point" thermodynamic functions; that is not the case for quantities resulting from integration of "inexact" differentials such as those of heat Q and of work W.

It is also observed that the form of the result, Eq. (A.11), is independent of the placement of the parameter a. True, the second bracket of Eq. (A.11) does contain the parameter a, but it is mixed in with the lower limit v of the second integral, resulting in a general constant of integration $C = (a^2 + av + 2v^2)$. Note also that the path A-B-C does not appear in the derivation.

Integration along a path

In this more elementary method of integration, a segmented zig-zag path is specified in the (x, y) plane in such a way that each corresponding integral is to be performed either along the x or y axis, along a path containing as many zigs and zags as desired. The general integration formula (A.11) is not used; instead we have the three integrals to compute, each over a single variable only,

$$f(x, y) = \overbrace{\int_u^a (2\alpha + v) \, d\alpha}^{A} + \overbrace{\int_v^y (a + 4\beta) \, d\beta}^{B} + \overbrace{\int_a^x (2\gamma + y) \, d\gamma}^{C} \tag{A.12}$$

yielding

$$f(x, y) = \left[\alpha^2 + v\alpha\right]_u^a + \left[a\beta + 2\beta^2\right]_v^y + \left[\gamma^2 + y\gamma\right]_a^x. \tag{A.13}$$

After cancellations and rearrangement we find

$$f(x,y) = (x^2 + xy + 2y^2) - (a^2 + av + 2v^2)\,;$$

just as in Eq. (A.11), illustrating once more, the path independence and the equivalence of the two methods, the algebraic and the graphical, so to speak. The former, relying on the necessary and sufficient proof given above and in Sec. 3.5, is actually the one which is simpler to carry out.

Reference

de La Vallée Poussin, Ch.-J. (1949). *Cours d'Analyse Infinitésimale*, Librairie Universitaire, Louvain.

Appendix B

The Second Law and the TdS Derivation

Recall that, with $U = U(\sigma, X_1, X_2, \ldots X_n)$, the differential of internal energy is given by expression (3.10). Comparing that expression with dU given by Eq. (3.8) leads to Eqs. (3.11) to (3.14). Going from Eqs. (3.13) to (3.15), i.e. going from the $\lambda\,d\sigma$ expression to $T\,dS$ requires a bit of additional work, however. Here is how it can be done, following Chandrasekhar (1957): Rewrite Eqs. (3.13) and (4.16) as

$$dQ_R = \tau\,d\sigma, \qquad \tau = \frac{\partial U}{\partial \sigma} \qquad (\text{B.1})$$

["τ" is more suitable than "λ" for reasons which will become apparent by comparing to the desired end result $dQ_R = T\,dS$].

Consider a system consisting of two subsystems "1" and "2". For simplicity, assume that the thermodynamic state of each (sub)system is determined by two independent variables. For each system take X_i and θ_i ($i = 1, 2$) as independent variables. Let systems 1 and 2 be mutually separated by an adiabatic partition but be separately in thermal equilibrium with a common "reservoir". Figure B.1 depicts the situation which is similar to that shown on the left-hand frame of Fig. 3.1. By the zeroth law, we may affirm that systems 1 and 2 are then also in thermal equilibrium with one another. Hence, there is but one empirical temperature to consider: $\theta_1 = \theta_2 = \theta$. Since part one of the proof (necessary condition) introduced a new state variable σ, we may take σ_i for X_i, so that the variables for the combined system are $(\sigma_1, \sigma_2, \theta)$. By Eq. (B.1), and for each system, sub-

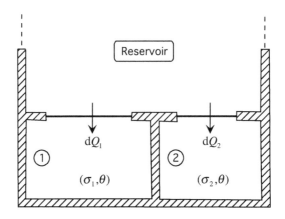

Figure B.1: Two sub-systems and a reservoir at same temperature, θ.

and combined, we may express the reversible differential of heat as

$$dQ_1 = \tau_1(\sigma_1, \theta)\, d\sigma_1 \tag{B.2a}$$

$$dQ_2 = \tau_2(\sigma_2, \theta)\, d\sigma_2 \tag{B.2b}$$

$$dQ = \tau(\sigma_1, \sigma_2, \theta)\, d\sigma \tag{B.2c}$$

with $\sigma = \sigma(\sigma_1, \sigma_2, \theta)$. By the "additive" nature of Q [all differentials in Eq. (B.2) pertain to extensive variables] $dQ = dQ_1 + dQ_2$, we have

$$\tau\, d\sigma = \tau_1\, d\sigma_1 + \tau_2\, d\sigma_2 \ . \tag{B.3}$$

By comparing to the total differential, and recalling the differentiation can only be done one way

$$d\sigma = \frac{\partial \sigma}{\partial \sigma_1}\, d\sigma_1 + \frac{\partial \sigma}{\partial \sigma_2}\, d\sigma_2 + \frac{\partial \sigma}{\partial \theta}\, d\theta \tag{B.4}$$

we obtain

$$\frac{\partial \sigma}{\partial \sigma_1} = \frac{\tau_1}{\tau}, \quad \frac{\partial \sigma}{\partial \sigma_2} = \frac{\tau_2}{\tau}, \quad \frac{\partial \sigma}{\partial \theta} = 0 \ . \tag{B.5}$$

Hence, σ turns out to be a function of σ_1 and σ_2 only. Therefore, by the first two equations of (5), the ratios of τ's are also independent θ:

$$\frac{\partial(\tau_1/\tau)}{\partial \theta} = \frac{\partial(\tau_2/\tau)}{\partial \theta} = 0 \tag{B.6}$$

or

$$\frac{1}{\tau_1}\frac{\partial \tau_1}{\partial \theta} = \frac{1}{\tau_2}\frac{\partial \tau_2}{\partial \theta} = \frac{1}{\tau}\frac{\partial \tau}{\partial \theta}\ . \tag{B.7}$$

Since τ_1 is a function of σ_1 and θ only, and τ_2 is a function σ_2 and θ only, each of the equated expressions in Eq. (B.7) must be a function of the empirical temperature θ only:

$$\frac{\partial \ln \tau_1}{\partial \theta} = \frac{\partial \ln \tau_2}{\partial \theta} = \frac{\partial \ln \tau}{\partial \theta} = \varphi(\theta). \tag{B.8}$$

This function φ is independent of the system considered: once a "thermometer" has been specified, then $\varphi(\theta)$ is a *universal* function of the empirical temperature, the same for all possible systems in the universe!

Partial differential equations (B.8) may be integrated, the "constant" of integration being expressed below as the logarithm of some function (Ψ) of the remaining variables:

$$\ln \tau_i = \int \varphi(\theta) \, d\theta + \ln \Psi_i(\sigma_i) \quad (i = 1, 2)$$

$$\ln \tau = \int \varphi(\theta) \, d\theta + \ln \Psi(\sigma_1, \sigma_2)$$

or

$$\tau = \Psi(\sigma_1, \sigma_2) \, e^{\int \varphi(\theta) \, d\theta} \tag{B.9a}$$

$$\tau_i = \Psi_i(\sigma_i) \, e^{\int \varphi(\theta) \, d\theta}. \tag{B.9b}$$

We now define the all-important *absolute temperature*

$$T = C e^{\int \varphi(\theta) \, d\theta} \tag{B.10}$$

where the constant C allows for different choices of temperature scales (the "size" of a "degree"). Note that, as a consequence of the integration process, the function T *cannot contain any arbitrary additive constant*; scaling is allowed (because of the arbitrary C), but *zero of the T scale is fixed*. It is the *absolute zero of temperature*. Note also that $T = 0$ can be reached only in the limit of $\int \varphi(\theta) \, d\theta$ tending to $-\infty$, a strong suggestion that there is something quite "singular" about the absolute zero.

Let us now incorporate C implicitly into the absolute temperature T so that Eq. (B.2) may be written, with the help of Eq. (B.8) as

$$dQ_1 = \tau_1 \, d\sigma_1 = T \, \Psi_1(\sigma_1) \, d\sigma_1 \tag{B.11a}$$

$$dQ_2 = \tau_2 \, d\sigma_2 = T \, \Psi_2(\sigma_2) \, d\sigma_2 \tag{B.11b}$$

$$dQ = \tau \, d\sigma = T \, \Psi_1(\sigma_1, \sigma_2) \, d\sigma. \tag{B.11c}$$

In the first two Eqs. (B.11) $\Psi_i(\sigma_i) \, d\sigma_i = dS_i$ $(i = 1, 2)$ is surely an exact differential, because only one independent variable appears. We do not need a special law to derive entropy S_i at this level. In his original papers, Carathéodory made that point very clearly, and Max Born emphasized it by stating that "proofs" of the integrability of the entropy differentials are just so much "empty verbiage" if they rest on discussions of systems of only one variable. So, the problem is: can we write, in the third Eq. (B.11) $\Psi(\sigma_1, \sigma_2)\, d\sigma$ as $\Psi(\sigma)\, d\sigma$? To show that it is possible to do so requires still a little more work.

By substituting the three expressions (B.11) into Eqs. (B.2) and eliminating the common factor T, we have:

$$\Psi(\sigma_1, \sigma_2)\, d\sigma = \Psi_1(\sigma_1)\, d\sigma_1 + \Psi_2(\sigma_2)\, d\sigma_2. \tag{B.12}$$

Equation (B.12) may be considered as an expression for the total differential of the function $\sigma(\sigma_1, \sigma_2)$ so that

$$\Psi_1(\sigma_1) = \Psi(\sigma_1, \sigma_2)\frac{\partial \sigma}{\partial \sigma_1}, \quad \Psi_2(\sigma_2) = \Psi(\sigma_1, \sigma_2)\frac{\partial \sigma}{\partial \sigma_2}. \tag{B.13}$$

These expressions may be differentiated once more, yielding

$$\frac{\partial \Psi_1}{\partial \sigma_2} = \frac{\partial \Psi}{\partial \sigma_2}\frac{\partial \sigma}{\partial \sigma_1} + \Psi \frac{\partial^2 \sigma}{\partial \sigma_2 \partial \sigma_1} = 0, \tag{B.14a}$$

$$\frac{\partial \Psi_2}{\partial \sigma_1} = \frac{\partial \Psi}{\partial \sigma_1}\frac{\partial \sigma}{\partial \sigma_2} + \Psi \frac{\partial^2 \sigma}{\partial \sigma_2 \partial \sigma_1} = 0, \tag{B.14b}$$

(recall that Ψ_1 depends only σ_1, and Ψ_2 on σ_2). Since the order of differentiation may be interchanged in the second derivative, it follows that the Jacobian (see Appendix E) of the transformation $(\sigma_1, \sigma_2) \to (\Psi, \sigma)$ vanishes:

$$\frac{\partial(\Psi, \sigma)}{\partial(\sigma_1, \sigma_2)} \equiv \begin{vmatrix} \frac{\partial \Psi}{\partial \sigma_1} & \frac{\partial \Psi}{\partial \sigma_2} \\ \frac{\partial \sigma}{\partial \sigma_1} & \frac{\partial \sigma}{\partial \sigma_2} \end{vmatrix} = \frac{\partial \Psi}{\partial \sigma_1}\frac{\partial \sigma}{\partial \sigma_2} - \frac{\partial \Psi}{\partial \sigma_2}\frac{\partial \sigma}{\partial \sigma_1} = 0. \tag{B.15}$$

That in turn means that there exists a relationship between the variables Ψ and σ; they are not independent. That relationship may be written $F(\Psi, \sigma) = 0$ for example, or $\Psi = \Psi(\sigma)$. In other words, Ψ is not a function of σ_1 and σ_2 independently, but a function of these variables in the particular combination $\sigma(\sigma_1, \sigma_2)$. Finally, we may write Eq. (B.11c) in the form of the desired result

$$dQ = \tau\, d\sigma = T\, dS \tag{B.16}$$

with the new function S, *entropy*, defined by its *exact* differential

$$dS = \Psi(\sigma)\, d\sigma. \tag{B.17}$$

The integral form of the entropy is thus

$$S = \int \Psi(\sigma)\, d\sigma + \text{const} \tag{B.18}$$

where "const" is a constant of integration which the third law attempts to deal with. From Eq. (B.3) we also have

$$dS = dS_1 + dS_2 \tag{B.19}$$

which exhibits the additive (extensive) nature of entropy. Note also that, unlike $T(\theta)$, entropy cannot be expressed by means of "universal functions" since, as seen in Eqs. (B.11) the functions Ψ_1, Ψ_2, Ψ are all different (in general) and depend on the system in question.

From Eq. (B.9b) we obtain the usual definition of entropy

$$dS = \frac{dQ_R}{T} \tag{B.20}$$

(the subscript R, for *reversible*, has been reinstated for emphasis). In mathematical terms we say that *the absolute temperature T is an integrating denominator for the (inexact) differential of heat dQ_R (reversible)*. In that sense, T converts the inexact dQ_R into the exact dS; thus S, the entropy, is a *state function* (or "point") function.

It is in that somewhat long-winded way that the notion of entropy in classical thermodynamics can be derived from a statement of the second law; not as it is sometimes stated, as part of the second law. The derivation is not straightforward, but the reward is one of the most important notions of classical science, that is to say: *entropy*.

Appendix C

Euler's Theorem on Homogeneous Functions

A function $f(x_1, x_2, \ldots x_n)$ of the variables x_i $(i = 1, 2 \ldots n)$ is said to be homogeneous of degree n if, by definition, it satisfies the identity

$$f(tx_1, tx_2, \ldots) = t^n f(x_1, x_2, \ldots), \qquad (C.1)$$

where t denotes an arbitrary constant. Euler's theorem states that for the homogeneous function $f(x_1, x_2, \ldots x_n)$ of degree n we have, identically,

$$x_1 f_{,1}(x_1, x_2, \ldots) + x_2 f_{,2}(x_1, x_2, \ldots) + \cdots = n\, f(x_1, x_2, \ldots), \qquad (C.2)$$

with obvious shorthand notation for the partial derivatives [see Eq. (3.34)]

$$f_{,i} = \frac{\partial f}{\partial x_i} \quad , \quad (i = 1, 2 \ldots n), \qquad (C.3)$$

and conversely, that expression (C.2) implies expression (C.1).

To prove the direct statement that Eq. (C.1) implies Eq. (C.2), simply take the derivative of Eq. (C.1)) to obtain

$$x_1 f_{,1}(tx_1, tx_2, \ldots) + x_2 f_{,2}(tx_1, tx_2, \ldots) + \cdots = n\, t^{n-1} f(x_1, x_2, \ldots), \qquad (C.4)$$

whence, by setting $t = 1$, one recovers Eq. (C.2). To prove the reciprocal statement that Eq. (C.2)) implies Eq. (C.1), substitute tx_i for x_i (for $i = 1, 2 \ldots n$) in Eq. (C.2). After division, there follows the expression

$$\frac{\sum_{i=1}^{n} x_i f_{,i}(tx_1, tx_2, \ldots)}{f(tx_1, tx_2, \ldots)} = \frac{n}{t}. \qquad (C.5)$$

It is seen that the latter expression is just the logarithmic derivative

$$\frac{d}{dt} \log f(tx_1, tx_2, \ldots) = \frac{d}{dt} \log t^n . \tag{C.6}$$

Integration then yields

$$\log f(tx_1, tx_2, \ldots) = \log t^n + \log C , \tag{C.7}$$

where C is a constant of integration, independent of t. Equating the arguments of the logarithms gives

$$f(tx_1, tx_2, \ldots) = Ct^n$$

and then with $t = 1$,

$$f(x_1, x_2, \ldots) = C .$$

Substituting for C then yields the required result

$$f(tx_1, tx_2, \ldots) = t^n f(x_1, x_2, \ldots)$$

which is just Eq. (C.1), the definition of a homogeneous function of degree n. The expressions (C.1) and (C.2) are thus completely equivalent.

Appendix D

Constrained Extrema and Lagrange Multipliers

Consider the function
$$F \equiv F(x_1, x_2, \ldots x_n)$$
suitably differentiable, for which we wish to find the coordinates of the extrema, subject to the $1 \leq q < n$ differentiable conditions, or constraints

$$G_i(x_1, x_2, \ldots x_n) = 0 \qquad (i = 1, 2, \ldots q). \tag{D.1}$$

Thus, among all the n variables, only $p = n - q$ are independent.

Direct Method

For the function F to have an extremum its value must be stationary for any small virtual variation of its coordinates dx_j, hence the total differential of F must vanish at that point:

$$dF = F_{,1}dx_1 + F_{,2}dx_2 + \ldots F_{,n}dx_n = 0. \tag{D.2}$$

Also, the total differentials of the G constraints must vanish, but for a different reason, which is that expressed by Eqs. (D.1):

$$dG_i = G_{i,1}dx_1 + G_{i,2}dx_2 + \ldots G_{i,n}dx_n = 0, \quad (i = 1, 2, \ldots q). \tag{D.3}$$

In both of these equations, and as in Eqs. (3.34), we have used the abbreviations

$$F_{,j} = \frac{\partial F}{\partial x_j}, \quad G_{i,j} = \frac{\partial G_i}{\partial x_j}, \quad (j = 1, 2 \ldots n). \tag{D.4}$$

If there were no constraints on F, then for arbitrary variations $\mathrm{d}x_i$, all partial derivatives $F_{,i}$ would have to vanish, yielding n equations in the n variables x_i, a soluble system. But if constraints are present, then the $\mathrm{d}x_i$ cannot be varied independently and the simple vanishing of all the partial derivatives is no longer the correct procedure. What must be done is to eliminate the dependent variations by solving the linear system (D.3) for these dependent differentials and inserting their values into the differential form (D.2). There remains a differential form in the independent differentials $\mathrm{d}x_i$ with $i = 1, 2, \ldots p$ and $p = n - q$ (the ordering of variables is immaterial), all the corresponding coefficients of which can now be zeroed out, yielding p equations in n variables. The required additional equations are the q equations of conditions (D.1). With $n = p + q$ the system of equations is solvable.

For the general case, the algebra can be daunting. For the case $q = 1$ (with $p = n - 1$) however, the procedure is quite simple: it suffices to solve for the differential $\mathrm{d}x_n$ in the only G–equation remaining in system (D.3)

$$\mathrm{d}x_n = -\frac{1}{G_{,n}}(G_{,1}\mathrm{d}x_1 + G_{,2}\mathrm{d}x_2 + \ldots G_{,n-1})$$

and insert back into Eq. (D.2), giving

$$\mathrm{d}F = \left[F_{,1} - \frac{G_{,1}}{G_{,n}}F_{,n}\right]\mathrm{d}x_1 + \ldots \left[F_{,n-1} - \frac{G_{,n-1}}{G_{,n}}F_{,n}\right]\mathrm{d}x_{n-1}. \qquad (\mathrm{D.5})$$

Now all $\mathrm{d}x_j$ differentials with $(j = 1, 2, \ldots n - 1)$ are independent and all $p = n - 1$ bracketed expressions may be set equal to zero, so that we have, equivalently

$$\begin{vmatrix} F_{,j} & F_{,n} \\ G_{,j} & G_{,n} \end{vmatrix} = 0 \qquad (j = 1, 2, \ldots n - 1). \qquad (\mathrm{D.6})$$

These $n - 1$ determinental equations (D.6) provide $n - 1$ equations in the variables x_j, the additional n^{th} one being given by the lone constraint of system (D.1).

For the general case of $2 \le q < n$, the equations corresponding to Eqs. (D.6) are quite similar, but with the elements of type G being replaced by determinants resulting from Cramer's rule applied the qth-order system (D.3); in particular, element $G_{,n}$ in (D.6) will be replaced by the determinant of system (D.3). The algebra then tends to be rather heavy, so that the method of undetermined multipliers, to which we now turn, is generally preferred.

Method of Lagrange Multipliers

In this clever method of seeking constrained extrema, we introduce as many *a priori* undetermined multipliers λ_j as there are equations of constraints (D.1), *i.e.* q in the present notation. We then multiply each total differential dG_i of the system (D.3) by the corresponding λ_i and add them all, term for term, to the differential dF of Eq. (D.2), grouping terms multiplied by the same differential, thus obtaining the differential form

$$K_1 dx_1 + K_2 dx_2 + \ldots K_n dx_n = 0 \tag{D.7}$$

with K-factors given by

$$K_j = F_j + \lambda_1 G_{1,j} + \lambda_2 G_{2,j} + \ldots \lambda_q G_{q,j}. \tag{D.8}$$

Let us now assume that the q undetermined multipliers λ_j are chosen so as to make those K_j factors vanish for just q of the differentials dx_j, so that the remaining total differential form (D.7) now contains only $p = n - q$ independent differentials. The p corresponding K-factors may now each be set to equal to zero at the extremum. It follows that *all* K_j factors may be set equal to zero, although for two different reasons. So, we have arrived at a system of n equations coupled to the q equations of constraints Eqs. (D.1) giving $n + q$ equations for the n variables x_j and the q constants λ_i.

And now for the clincher: The resulting system of equations which defines the extremum conditions, is exactly the same as that which would be obtained by seeking the *unconstrained* extrema of the function Φ given by

$$\Phi = F + \lambda_1 G_1 + \lambda_2 G_2 + \ldots \lambda_n G_n. \tag{D.9}$$

It is as if the introduction of the constants λ_i had "liberated" the function to be maximized, allowed it, or at least its replacement Φ, to be treated as a problem of unconstrained extrema. The multipliers also allow the problem a greater symmetry, and moreover, it turns out in many cases that these multipliers have important practical significance, as will be seen in the section on statistical thermodynamics (Sec. 8.2, for example).

To summarize: the general procedure for finding the extrema of a function F of the variables $(x_1, x_2, \ldots x_n)$ subject to the constraints G is to differentiate totally the function F, then add to it the total differentials of the constraints G_i multiplied respectively by unknown constants λ_i to produce the differential of a new function, Φ. That function is then to be treated as not subjected to constraints, so that the coefficients of all the differentials dx_j should be set to zero, and the resulting n equations plus the equation of constraints (D.1) should be solved for the maximizing unknown and the Lagrange multipliers. Written out in full, the resulting system of n equations in $n + q$ variables should look like this:

$$\begin{cases} \frac{\partial F}{\partial x_1} + \lambda_1 \frac{\partial G_1}{\partial x_1} + \lambda_2 \frac{\partial G_2}{\partial x_1} + \ldots + \lambda_q \frac{\partial G_q}{\partial x_1} = 0 \\[2mm] \frac{\partial F}{\partial x_2} + \lambda_1 \frac{\partial G_1}{\partial x_2} + \lambda_2 \frac{\partial G_2}{\partial x_2} + \ldots + \lambda_q \frac{\partial G_q}{\partial x_2} = 0 \\[2mm] \cdots\cdots\cdots\cdots\cdots\cdots\cdots\cdots\cdots\cdots\cdots\cdots\cdots \\[2mm] \frac{\partial F}{\partial x_q} + \lambda_1 \frac{\partial G_1}{\partial x_q} + \lambda_2 \frac{\partial G_2}{\partial x_q} + \ldots + \lambda_q \frac{\partial G_q}{\partial x_q} = 0 \\[2mm] \cdots\cdots\cdots\cdots\cdots\cdots\cdots\cdots\cdots\cdots\cdots\cdots\cdots \\[2mm] \frac{\partial F}{\partial x_n} + \lambda_1 \frac{\partial G_1}{\partial x_n} + \lambda_2 \frac{\partial G_2}{\partial x_n} + \ldots + \lambda_q \frac{\partial G_q}{\partial x_n} = 0 \end{cases} \qquad (D.10)$$

which must be augmented by the q constraints (D.1)

$$G_i(x_1, x_2, \ldots x_n) = 0 \qquad (i = 1, 2, \ldots q)$$

to obtain the complete solution. The present derivation, which follows that of de La Vallée Poussin (1949), may appear circuitous, but how else can one demonstrate that the introduction of Lagrange multipliers almost magically transforms a problem of constrained extrema into one of free extrema, despite the cost of additional equations to be solved?

References

de La Vallée Poussin, Ch.-J. (1949). *Cours d'Analyse Infinitésimale*, Librairie Universitaire, Louvain.

Appendix E

Jacobians in Thermodynamics

Consider n independent variables $y_i (i = 1, 2, \ldots n)$ and $x_j (j = 1, 2, \ldots n)$ differentiable functions of these variables:

$$x_j = f_j(y_1, y_2, \ldots y_n).$$

The Jacobian of the transformation of the y to the x variables is, by definition, the determinant

$$J \equiv \begin{vmatrix} \frac{\partial x_1}{\partial y_1} & \frac{\partial x_1}{\partial y_2} & \cdots & \frac{\partial x_1}{\partial y_n} \\ \frac{\partial x_2}{\partial y_1} & \frac{\partial x_2}{\partial y_2} & \cdots & \frac{\partial x_2}{\partial y_n} \\ \cdots & \cdots & \cdots & \cdots \\ \frac{\partial x_n}{\partial y_1} & \frac{\partial x_n}{\partial y_2} & \cdots & \frac{\partial x_n}{\partial y_n} \end{vmatrix} \overset{\text{def}}{=} \frac{\partial(x_1, x_2, \ldots, x_n)}{\partial(y_1, y_2, \ldots, y_n)}. \tag{E.1}$$

Properties of Jacobians can be derived from the known properties of determinants. Here are some of the most important ones, reminiscent of the analogous properties of partial derivatives:

Property 1.

If all y variables are the same as all x (*i.e.* $x_j = y_j$), then we have

$$J = \frac{\partial(x_1, \ldots, x_n)}{\partial(x_1, \ldots, x_n)} = 1. \tag{E.2}$$

Property 2.

If the y variables are themselves functions of an equal number of z variables, then we have, by the property of multiplication of determinants,

$$
\begin{vmatrix} \frac{\partial x_1}{\partial z_1} & \cdots & \frac{\partial x_1}{\partial z_n} \\ \cdots\cdots\cdots\cdots\cdots \\ \frac{\partial x_n}{\partial z_1} & \cdots & \frac{\partial x_n}{\partial z_n} \end{vmatrix}
$$

$$
= \begin{vmatrix} \left(\frac{\partial x_1}{\partial y_1}\frac{\partial y_1}{\partial z_1} + \cdots + \frac{\partial x_1}{\partial y_n}\frac{\partial y_n}{\partial z_1} \right) & \cdots & \left(\frac{\partial x_1}{\partial y_1}\frac{\partial y_1}{\partial z_n} + \cdots + \frac{\partial x_1}{\partial y_n}\frac{\partial y_n}{\partial z_n} \right) \\ \cdots\cdots\cdots\cdots\cdots\cdots\cdots\cdots\cdots\cdots\cdots\cdots\cdots \\ \left(\frac{\partial x_n}{\partial y_1}\frac{\partial y_1}{\partial z_1} + \cdots + \frac{\partial x_n}{\partial y_1}\frac{\partial y_1}{\partial z_1} \right) & \cdots & \left(\frac{\partial x_n}{\partial y_1}\frac{\partial y_1}{\partial z_1} \cdots + \cdots \frac{\partial x_n}{\partial y_n}\frac{\partial y_n}{\partial z_n} \right) \end{vmatrix}
$$

$$
= \begin{vmatrix} \frac{\partial x_1}{\partial y_1} & \cdots & \frac{\partial x_1}{\partial y_n} \\ \cdots\cdots\cdots\cdots \\ \frac{\partial x_n}{\partial y_1} & \cdots & \frac{\partial x_n}{\partial y_n} \end{vmatrix} \times \begin{vmatrix} \frac{\partial y_1}{\partial z_1} & \cdots & \frac{\partial y_1}{\partial z_n} \\ \cdots\cdots\cdots\cdots \\ \frac{\partial y_n}{\partial z_1} & \cdots & \frac{\partial y_n}{\partial z_n} \end{vmatrix}, \quad \text{(E.3)}
$$

so that

$$
\frac{\partial(x_1, \ldots, x_n)}{\partial(z_1, \ldots, z_n)} = \frac{\partial(x_1, \ldots, x_n)}{\partial(y_1, \ldots, y_n)} \frac{\partial(y_1, \ldots, y_n)}{\partial(z_1, \ldots, z_n)}. \quad \text{(E.4)}
$$

Property 3

It follows that

$$
\frac{\partial(x_1, \ldots, x_n)}{\partial(y_1, \ldots, y_n)} = \frac{1}{\frac{\partial(y_1, \ldots, y_n)}{\partial(x_1, \ldots, x_n)}}. \quad \text{(E.5)}
$$

Property 4

If two columns — or two rows — of a Jacobians (J) are interchanged, its absolute value stays the same, but with a change of sign. If a Jacobian has two identical columns — or two rows — it vanishes. Indeed, by the preceding property, interchanging those two columns — or rows — changes the sign of J, but since the interchanging columns (rows) are identical, the value of J does not change, so that $J = -J$, hence $J = 0$.

Property 5

In thermodynamics it is often useful to express partial derivatives in terms of Jacobians, in the following manner:

$$
J = \left(\frac{\partial x_k}{\partial y_k} \right)_{y_{i \neq k}} = \frac{\partial(y_1, \ldots, x_k, \ldots, y_n)}{\partial(y_1, \ldots, y_k, \ldots, y_n)}. \quad \text{(E.6)}
$$

This result is formally correct since all off-diagonal elements of the Jacobian $(\partial x_i/\partial y_j)$ for $i \neq j$ vanish and all diagonal elements $(\partial y_i/\partial y_i)$ are equal to unity, except for the one of interest.

Property 6

One property, of interest to Appendix B in the TdS derivation, is the following: if there exists a functional relationship between the variables x_i, $F(x_1, \ldots, x_n) = 0$, then the Jacobian (E.1) vanishes.

Property 7

The following property follows directly from Property 5. Start with the Jacobian (E.1) (or rather its inverse written with y variables instead of the x variables, and conversely) and express it as a product of other Jacobians, related to the unit Jacobian (E.2) by performing a set of operations successively replacing the y's by x's in the "denominators" of the "fractions" formally representing the Jacobians, as indicated in the first line of the equation

$$\frac{\partial(y_1, y_2, \ldots, y_n)}{\partial(x_1, x_2, \ldots, x_n)} = \frac{\partial(y_1, y_2, \ldots, y_n)}{\partial(x_1, y_2, \ldots, y_n)} \frac{\partial(x_1, y_2, \ldots, y_n)}{\partial(x_1, x_2, \ldots, y_n)} \cdots \frac{\partial(x_1, x_2, \ldots, y_n)}{\partial(x_1, x_2, \ldots, x_n)}$$

$$= \left(\frac{\partial y_1}{\partial x_1}\right)_{y_2 \cdots y_n} \times \left(\frac{\partial y_2}{\partial x_2}\right)_{x_1, y_3 \cdots y_n} \cdots \times \left(\frac{\partial y_n}{\partial x_n}\right)_{x_1, x_2 \ldots, x_{n-1}} \tag{E.7}$$

and as used in Sec. 16. Then in the second line of Eq. (E.7), the individual Jacobians are expressed as partial derivatives. The variables maintained constant are indicated by subscripts, as usual, but here they can be "mixed", partly x, partly y.

Example

Recall the definitions of some useful thermodynamic properties:

$$\beta_p = \frac{1}{V}\left(\frac{\partial v}{\partial T}\right)_P \qquad \text{isobaric expansivity,} \tag{E.8a}$$

$$\kappa_T = -\frac{1}{V}\left(\frac{\partial v}{\partial P}\right)_T \qquad \text{isothermal compressibility,} \tag{E.8b}$$

$$c_p = T\left(\frac{\partial s}{\partial T}\right)_P \qquad \text{constant pressure heat capacity,} \tag{E.8c}$$

$$c_v = T\left(\frac{\partial s}{\partial T}\right)_V \qquad \text{constant volume heat capacity.} \tag{E.8d}$$

For the latter partial derivative we find, from the properties of Jacobian determinants,[1]

$$\left(\frac{\partial s}{\partial T}\right)_V = \frac{\partial(s,v)}{\partial(T,v)} = \frac{\partial(s,v)}{\partial(T,v)} = \frac{\partial(s,v)}{\partial(T,P)}\frac{\partial(T,P)}{\partial(T,v)}. \tag{E.9}$$

Thus we have, from the appropriate Eqs. (E.8) and (E.9),

$$c_v = -\frac{T}{v\kappa_T}\frac{\partial(s,v)}{\partial(T,P)} = -\frac{T}{v\kappa_T}\begin{vmatrix}\frac{\partial s}{\partial T} & \cdots & \frac{\partial s}{\partial P} \\ \cdots\cdots\cdots\cdots \\ \frac{\partial v}{\partial P} & \cdots & \frac{\partial v}{\partial P}\end{vmatrix}.$$

Expanding the determinant yields

$$c_v = -\frac{T}{v\kappa_T}\left[\left(\frac{\partial s}{\partial T}\right)_P\left(\frac{\partial v}{\partial T}\right)_T - \left(\frac{\partial s}{\partial P}\right)_T\left(\frac{\partial v}{\partial T}\right)_P\right]. \tag{E.10}$$

The last two partial derivatives are equal by a Maxwell equation. No need to memorize the Maxwell equations: it suffices to recognize that the independent variables are T and P, so that the Gibbs free energy differential is the one required: $dg = -s dT + v dP$. The corresponding Maxwell equation is then obtained from the condition of integrability (3.40), here given by

$$\left(\frac{\partial s}{\partial P}\right)_T = \left(\frac{\partial v}{\partial T}\right)_P,$$

which, inserted into Eq. (E.10) gives

$$c_v = -\frac{T}{v\kappa_T}\left\{-\frac{c_P}{v\kappa_T} + \left[\frac{\partial v}{\partial T}\right]^2\right\}. \tag{E.11}$$

The difference in heat capacities is obtained by rearranging Eq. (E.11) and using Eq. (E.8a)

$$c_P - c_V = T v \beta_P^2/\kappa_T. \tag{E.12}$$

Recall that for an ideal gas, the difference in (molar) heat capacities is simply given by $C_P - C_V = R$, a special case of Eq. (E.12).

---

[1] The derivation follows that in Callen, H. (1960), Sec. 7.5, *Thermodynamics*, J. Wiley & Sons.

Appendix F

The Third Law in Statistical Mechanics

The following is a proof for the expression of the entropy as T approaches absolute zero.[1] Start with the partition function given by Eq. (8.27):

$$\mathcal{Z} = \sum_{j=0}^{m} e^{-E_j/(k_B T)} , \qquad (F.1)$$

with Boltzmann's constant written explicitly. Let the energy level of the ground state (state of lowest energy at T=0) be written as E_o (with $E_o < E_j$, $j = 1, 2, \ldots m$), and take the exponential of the E_o contribution outside the summation to obtain

$$\mathcal{Z} = e^{-E_o/(k_B T)} \sum_{j=0}^{m} e^{-(E_j - E_o)/(k_B T)} .$$

For $j = 0$ the first term in the sum is unity, and as $T \to 0$ the other terms vanish exponentially ($e^{-\infty} \to 0$). We thus end up with

$$\lim_{T \to 0} \mathcal{Z} = g_o \, e^{-E_o/(k_B T)} .$$

Taking the logarithm of this limiting value and multiplying by k_B gives

$$k_B \ln \mathcal{Z} = k_B \ln g_o - \frac{E_o}{T} ,$$

---

[1]The proof is from Diu, B. *et al.* (1989), *Physique Statistique*, Herman, Paris.

whence, by rearranging,

$$F_o = E_o - k_B T \ln g_o \qquad \text{(F.2)}$$

where the subscript "o" indicates limiting values at $T = 0$, which shall call "residual". Since we have the identity

$$F = E - TS \qquad \text{(F.3)}$$

we may equate the last terms of the latter two equations to obtain the desired residual entropy

$$S_{res} = k_B \ln g_o . \qquad \text{(F.4)}$$

Appendix G

Proof of Triangle Equalities

Consider triangle $\overline{ABC}$ with opposing sides a, b, c. The circumcircle of the triangle is centered at point O, where the three perpendicular bisectors meet, one of which is shown as the dotted line $\overline{OP}$ in Fig. G.1. It is desired to prove the well-known relationship between angles and edges

$$\frac{a}{\sin A} = \frac{b}{\sin B} = \frac{c}{\sin C}. \qquad (G.1)$$

Three radii $(R), \overline{OA}, \overline{OB}, \overline{OC}$, shown as dashed lines, split the original triangle into three isosceles triangles $\overline{BOC}, \overline{COA},$ and $\overline{AOB}$, whose angles around point O are $\angle\alpha'\ \angle\beta'$ and $\angle\gamma'$, with corresponding angles at edges $\alpha, \beta,$ and γ. We have the following relations between angles:

$$\beta + \gamma = A$$
$$\beta' + \gamma' = 2\pi - 2(\beta + \gamma)$$
$$\alpha' = 2\pi - (\beta' + \gamma') \qquad \text{so that}$$
$$\alpha' = 2A. \qquad (G.2)$$

It has therefore been demonstrated (1) that the internal angle $\frac{1}{2}\alpha'$ at O is twice the corresponding angle at vertex A (that property being emphasized by the double arc at points A and O), and (2) that the angle that "sees" a chord, such as $\overline{BC}$, is always the same regardless of the position of the point A on the circumcircle.

It remains to prove that the expressions (G.1) always hold. For that, we appeal to the proof given by Coxeter (1969, Sec. 1.531).[1] It is seen that, in the right triangle $\overline{BOP}$ (or $\overline{COP}$), we have $\frac{1}{2}a = R \sin \frac{1}{2}\alpha'$, or equivalently, by Eq. (G.2),

$$\frac{a}{\sin A} = 2R. \tag{G.3}$$

Since the same argument holds for the other two sides of the triangle $\overline{ABC}$, those three ratios of side–to–sine of the opposite angle are the same and equal to the diameter of the circle. Thus Eqs. (G.1) are proved; *q. e. d.*

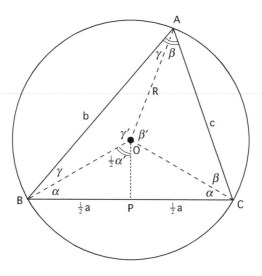

Figure G.1: Triangle $\overline{ABC}$ and circumcircle centered at O.

Reference

Coxeter, H. S. M., (1961), *Introduction to Geometry*, John Wiley & Sons, Inc., New York.

---

[1] Coxeter's book is remarkable both for the choice of subjects treated and for the elegance of the proofs. His textbook was already quoted in the present book in conjunction with the problem of close packing in Sec. 10.2.

Appendix H

Variational Calculus

The calculus of variations is an essential mathematical tool introduced in classical mechanics, and is associated with the names of Euler (1707–1783), Lagrange (1736–1813), Hamilton (1805–1865) and others. Many books treat the subject, but one of the most readable, *i.e.* most user-friendly, is that by Herbert Goldstein, entitled, appropriately enough, *Classical Mechanics* (1950), perfect for first-year graduate studies, and used throughout the world, having been translated in nine languages. Here however, we shall mostly follow the textbook by Margenau and Murphy (*The Mathematics of Physics and Chemistry*, 1955) because that text mentions specifically an alternate formulation of the Euler–Lagrange equation, useful in Chapter 14 and quoted by Cahn and Hilliard (1958–I).

The general problem of variational calculus is that of finding the path $y = y(x)$ between specified points x_1 and x_2 which will give the functional $F[f]$ its extremal value, equivalently which will make the following integral stationary:

$$\delta F[f] = C \int_{x_1}^{x_2} \delta f(x, y, y') \, \mathrm{d}x = 0 \tag{H.1}$$

where C is a suitable normalization constant, with integrand f function of three variables: the independent one x (only one, for simplicity) and $y' = dy/dx$. By carrying out the variation on f we obtain

$$\int_{x_1}^{x_2} \left[\frac{\partial f}{\partial y} \delta y + \frac{\partial f}{\partial y'} \frac{\mathrm{d}}{\mathrm{d}x}(\delta y) \right] \mathrm{d}x = 0 \, . \tag{H.2}$$

Integration by parts on the second term gives

$$\left[\frac{\partial f}{\partial y'} \right]_{x_1}^{x_2} - \int_{x_1}^{x_2} \left(\frac{\mathrm{d}}{\mathrm{d}x} \frac{\partial f}{\partial y'} \right) \delta y \, \mathrm{d}x \, ; \tag{H.3}$$

341

but the integrated term vanishes since the variations of y vanish at the two fixed points, so we are left with

$$\int_{x_1}^{x_2} \left(\frac{\partial f}{\partial y} - \frac{\mathrm{d}}{\mathrm{d}x} \frac{\partial f}{\partial y'} \right) \delta y \, \mathrm{d}x = 0 \,. \tag{H.4}$$

Hence, since the variations δy are arbitrary, the only way that the integral can vanish is by setting the bracket itself equal to zero

$$\frac{\partial f}{\partial y} - \frac{\mathrm{d}}{\mathrm{d}x} \frac{\partial f}{\partial y'} = 0 \,. \tag{H.5}$$

Such is the famous Euler–Lagrange condition for the integral $F[f]$ to reach stationary value, denoting an equilibrium state. The Landau and Lifshitz, and Cahn and Hilliard approaches are thus seen as applications of the mathematical work of Leonhard Euler, one of the greatest and most prolific mathematicians of all time.

It turns out that there is an alternate way to express the Euler condition. To obtain this formulation we first write down the total derivative of the function $f(x, y, y')$

$$\frac{\mathrm{d}f}{\mathrm{d}x} = \frac{\partial f}{\partial x} + \frac{\partial f}{\partial y} \frac{\mathrm{d}y}{\mathrm{d}x} + \frac{\partial f}{\partial y'} \frac{\mathrm{d}y'}{\mathrm{d}x} \,, \tag{H.6}$$

in which we must clearly distinguish between total (d) and partial (∂) derivatives. We now use the Euler–Lagrange equation to perform the following substitution

$$\frac{\partial f}{\partial y} \implies \frac{\mathrm{d}}{\mathrm{d}x} \frac{\partial f}{\partial y'}$$

into Eq.(H.6) to obtain

$$\frac{\mathrm{d}f}{\mathrm{d}x} = \frac{\partial f}{\partial x} + y' \frac{\mathrm{d}}{\mathrm{d}x} \frac{\partial f}{\partial y'} + \frac{\partial f}{\partial y'} \frac{\mathrm{d}y'}{\mathrm{d}x} \,. \tag{H.7}$$

The last two terms of this equation are just the total derivative with respect to x of the product $y' \frac{\partial f}{\partial y'}$, thereby simplifying the expression on the right hand side of the equation. If it turns out that the function f is not an explicit function of x, then the *partial* derivative $\partial f / \partial x$ vanishes and we are left with

$$\frac{\mathrm{d}}{\mathrm{d}x} \left(f - y' \frac{\partial f}{\partial y'} \right) = 0 \,. \tag{H.8}$$

Hence, when $\partial f / \partial x = 0$, we find that

$$f - y' \frac{\partial f}{\partial y'} = \text{constant} \,. \tag{H.9}$$

So, in this special case, the expression on the left of the latter equation must be a constant, and we have therefore eliminated one integration of the Euler–Lagrange equation, which is in fact a second-order differential equation. This trick will be very useful in Sec. 14.1, describing a system with a one-dimensional compositional inhomogeneity.

Of course, three dimensional problems can be treated, in which cases the derivation of the corresponding Euler–Lagrange equation is very similar to the one given here, but expressions such as $\frac{dy}{dx}$ must be replaced by ∇y (gradient of y) to yield

$$\frac{\partial f}{\partial y} - \nabla y \, \frac{\partial f}{\partial \nabla y} = 0 \,. \tag{H.10}$$

The expression on the left can be considered as a generalized (chemical) potential, the gradient of which is a corresponding flux whose divergence times a mobility term will enter into the master diffusion equation used in phase field kinetic models.

References

Cahn, J. W. and Hilliard, J. E. (1958-I). *J. Chem. Phys.* **28**, 258.

Goldstein, H. (1950). *Classical Mechanics* Addison-Wesley Publishing Co., Reading, MA.

Margenau, H. and Murphy, G. M. (1955). *The Mathematics of Physics and Chemistry* D. Van Nostrand Co., New York.

Appendix I

Discrete Lattice Fourier Series

Most solid matter is of crystalline nature, *i.e.* it consists of periodically repeating elements, or *unit cells*. Periodicity almost immediately suggests the mathematical technique of Fourier series, to be applied here to spatially *discrete* systems, from which continuum space is banned. Yet, as we saw earlier, the thermodynamics of phase transformation often requires the continuum hypothesis. This apparently contradictory requirement can be met, as was mentioned, by resorting to *coarse graining*. That means, in essence, that we shall be dealing with discrete periodic arrangements of *average* units, in the limit, actually *average atoms* (or molecules), a physical nonsense, but a useful one within the framework of mean field thermodynamic models. Classical Fourier series, and Fourier transforms, can easily be adapted to these semi-continuum spaces, as will now be shown. For simplicity, we shall consider only binary solutions, and to begin with, only one-dimensional variations of concentrations, or of order parameters ξ on simple lattices, therefore consisting of N average point units (atoms), placed periodically on a line, with average ξ (in the sense described in the text) located at each of these p points ($p = 0, 1, 2, \ldots, N-1$).

Discrete Fourier Transform

By definition, the discrete function ξ has the Fourier series expansion

$$\xi(p) \stackrel{\text{def}}{=} \sum_h X(h) \, e^{2\pi i h p}, \qquad (\text{I.1})$$

where the summation is over the index h which has formula $h = m/N$ with $m = 0, 1, 2, \ldots$. Thus the index runs over fractional numbers which refer to points in a linear space of N points, which are coordinates of so-called "reciprocal space" in terms of diffraction (by X-rays, neutrons, electrons) or "k-space" in condensed matter physics. Both reciprocal (h) and direct (p) spaces in one dimension have N points which repeat periodically. By definition then, $X(h)$ is the (discrete) Fourier transform of $\xi(p)$.

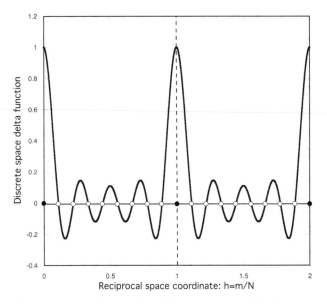

Figure I.1: One-dimensional normalized discrete space delta function vanishes at all integer points $m = 0, 1, 2, \ldots$ (open circles), except when m is a multiple of N (here with $N = 9$) where it equals unity (filled circles).

Let us multiply both sides of Eq. (I.1) by the complex exponential $e^{-2\pi i h' p}$ and sum over all p in the direct space period (N) to obtain

$$\sum_{p=0}^{N-1} \xi(p)\, e^{-2\pi i h' p} = \sum_{h} X(h) \left[\sum_{p=0}^{N-1} e^{-2\pi i (h'-h)p} \right] \qquad (I.2)$$

where the expression inside brackets is N times the discrete space (or lattice) *delta* function $\delta(h' - h)$. A representation of this delta function, for the case $N = 9$, is shown in Fig. I.1. Recall that in discrete space only those values of a continuous function, such as ξ, that are meaningful are those with integer values of its argument, *i.e.* those on the direct lattice sites. The reciprocal lattice is made up of peaks (at the location of filled circles in Fig. I.1, the fundamental Bragg peaks dear to the heart of diffraction-ists. Indeed, the delta function is seen to vanish at all h values for which $h = m/N$ with $m =$ integer, except those for which h is a multiple of the period N, including zero, for which the normalized (by $1/N$) delta function has value unity (see Fig. I.1).

Lattice Delta Function

To prove that the bracketed expression in Eq. (I.2) acts as a discrete space delta function, we first recognize that the expression in question is a power series for which well-known formulas for the sum of the terms do exist. For the moment, let us write the terms of this series as e^{ipy} with, temporarily, $y = -2\pi p$, with p integer. Let us denote the sum of the first N terms of this series by S, as in the first of Eqs. (I.3):

$$\begin{cases} S = 1 + e^{iy} + e^{2iy} + \ \dots \ + e^{(N-1)iy} \\ e^{iy} S = e^{iy} + e^{2iy} + \dots + e^{(N-1)iy} + e^{Niy}. \end{cases} \tag{I.3}$$

In the second equation (I.3) we have multiplied all the terms of the first equation by e^{iy}, then subtract the second sum from the first, noting that all terms cancel except the first one (unity) from the first equation and the last term from the second. There results the expression

$$S\left(1 - e^{iy}\right) = 1 - e^{Niy}$$

so that

$$S = \frac{1 - e^{Niy}}{1 - e^{iy}} = \frac{e^{Ni\frac{y}{2}}}{e^{i\frac{y}{2}}} \frac{e^{-Ni\frac{y}{2}} - e^{Ni\frac{y}{2}}}{e^{-i\frac{y}{2}} - e^{i\frac{y}{2}}},$$

or, after normalization by N and returning to the original notation,

$$\delta(h' - h \mid m) = \frac{S}{N} = \frac{1}{N} \frac{\sin N\pi(h' - h)}{\sin \pi(h' - h)} e^{-i\pi(N-1)(h'-h)}. \tag{I.4}$$

Thus the bracketed expression in Eq. (I.2) picks out of the sum over p only those values for which $(h' - h) = m =$ integer, including zero. So, with $h = h'$, equation (I.2) reduces to (after dropping the *prime*)

$$X(h) = \frac{1}{N} \sum_{p=0}^{N-1} \xi(p) e^{-2\pi i h p}, \tag{I.5}$$

which determines the expression of the *Fourier coefficients*. For $h - h' = m \neq 0$ the same result is obtained, of course, so no new information is gained. Equations (I.1) and (I.5) are (discrete) Fourier transform pairs, transforming functions from direct (p) to reciprocal (h) space and conversely. Unlike the continuum case, the transformation is *exact*; there are no arbitrary cutoffs of the series, no convergence problems. Note that we have used lower-case symbols for direct space functions and upper-case symbols for reciprocal space functions, a convention that we shall generally adhere to in what follows.

Plancherel and Parseval Theorems

A useful application of the Fourier series is *Plancherel's theorem* which states that the sum of the product of two functions in direct space is equal to the sum of the products of the Fourier transforms of these functions in reciprocal space, that is

$$\sum_p \psi(p)\, q(p) = \frac{1}{N} \sum_h \Psi(h)\, Q^*(h)\,. \tag{I.6}$$

The proof is obtained by substituting the expressions, such as Eq. (I.1), for the Fourier expansions of functions $\psi(p)$ and $q(p)$ into the left side of (I.6) to obtain

$$\sum_p \left(\frac{1}{N} \sum_h Q(h) e^{2\pi i h p} \right) \left(\frac{1}{N} \sum_h \Psi^*(h) e^{-2\pi i h p} \right) \tag{I.7}$$

which, by rearranging, gives

$$\frac{1}{N^2} \sum_{h,h'} Q(h)\, \Psi^*(h') \underbrace{\sum_p e^{2\pi i (h-h'')p}}_{\delta \text{ function}} = \frac{1}{N} \sum_h Q(h)\, \Psi^*(h)\,, \tag{I.8}$$

proving the relation (I.6). If the two (real) functions q and ψ are the same, we have *Parseval's theorem*

$$\sum_p [q(r)]^2 = \sum_h |Q(h)|^2\,, \tag{I.9}$$

which states that the sum of the squares of a function (over its period) is equal to the sum of its squared Fourier transforms.

Convolution Theorem

Let us prove the *convolution theorem* which states that the Fourier transform $(\mathcal{F})$ of the convolution of two functions, ψ and q (not necessarily the ones used above, of course) is equal to the product of the Fourier transforms of those two functions. First define the convolution $\varphi(p)$ as

$$\varphi(p) \equiv (\psi * q) \overset{\text{def}}{=} \frac{1}{N} \sum_{p'} \psi(p') \, q(p - p') \tag{I.10}$$

so that its Fourier transform is

$$\mathcal{F}(\psi * q) = \frac{1}{N^2} \sum_{p} e^{-2\pi i h p} \sum_{p'} \psi(p') \, q(p - p') \tag{I.11a}$$

$$= \Big(\frac{1}{N} \sum_{p'} \psi(p') \, e^{-2\pi i h p'} \Big) \Big(\frac{1}{N} \sum_{r} q(r) \, e^{-2\pi i h r} \Big), \tag{I.11b}$$

where in Eq. (I.11b) we have set $r = p - p'$ in the argument of the function q. Thus, we have proven the convolution theorem,

$$\mathcal{F}(\varphi * q) = (\mathcal{F}\psi)(\mathcal{F}q) = \Phi(h) \, Q(h). \tag{I.12}$$

with Fourier transforms

$$\Psi(h) = \frac{1}{N} \sum_{p} \psi(p) \, e^{-2\pi i h p} \tag{I.13}$$

$$Q^*(h) = \frac{1}{N} \sum_{r} q(p) \, e^{2\pi i h r}. \tag{I.14}$$

If the two functions are the same, we are then dealing with the auto-correlation function

$$\omega(p) = \frac{1}{N} \sum_{p'} q(p') \, q(p - p'), \tag{I.15}$$

and the (auto-) convolution theorem now reads

$$\Omega(h) = N Q(h) Q^*(h) = |Q^*(h)|^2, \tag{I.16}$$

with the usual convention regarding the upper and lower-case symbols.

Diagonalization of Quadratic Forms

Fourier transforms are particularly useful in the context of quadratic forms, such as

$$F = \sum_{p} \sum_{p'} q(p, p') \, \xi(p) \, \xi(p') \tag{I.17}$$

particularly when the coefficients of the products are of the type $q(p - p')$, a special case. Then we have successively (taking lattice periodicity into account)

$$F = \sum_{p} \sum_{p'} q(p' - p)\, \xi(p)\, \xi(p') \tag{I.18a}$$

$$= \sum_{r} q(r) \sum_{p} \xi(p)\, \xi(p + r) \tag{I.18b}$$

$$= N \sum_{r} q(r)\, \varphi(r) \tag{I.18c}$$

where $\varphi(r)$ is a version of the *autocorrelation function*

$$\varphi(r) \stackrel{\text{def}}{=} \frac{1}{N} \sum_{p} \xi(p)\, \xi(p + r)\,, \tag{I.19}$$

with Fourier expansion ($\mathcal{F}$) and its inverse ($\mathcal{F}^{-1}$), respectively:

$$\varphi(r) = \sum_{h} \Phi(h)\, e^{2\pi i h r}, \quad \text{and} \tag{I.20a}$$

$$\Phi(h) = \frac{1}{N} \sum_{r} \varphi(r)\, e^{-2\pi i h r}\,. \tag{I.20b}$$

If the order parameter ξ is taken to be $(c - \bar{c})$, then the autocorrelation function (I.19) at the origin ($r = 0$) is the *variance*

$$\varphi_o = \overline{c^2} - \bar{c}^2 \tag{I.21}$$

in statistics.[1]

These and other considerations are handy in the study of order-disorder reactions. In particular, in the Landau expansion it is advantageous to reduce the second-order term, which is actually a quadratic form in the direct space of pair interactions [see Eq. (I.18a or b)], to a sum of squares in reciprocal space, as shown below. To show that, start from Eq. (I.18c) and use the Plancherel theorem (I.6) to transform that expression into

$$F = \frac{1}{N} \sum_{r} q(r)\, \varphi(r) = \frac{1}{N} \sum_{h} Q(h)\, \Phi^*(h). \tag{I.22}$$

We also have, by Eq. (I.19),

$$q(r) = \sum_{h} |X(h)|^2\, e^{2\pi i h r}\,. \tag{I.23}$$

---

[1] If the variable c is replaced by f, the atomic scattering factor, the variance becomes the *diffuse Laue monotonic scattering* in scattering theory.

But $q(r)$ has Fourier transform $Q(h)$, hence $|X(h)|^2 = Q(h)$. Also, by Eq. (I.20b), we have

$$\sum_r q(r)\,\varphi(r) = \sum_h Q(h) \sum_r \varphi(r)\,e^{2\pi ihr}$$
$$= N \sum_h Q(h)\,\Phi^*(h) = N \sum_h |X(h)|^2 \Phi^*(h)\,, \qquad (I.24)$$

and the quadratic forms of Eqs. (I.18 a and b) have been fully diagonalized.

Appendix J

Symmetry of Tensors

Definition of tensors was given in Chapter 14, in particular with reference to Table 16.2. The present Appendix briefly shows how the structure of tensors depends on the symmetry of the crystalline material being considered. A much more complete treatment is given in the excellent text by Nye (1957, and many reprints). Here only a summary of the essentials will be given. As in the main text, only Cartesian tensors will be considered.

Voigt notation

First the Voigt notation should be introduced. Normally, the range of tensor indices should be 1, 2, 3, reflecting their use in three-dimensional space. To exhibit all the tensor components of a second-rank tensor it is convenient to represent it as a square symmetric matrix of $n(n+1)/2$ components ($=6$ for $n=3$). The symmetry of the components about the main diagonal is dictated by the thermodynamics of the problem, giving rise to the equality of second derivatives as occurs with the Maxwell relations. For tensors of rank higher than 2, a change of notation is required, accomplished by the Voigt notation. In that scheme, a pair of indices, such as $\{ij\}$ ($i,j = 1,2,3$) is replaced by the single index $\{I=1, 2 \ldots 6\}$ in the following manner

$$11 \rightarrow 1 \ , \ 22 \rightarrow 2 \ , \ 33 \rightarrow 3 \ , \ 23 = 32 \rightarrow 4 \ , \ 31 = 13 \rightarrow 5 \ , \ 12 = 21 \rightarrow 6,$$

this assignment being valid thanks to the tensor symmetry mentioned above. In this new notation, the components are not tensors, but may be arranged as elements of matrices as in the following examples of elastic constants C

and piezoelectric parameters d, respectively

$$
\begin{pmatrix}
C_{11} & C_{12} & C_{13} & C_{14} & C_{15} & C_{16} \\
C_{21} & C_{22} & C_{23} & C_{24} & C_{25} & C_{26} \\
C_{31} & C_{32} & C_{33} & C_{34} & C_{35} & C_{36} \\
C_{41} & C_{42} & C_{43} & C_{44} & C_{45} & C_{46} \\
C_{51} & C_{52} & C_{53} & C_{54} & C_{55} & C_{56} \\
C_{61} & C_{62} & C_{63} & C_{64} & C_{65} & C_{66}
\end{pmatrix}
,
\begin{pmatrix}
d_{11} & d_{12} & d_{13} & d_{14} & d_{15} & d_{16} \\
d_{12} & d_{22} & d_{23} & d_{24} & d_{25} & d_{26} \\
d_{13} & d_{32} & d_{33} & d_{34} & d_{35} & d_{36}
\end{pmatrix}.
$$

$$(\mathrm{J}.1)$$

For square matrices, the symmetry about the main diagonal is still preserved. If this notation appears cumbersome, it is still much more compact than that of the full tensor notation. Indeed, without taking symmetry into account, the number of fourth rank tensor components would be $3^4 = 81$, but the number of symmetric 6×6 matrix elements in the Voigt notation is only $(6 \times 7)/2 = 21$, a gain of almost a factor of 4. Crystal symmetry can reduce the number of independent elements even more, as will now be shown.

Crystal Symmetry

Since tensors represent physical properties, the components must remain invariant with respect to symmetry operations which leave the crystal structure unchanged. The tensor must then have the symmetry of the corresponding *crystallographic point group*, the point group is the set (having the mathematical structure of a group) of all symmetry operations leaving a least one lattice point fixed. The symmetry operations are therefore: inversion (center of symmetry), mirror planes, rotation axes, and excluding translations. As shown in texts on crystallography (see *International Tables for Crystallography* for example) there are 32 distinct point groups of symmetry operations which leave a lattice invariant, grouped into classes.

From the definition of tensors, it is seen that tensor components transform as the product of the coordinates of an arbitrary point x_i, so that $t_{ijk...}$ transforms as $x_i x_j x_k \ldots$, the direct products of the coordinates. That property allows us to use the *direct inspection method* due to Fumi (1951, 1953), as illustrated in Nye's textbook (1962). The general idea is the following: first, consider a second rank tensor t_{ij}. Its components transform like the product $x_i x_j$. Under inversion, the coordinates of a point transform by changing sign: $x_i \to -x_i$. The product therefore transforms as $x_i x_j \to (-x_i)(-x_j) = x_i x_j$, meaning that a second rank tensor in invariant under inversion. A third rank tensor however transforms under inversion as the product $x_i x_j x_k \to (-x_i)(-x_j)(-x_k)$ thus $t_{ijk} = -t'_{ijk}$, the prime indicating the transformed component. But since the inversion was an allowed symmetry operation, we assumed, the operation must leave the tensor invariant, we must therefore have $t_{ijk} = 0$. Conclusion: all third rank tensors must vanish if a center of inversion is an allowed operation in a given crystal. For a fourth rank tensor, $t_{ijkl} \rightarrow t_{ijkl}$ so there is no problem, there is no restriction in this case.

We now turn to rotation axes of symmetry; first a binary one (*diad*) along the x_3 direction. Then we have $x_1 \rightarrow -x_1$, $x_2 \rightarrow -x_2$ and $x_3 \rightarrow x_3$. Hence for t_{12} there is no restriction; but for t_{13} there is since $x_1 x_3 \rightarrow (-x_1)x_3$ and thus $t_{13} = -t'_{13}$, that is the component (13) must vanish, as is the case also for component (23) and the symmetric counterparts. Let us now turn to a four-fold axis (*tetrad*). A tetrad along the x_3 axis changes the coordinates of an arbitrary point as follows: $x_1 \rightarrow x_2, x_2 \rightarrow -x_1, x_3 \rightarrow x_3$. By similar reasoning we find: $t_{11} = t_{22} \neq t_{33}$ and $t_{12} = t_{23} = t_{31} = 0$ and the same for symmetric partners. Hence in the tetragonal point symmetry class, characterized by just one tetrad, the second rank tensor in the matrix representation is limited to elements along the main diagonal with two of them equal (matrix on the left side of Eq. (J.2))

$$\text{Tetragonal} \equiv \begin{pmatrix} t_{11} & 0 & 0 \\ 0 & t_{11} & 0 \\ 0 & 0 & t_{33} \end{pmatrix}, \quad \text{Cubic} \equiv \begin{pmatrix} t_{11} & 0 & 0 \\ 0 & t_{11} & 0 \\ 0 & 0 & t_{11} \end{pmatrix}. \quad \text{(J.2)}$$

The cubic class (3 orthogonal tetrads) is similar but with all diagonal elements equal (matrix on the right side of Eq. (J.2)); it is therefore the unit matrix multiplied by a scalar.

For fourth rank tensors we proceed the same way, but after determining the symmetry conditions on tensors c_{ijkl} $(i, j, k, l = 1, 2, 3)$ for lattice constants let us say, there remains the conversion to the Voigt notation C_{IJ} $(I, J = 1, 2, \ldots 6)$ to be carried out. For example, the results for all but one of the tetragonal systems — characterized by a single 4-fold axis in the x_3 direction — is

$$\text{Tetragonal} \equiv \begin{pmatrix} C_{11} & C_{12} & C_{13} & 0 & 0 & 0 \\ C_{21} & C_{11} & C_{13} & 0 & 0 & 0 \\ C_{31} & C_{31} & C_{33} & 0 & 0 & 0 \\ 0 & 0 & 0 & C_{44} & 0 & 0 \\ 0 & 0 & 0 & 0 & C_{44} & 0 \\ 0 & 0 & 0 & 0 & 0 & C_{66} \end{pmatrix}, \text{with } C_{IJ} = C_{JI}.$$

$$\text{(J.3)}$$

In similar ways, we can use Fumi's method to construct appropriate matrices for second, third, and fourth rank tensors in the Voigt notation for all 32 point groups. The results are tabulated, for instance, in Nye's textbook, cited several times previously. The Voigt matrix for a fourth rank tensor in the all-important cubic system is given in Eq. (16.16) of the main text.

References

Fieschi, R. and Fumi, F. G. (1953) *Il Nuovo Cimento*, **10**, 865-82 (in English).

Fumi, F. G. (1951) *Phys. Rev.*, **83**, 1274-5.

International Tables for Crystallography (see 8 Volumes on the Web, URL http://it.iucr.org/).

Nye, J. F. (1962), *Physical properties of Crystals*, Oxford, Clarendon Press.

Appendix K

Diagonalization of Quadratic Forms

This Appendix briefly recalls known features of linear algebra pertaining to subjects treated in this book. Proofs are merely sketched and special cases are not covered. For more complete and rigorous treatments the reader is referred to textbooks on linear algebra, of which there are many.[1]

Eigenvalues and Eigenvectors

Let the quadratic form Q be given in explicit form, then in matrix notation, by[2]

$$Q = \sum_{i,j=1}^{n} A_{i,j} x_i x_j = \mathbf{X'AX} \tag{K.1}$$

where the $n \times n$ quadratic form matrix $\mathbf{A}$ of elements a_{ij} can always be taken as symmetric, $\mathbf{X}$ is a column vector of n elements x_i, and $\mathbf{X'}$ its corresponding (row) transpose. To express Q into a sum of squares we seek a linear transformation $x_i \to z_k$

$$\mathbf{X = BZ} \tag{K.2}$$

---

[1]The mathematical methods employed are mostly due to mathematicians of the past such as Lagrange (1736–1813), Gauss (1777–1855), Jacobi (1804–1851) and Sylvester (1814–1897), the dates listed indicating that the relevant mathematics were often created much in advance of their applications. For a more modern treatment, see for example Parlett (1980).

[2]The present treatment follows that of Acad. V. I. Smirnov's in *Linear Algebra and Group Theory*, translated from Russian by Richard Silverman, Mc.Graw Hill, N.Y. (1961).

with $\mathbf{B}$ taken as an $n \times n$ real orthogonal matrix of elements b_{ij}.[3] Substituting from Eq. (K.2) into Eq. (K.1) gives

$$Q = \mathbf{Z}'\mathbf{B}'\mathbf{A}\mathbf{B}\mathbf{Z} = \mathbf{Z}'(\mathbf{B}^{-1}\mathbf{A}\mathbf{B})\mathbf{Z}. \tag{K.3}$$

It is always possible to find a *similarity transformation*

$$\mathbf{B}^{-1}\mathbf{A}\,\mathbf{B} = \mathbf{\Lambda} \tag{K.4}$$

where $\mathbf{\Lambda}$ is a diagonal matrix of elements $\lambda_k(k = 1, \ldots, n)$. It is now necessary to determine the elements of the matrix $\mathbf{B}$ and the n parameters λ_k. This shall be done one column of $\mathbf{B}$ at a time: Eqs.(K.4), expanded out, are a system of n linear homogeneous equations which, written out in full for the elements b_{jk} of column $k(k = 1, \ldots, n)$, is:

$$\begin{cases} (a_{11} - \lambda_k)b_{1k} + & a_{12}b_{2k} & + \ldots + & a_{1n}b_{nk} & = 0 \\ a_{21}b_{1k} & + (a_{22} - \lambda_k)b_{2k} + \ldots + & a_{2n}b_{nk} & = 0 \\ \cdots\cdots\cdots\cdots\cdots\cdots\cdots\cdots\cdots\cdots\cdots\cdots\cdots \\ a_{n1}b_{1k} & + a_{n2}b_{2k} & + \ldots + (a_{nn} - \lambda_k)b_{nk} & = 0 \end{cases} . \tag{K.5}$$

This system, let's call it *subsystem k*, can only have trivial (all zero) solutions *unless* the determinant of the system vanishes:

$$\begin{vmatrix} a_{11} - \lambda & a_{12} & \ldots & a_{1n} \\ a_{21} & a_{22} - \lambda & \ldots & a_{2n} \\ \cdots\cdots\cdots\cdots\cdots\cdots\cdots\cdots\cdots\cdots \\ a_{n1} & a_{n2} & \ldots & a_{nn} - \lambda \end{vmatrix} = 0 . \tag{K.6}$$

In this determinant, the index k on the λ parameters has been left out because the determinant of $\mathbf{A}$ is the same for all subsystems k and can serve for all of them. Equation (K.9) is an nth-degree polynomial in λ whose zeros are all real because the matrix $\mathbf{A}$ is symmetric. These n roots λ_k, called *eigenvalues* of the matrix $\mathbf{A}$, are to be inserted into the linear system (K.5) in turn, thus obtaining n such linear systems, each differing only by the value of the eigenvalue λ_k, thereby obtaining all b_{jk} $(j, k = 1, \ldots, n)$ elements of the $\mathbf{B}$ matrix. The n columns of this matrix can be regarded as vectors, one for each eigenvalue, and are called *eigenvectors* or *characteristic vectors* of $\mathbf{A}$. Note that Eq. (K.9) can have multiple roots, which can modify the procedure of diagonalization somewhat, but will not be covered here.

---

[3]An orthogonal matrix is a square matrix whose columns and rows are orthogonal unit vectors (i.e. orthonormal vectors); its transpose $\mathbf{B}'$ is then the same as its inverse $\mathbf{B}^{-1}$.

Having thus determined the eigenvalues and eigenvectors of the original matrix, one can now diagonalize the quadratic form Q and rewrite Eq. (K.1) as:

$$Q = \sum_{i,j=1}^{n} A_{i,j} x_i x_j \Rightarrow \sum_{k=1}^{n} \lambda_k z_k^2 = \mathbf{Z}'\mathbf{\Lambda Z}, \tag{K.7}$$

where, from Eq. (K.2),

$$\mathbf{Z} = \mathbf{B}'\mathbf{X}. \tag{K.8}$$

One application of the method is given here:

Gaussian Reduction Algorithm

Apparently this algorithm was known to the Chinese in the fourth century A.D. and to later to European mathematicians, but Gauss' name has stuck to it. Here we shall apply the algorithm to the diagonalization of quadratic forms. To simplify, we shall look first at a 3×3 system for the determinant of the matrix $\mathbf{A}$, featured in the linear system (K.5):

$$\Delta_3 \equiv |\mathbf{A}| = \begin{vmatrix} a_{11} & a_{12} & a_{13} \\ a_{21} & a_{22} & a_{23} \\ a_{31} & a_{32} & a_{33} \end{vmatrix}. \tag{K.9}$$

The algorithm in question proposes to reduce this matrix to a "triangular" one, *i.e.* to one where all the elements below the main diagonal are zero.

Here is how it's done: we first make sure that the first element a_{11} is occupied by the non-zero diagonal element of $|\mathbf{A}|$ with the lowest index. We then perform successively the following row operations which leave the determinant invariant: as a first step add to the elements of the second row of $|\mathbf{A}|$ the corresponding elements of the first row multiplied by the ratio $-\frac{a_{21}}{a_{11}}$, thereby getting rid of the element a_{21} in the new row number two, through the use of Eq. (K.10):

$$a_{22}^{(1)} = a_{22} - a_{12}\frac{a_{21}}{a_{11}}, \tag{K.10}$$

the superscript (1) indicating "level 1" of the reduction operation. This operation (Eq. (K.10)) clearly indicates that element $a_{21}^{(1)}$ vanishes. This procedure must be continued until the whole second row elements have been similarly treated, and the third row handled in a like manner. The resultant determinant looks like this

$$\Delta_3 \equiv |\mathbf{A}| = \begin{vmatrix} a_{11} & a_{12} & a_{13} \\ 0 & a_{22}^{(1)} & a_{23}^{(1)} \\ 0 & a_{32}^{(1)} & a_{33}^{(1)} \end{vmatrix}. \tag{K.11}$$

There results a sub-determinant which is that of a 2×2 system of elements, recognized by the superscript (1). In turn that subsystem can be reduced to a triangular one by similar row operation [level (2)], producing the desired triangular determinant

$$\Delta_3 \equiv |\mathbf{A}| = \begin{vmatrix} a_{11} & a_{12} & a_{13} \\ 0 & a_{22}^{(1)} & a_{23}^{(1)} \\ 0 & 0 & a_{33}^{(2)} \end{vmatrix}. \tag{K.12}$$

Clearly, the process can be continued to the arbitrary $n \times n$ case, with the resulting determinant

$$\Delta_n = \begin{vmatrix} a_{11} & a_{12} & a_{13} & \cdots & a_{1n} \\ 0 & a_{22}^{(1)} & a_{23}^{(1)} & \cdots & a_{2n}^{(1)} \\ 0 & 0 & a_{33}^{(2)} & \cdots & a_{3n}^{(2)} \\ \cdots\cdots\cdots\cdots\cdots\cdots\cdots \\ 0 & 0 & 0 & \cdots & a_{nn}^{(n-1)} \end{vmatrix} = a_{11}\, a_{22}^{(1)} a_{33}^{(2)} \ldots a_{nn}^{(n-1)}, \tag{K.13}$$

where the superscripted variables are given by

$$a_{ij}^{(i-1)} = \begin{cases} 0 & \text{if} \quad j < i \\ a_{ij}^{(m-1)} - \dfrac{a_{im}^{(m-1)}}{a_{mm}} a_{mj}^{(m-1)} & \text{if} \quad j \geq i, \end{cases} \tag{K.14}$$

that equation containing the generalized form of Eq. (K.10). The great advantage of working with triangular determinants is that their value is given directly by the product of their main diagonal elements, as indicated in Eq. (K.13).

Jacobi Formulation

From Eq. (K.13) we also have

$$\begin{cases} \Delta_1 = a_{11} \\ \Delta_2 = a_{11}\, a_{22}^{(1)} \\ \Delta_3 = a_{11}\, a_{22}^{(1)} a_{33}^{(2)} \\ \cdots\cdots\cdots\cdots\cdots\cdots \\ \Delta_k = a_{11}\, a_{22}^{(1)} a_{33}^{(2)} \ldots a_{kk}^{(k-1)} \\ \cdots\cdots\cdots\cdots\cdots\cdots \end{cases} \quad \text{hence} \quad \begin{cases} a_{11} = \Delta_1/\Delta_0 \\ a_{22}^{(1)} = \Delta_2/\Delta_1 \\ a_{33}^{(2)} = \Delta_3/\Delta_2 \\ \cdots\cdots\cdots \\ a_k^{(k-1)} = \Delta_k/\Delta_{k-1} \\ \cdots\cdots\cdots \end{cases} \tag{K.15}$$

from which we may immediately obtain the neat formula for the diagonalized quadratic form Q in terms of the transformed variables z_k:

$$Q(z_1, \ldots, z_n) = \frac{\Delta_1}{\Delta_0} z_1^2 + \frac{\Delta_2}{\Delta_1} z_2^2 + \cdots + \frac{\Delta_n}{\Delta_{n-1}} z_n^2 . \tag{K.16}$$

Such is the *Jacobi formula* for diagonalization of a quadratic form: the original quadratic form now contains only square terms in the new variables, and it is easy to determine the stability conditions of Q from this sum of squares.

$$(1) \quad \frac{\partial z_1}{\partial x_1} = \frac{\partial(z_1, x_2, x_3 \ldots x_n)}{\partial(x_1, x_2, x_3 \ldots x_n)} > 0 \tag{K.17}$$

$$\Rightarrow \left(\frac{dz_1}{dx_1} \right)_{x_2, x_3, \ldots x_n} > 0$$

$$(2) \quad \frac{\partial(z_1, z_2)}{\partial(x_1, x_2)} = \frac{\partial(z_1, z_2, x_3 \ldots x_n)}{\partial(x_1, z_2, x_3 \ldots x_n)} \frac{\partial(x_1, z_2, x_3 \ldots x_n)}{\partial(x_1, x_2, x_3 \ldots x_n)} > 0$$

$$\Rightarrow \left(\frac{dz_2}{dx_2} \right)_{z_1, x_3, \ldots x_n} > 0$$

$$(3) \quad \frac{\partial(z_1, z_2, z_3)}{\partial(x_1, x_2, x_3)} = \frac{\partial(z_1, z_2, z_3 \ldots)}{\partial(x_1, z_2, z_3 \ldots)} \frac{\partial(x_1, z_2, z_3 \ldots)}{\partial(x_1, x_2, z_3 \ldots)} \frac{\partial(x_1, x_2, z_3 \ldots)}{\partial(x_1, x_2, x_3 \ldots)} > 0$$

$$\Rightarrow \left(\frac{dz_3}{dx_3} \right)_{z_1, z_2, x_4 \ldots} > 0$$

in general, for the kth partial:

$$(k) \quad \Rightarrow \left(\frac{dz_k}{dx_k} \right)_{z_1, z_2, \ldots z_{k-1}, x_{k+1} \ldots x_n} > 0 . \tag{K.18}$$

An application of these stability conditions is given in Sec. 16.2.

Index